OXFORD MATHEMATICAL MONOGRAPHS

Series Editors

J. M. BALL   E. M. FRIEDLANDER   I. G. MACDONALD
L. NIRENBERG   R. PENROSE   J. T. STEWART

OXFORD MATHEMATICAL MONOGRAPHS

A. Belleni-Moranti: *Applied semigroups and evolution equations*
A. M. Arthurs: *Complementary variational principles* 2nd edition
M. Rosenblum and J. Rovnyak: *Hardy classes and operator theory*
J. W. P. Hirschfeld: *Finite projective spaces of three dimensions*
A. Pressley and G. Segal: *Loop groups*
D. E. Edmunds and W. D. Evans: *Spectral theory and differential operators*
Wang Jianhua: *The theory of games*
S. Omatu and J. H. Seinfeld: *Distributed parameter systems: theory and applications*
J. Hilgert, K. H. Hofmann, and J. D. Lawson: *Lie groups, convex cones, and semigroups*
S. Dineen: *The Schwarz lemma*
S. K. Donaldson and P. B. Kronheimer: *The geometry of four-manifolds*
D. W. Robinson: *Elliptic operators and Lie groups*
A. G. Werschulz: *The computational complexity of differential and integral equations*
L. Evens: *Cohomology of groups*
G. Effinger and D. R. Hayes: *Additive number theory of polynomials*
J. W. P. Hirschfeld and J. A. Thas: *General Galois geometries*
P. N. Hoffman and J. F. Humphreys: *Projective representations of the symmetric groups*
I. Györi and G. Ladas: *The oscillation theory of delay differential equations*
J. Heinonen, T. Kilpelainen, and O. Martio: *Non-linear potential theory*
B. Amberg, S. Franciosi, and F. de Giovanni: *Products of groups*
M. E. Gurtin: *Thermomechanics of evolving phase boundaries in the plane*
I. Ionescu and M. Sofonea: *Functional and numerical methods in viscoplasticity*
N. Woodhouse: *Geometric quantization* 2nd edition
U. Grenander: *General pattern theory*
J. Faraut and A. Koranyi: *Analysis on symmetric cones*
I. G. Macdonald: *Symmetric functions and Hall polynomials* 2nd edition
B. L. R. Shawyer and B. B. Watson: *Borel's methods of summability*
M. Holschneider: *Wavelets: an analysis tool*
Jacques Thévenaz: *G-algebras and modular representation theory*
Hans-Joachim Baues: *Homotopy type and homology*
P. D. D'Eath: *Black holes: gravitational interactions*
R. Lowen: *Approach spaces: the missing link in the topology–uniformity–metric triad*
Nguyen Dinh Cong: *Topological dynamics of random dynamical systems*
J. W. P. Hirschfeld: *Projective geometries over finite fields* 2nd edition
K. Matsuzaki and M. Taniguchi: *Hyperbolic manifolds and Kleinian groups*
David E. Evans and Yasuyuki Kawahigashi: *Quantum symmetries on operator algebras*
Norbert Klingen: *Arithmetical similarities: prime decomposition and finite group theory*
Isabelle Catto, Claude Le Bris, and Pierre-Louis Lions: *The mathematical theory of thermodynamic limits: Thomas–Fermi type models*
D. McDuff and D. Salamon: *Introduction to symplectic topology* 2nd edition
William M. Goldman: *Complex hyperbolic geometry*
Charles J. Colbourn and Alexander Rosa: *Triple systems*
V. A. Kozlov, V. G. Maz'ya and A. B. Movchan: *Asymptotic analysis of fields in multi-structures*
Gérard A. Maugin: *Nonlinear waves in elastic crystals*
George Dassios and Ralph Kleinman: *Low frequency scattering*

Asymptotic Analysis of Fields in Multi-Structures

VLADIMIR KOZLOV
*Department of Mathematics, Linkoeping University,
Sweden*

VLADIMIR MAZ'YA
*Department of Mathematics, Linkoeping University,
Sweden*

ALEXANDER MOVCHAN
*Department of Mathematical Sciences, University of Liverpool,
United Kingdom*

CLARENDON PRESS · OXFORD
1999

OXFORD
UNIVERSITY PRESS

Great Clarendon Street, Oxford OX2 6DP
Oxford University Press is a department of the University of Oxford.
It furthers the University's objective of excellence in research, scholarship,
and education by publishing worldwide in
Oxford New York
Athens Auckland Bangkok Bogotá Buenos Aires Calcutta
Cape Town Chennai Dar es Salaam Delhi Florence Hong Kong Istanbul
Karachi Kuala Lumpur Madrid Melbourne Mexico City Mumbai
Nairobi Paris São Paulo Singapore Taipei Tokyo Toronto Warsaw
with associated companies in Berlin Ibadan

Oxford is a registered trade mark of Oxford University Press
in the UK and in certain other countries

Published in the United States
by Oxford University Press Inc., New York

© V. A. Kozlov, V. G. Maz'ya, and A. Movchan, 1999

The moral rights of the authors have been asserted
Database right Oxford University Press (maker)

First published 1999

All rights reserved. No part of this publication may be reproduced,
stored in a retrieval system, or transmitted, in any form or by any means,
without the prior permission in writing of Oxford University Press,
or as expressly permitted by law, or under terms agreed with the appropriate
reprographics rights organization. Enquiries concerning reproduction
outside the scope of the above should be sent to the Rights Department,
Oxford University Press, at the address above

You must not circulate this book in any other binding or cover
and you must impose this same condition on any acquirer

A catalogue record for this book is available from the British Library

Library of Congress Cataloging in Publication Data
(Data available)

ISBN 0 19 851495 6 (Hbk)

Typeset by the authors
Printed in Great Britain by
Biddles Ltd., Guildford and King's Lynn

PREFACE

The main objective of the present text is to describe methods of asymptotic analysis of boundary value problems in multi-structures, i.e. domains dependent on a small parameter ε in such a way that the limit region, as $\varepsilon \to 0$, consists of subsets of different space dimensions. At present, the asymptotic analysis of boundary value problems in parameter-dependent domains is a rapidly developing area in the theory of partial differential equations.

In general, mathematical problems connected with perturbations of domains occur naturally in electrostatics, hydrodynamics, fracture mechanics etc. (see the monographs of Van Dyke (1964), Cole (1968) and Cherepanov (1974)). Asymptotic engineering theories of thin elastic rods, plates and shells are well established in the mechanics literature. However, a rigorous asymptotic analysis of such problems began to develop only about 20 years ago. Of course, there were earlier roots such as, for example, a variational formula for Green's function by Hadamard (1908), or the asymptotic estimates for the capacity of a thin condenser due to Pólya and Szegö (1951).

Boundary value problems in multi-structures arise in many engineering applications, in particular in the analysis of dynamics and strength of mechanical structures and machines and forecasting their reliability. In general, the geometry of multi-structures is quite complex, which makes the effective solution of corresponding boundary value problems rather difficult. In the engineering literature one can find simplified methods of description of fields in multi-structures. One of the simplest approaches involves the finite-dimensional model, where the structure is represented as a system of discrete masses with weightless elastic constraints. However, this model is not always the best one with regard to simplicity of the algorithm, the amount of computation and accuracy obtained.

A more adequate model involves a description of a structure as a system of connected elastic bodies with distributed elastic inertial and dissipative parameters. These bodies may be one dimensional (rods), two dimensional (plates, shells) as well as three dimensional (thick disks, thick hollow cylinders, etc.). The motion of this structure is described by the equations of dynamics of elastic bodies complemented by junction conditions on the interfaces and the exterior boundary.

In engineering practice the entire boundary value problem in a multi-structure is formulated on the basis of physical assumptions, and often it is intuitive in some instances. Hence it is important to develop rigorous mathematical models of complicated mechanical structures to provide efficient methods for their theoretical and numerical analysis.

Whereas asymptotic theories for thin rods, plates and shells attracted much attention from researchers during the whole history of elasticity theory, the problem of asymptotic analysis of fields in multi-structures is relatively young. This problem was explicitly formulated by Ciarlet and Destuynder (1979) and still generates many interesting questions. Numerous results on boundary value problems in multi-structures obtained by a variational approach are surveyed in the comprehensive monograph by Ciarlet (1997). In the present book we use a new approach based upon the so-called method of compound asymptotic expansions (see Maz'ya, Nazarov and Plamenevskii (1991, 1992)). In this respect in particular, the text differs from the other monographs dealing with boundary value problems for parameter-dependent domains by Marchenko and Khruslov (1974), Bakhvalov and Panasenko (1989), Sanchez-Palencia (1980), Il'in (1992), Ciarlet (1990), Ciarlet (1997) and Le Dret (1991).

We are concerned with an asymptotic theory of boundary value problems in multi-structures with junctions of one-dimensional and three-dimensional elements, as well as with a spectral analysis of boundary value problems in multi-structures. This selection of topics was strongly influenced by our interests and the presented material reflects mainly results of our work during the last five years.

The volume is organised in six chapters as follows.

Chapter 1 can serve as an introduction to the method of compound asymptotic expansions illustrated by examples of boundary value problems in comparatively simple multi-structures. This material was used in graduate courses on asymptotic methods.

In Chapters 2–5 we are concerned with the solutions in multi-structures of more complicated geometries, first for the Laplace equation and then for the Lamé system of linear isotropic elasticity. We deal with the asymptotic analysis of junctions of three-dimensional and one-dimensional (3D–1D) elements, which was an open question until recently.

In Chapter 2 we obtain a uniform asymptotic expansion for a solution of a mixed boundary value problem for the Laplace operator in a 3D–1D multi-structure consisting of a large three-dimensional domain Ω connected with a number of thin cylinders. We prescribe the Dirichlet data on the bases of thin cylinders; on the remaining part of the surface we have the Neumann boundary condition. The solution is represented by an asymptotic series whose coefficients are linear combinations of solutions to certain model problems posed in canonical domains independent of ε. We obtain the remainder estimate for the asymptotic approximation and discuss some examples.

In Chapter 3 we collect some information to be used in the sequel, such as Korn's inequalities and special solutions of Lamé's system in canonical domains.

Chapter 4 deals with elastic fields in a 3D–1D multi-structure, which has the same geometry as in Chapter 2. Similar to Chapter 2, we obtain an asymptotic series which approximates the solution uniformly in the entire multi-structure. As a by-product of our asymptotic analysis, we show that the classical engineering approach needs modification in order to take into account the elastic interaction between Ω and the rods.

In Chapter 5 we allow the thin rods to have different orientations and we define degenerate and non-degenerate 3D–1D multi-structures in accordance with degeneracy or non-degeneracy of the stiffness matrix. From the point of view of this classification the multi-structure considered in Chapter 4 is degenerate. For the case of a non-degenerate multi-structure our asymptotic analysis gives a rigorous justification of the engineering pile structure model.

Chapter 6 is devoted to the asymptotics of eigenvalues of Laplace's and Lamé's operators in the same multi-structure as in Chapter 2 and Chapter 4.

This volume is addressed to mathematicians who work in partial differential equations, asymptotic methods and their applications. Also, we hope that it will be useful for those who are interested in problems in elasticity, thermoconductivity and structural mechanics. The reader is assumed to have undertaken undergraduate courses in partial differential equations and functional analysis.

Acknowledgements

We are grateful to Jan Åslund and Özgür Selsil for reading the manuscript and valuable comments and to Nadya Movchan for help with preparation of figures. V. Kozlov and V. Maz'ya acknowledge the support of Linköping University, the Swedish Natural Science Research Council (NFR) and the Swedish Research Council for Engineering Sciences (TFR). V. Kozlov was supported by the UK Engineering and Physical Science Research Council (EPSRC) and by the London Mathematical Society during his time at Bath University. A. Movchan would like to thank the Department of Mathematics at Linköping University for hospitality.

CONTENTS

List of symbols   xiv

1 Introduction to compound asymptotic expansions   1
    1.1 Elementary examples of perturbation problems for ordinary differential equations   1
    1.2 A one-dimensional singularly perturbed problem   5
    1.3 Neumann boundary value problem in a domain with small cavity   10
       1.3.1 Formulation of the problem   10
       1.3.2 The leading order approximation   12
       1.3.3 Remainder estimate   13
       1.3.4 Complete asymptotic expansion   16
       1.3.5 Asymptotic formula for the energy   20
    1.4 Dirichlet boundary value problem in a domain with small inclusion   22
       1.4.1 The leading order approximation   22
       1.4.2 The next approximation   24
       1.4.3 The complete asymptotic expansion   25
    1.5 Mixed boundary value problem for the Laplacian in a thin rectangle   30
       1.5.1 Formulation of the boundary value problem   30
       1.5.2 Two-term approximation   31
       1.5.3 The next approximation   34
       1.5.4 Higher-order approximation   36
    1.6 Problem of junction between thin bodies   38
       1.6.1 Model problems   40
       1.6.2 The leading order approximation   47
       1.6.3 The next-order approximation   49
       1.6.4 The complete asymptotic expansion   51
       1.6.5 The remainder estimate   53

2 A boundary value problem for the Laplacian in a multi-structure   56
    2.1 Formulation of the problem   58
    2.2 Model problems   59
       2.2.1 Limit domains   59
       2.2.2 Model problem in Ω 60
       2.2.3 Model problem for the junction region   63
       2.2.4 Junction layer   70

	2.2.5	Model problem for the bottom region	71
	2.2.6	Two model problems for a thin cylinder	75
	2.2.7	Algebraic system for the skeleton	76
2.3	Right-hand sides	77
	2.3.1	Local coordinates and limit domains	78
	2.3.2	Cut-off functions	78
	2.3.3	Asymptotic representations of the right-hand sides	79
2.4	The leading term of the asymptotic solution	81
	2.4.1	Domain Ω	81
	2.4.2	Junction layer	83
	2.4.3	Thin cylinder	85
	2.4.4	Bottom layer	87
	2.4.5	Evaluation of $W_0^{(j)}$	88
	2.4.6	Evaluation of the constants $T_0^{(j)}$ and C_0	88
	2.4.7	Concluding remarks on formal algorithm	89
2.5	Complete asymptotic expansion	89
	2.5.1	Structure of the asymptotic expansion	89
	2.5.2	The asymptotic algorithm	91
2.6	Justification of the asymptotic expansion	92
	2.6.1	Auxiliary estimates for functions in $H^1(\Omega_\varepsilon)$	92
	2.6.2	Estimate for solutions	94
	2.6.3	Estimate for the remainder term	97
2.7	A constant right-hand side	98
2.8	Application to the asymptotics of the energy integral	100
	2.8.1	The case of the right-hand sides concentrated in $\overline{\Omega}$	101
	2.8.2	The case of the Dirichlet data at the bases of thin cylinders	103
2.9	On a general 1D–3D multi-structure	105
2.10	A multi-structure with a thin-walled tube	108

3 Auxiliary facts from mathematical elasticity — 115

3.1	Basic formulae of linear elasticity	115
	3.1.1	Stress and strain	115
	3.1.2	Equations of equilibrium and boundary conditions	116
3.2	Two-dimensional problems of linear elasticity	118
	3.2.1	Plane strain	118
	3.2.2	Anti-plane shear	119
3.3	Differential equations for engineering models of elastic rods	120

3.4	Classical solutions of linear elasticity for a half-space		121
	3.4.1	Boussinesq–Cerruti's solution	121
	3.4.2	Mindlin's solution	123
	3.4.3	Connection between the Boussinesq–Cerruti and Mindlin solutions	124
3.5	Special solutions for a bounded two-dimensional domain		126
	3.5.1	The torsion potential	127
	3.5.2	The bending potentials	128
	3.5.3	Example	128
3.6	Special solutions of linear elasticity for an infinite cylinder		129
	3.6.1	Representation of differential operators	129
	3.6.2	The spectral problem	130
	3.6.3	X_3-polynomial solutions	132
	3.6.4	Biorthogonality conditions	133
	3.6.5	The normalised stiffness coefficients	135
	3.6.6	Biorthogonality relations for eigenvectors and generalised eigenvectors	136
	3.6.7	There are no other polynomials	138
3.7	Green's matrix in Ω		141
	3.7.1	Definition	141
	3.7.2	Asymptotics	141
3.8	Korn's inequalities		143
	3.8.1	The case of bounded Lipschitz domains	143
	3.8.2	Half-space and a cylinder	146
3.9	Asymptotics at infinity for solutions to the traction problem for a half-cylinder		151

4 Elastic multi-structure — 155

4.1	Multi-structure and boundary value problem		156
4.2	Model problems		158
	4.2.1	Limit domains	158
	4.2.2	Model problem for the body Ω	158
	4.2.3	Junction layer	160
	4.2.4	Model problem for the bottom layer	167
	4.2.5	Model problem for a bounded two-dimensional domain	171
	4.2.6	Model problems on the axis of an elastic rod	172
	4.2.7	Model matrices and the pile structure	174
	4.2.8	Special cases	179
4.3	Asymptotic expansion of the solution		180

		4.3.1	Asymptotic representation of the right-hand sides	180
		4.3.2	Description of the asymptotic series for the solution	182
		4.3.3	Auxiliary solutions of the Lamé system in a thin elastic rod	184
		4.3.4	Expansions for displacement in a thin rod	187
		4.3.5	Junction layer	189
		4.3.6	Displacement in Ω	190
		4.3.7	Bottom layer	191
		4.3.8	Functions $v_k^{(m,j)}$	193
		4.3.9	The recurrent procedure for the asymptotic expansion	197
	4.4	Justification of the asymptotic expansion		198
		4.4.1	Korn's inequality in Ω_ε	198
		4.4.2	An estimate for the solution	199
	4.5	The leading order approximation		203
		4.5.1	The term $\mathbf{u}^{(-1)}$	203
		4.5.2	The term $\mathbf{u}^{(0)}$	204
	4.6	Physical interpretation of the results		208
		4.6.1	The case $|\mathbb{M}_3| + |\mathbb{F}_1| + |\mathbb{F}_2| \neq 0$	208
		4.6.2	The case $\mathbb{F}_1 = \mathbb{F}_2 = \mathbb{M}_3 = 0$	210

5 Non-degenerate elastic multi-structures 213

	5.1	Pile structure model		214
		5.1.1	Skeleton of the multi-structure	214
		5.1.2	The pile structure	216
		5.1.3	Mathematical model of the pile structure	216
		5.1.4	Solution of the pile structure equations	217
		5.1.5	Algebraic system for the pile structure model	219
		5.1.6	Non-degenerate and degenerate pile structures	221
		5.1.7	Examples	222
	5.2	Multi-structure and the boundary value problem		224
		5.2.1	Description of the multi-structure	224
		5.2.2	Formulation of the boundary value problem	226
	5.3	Model problems		227
		5.3.1	Junction layer	227
		5.3.2	Remaining model problems	230
	5.4	Asymptotic expansion of the solution		231
		5.4.1	Cut-off functions	231
		5.4.2	Asymptotic representation of the right-hand sides for the case of a non-degenerate multi-structure	233

		5.4.3	Structure of the asymptotic series for the displacement field in Ω_ε	233

		5.4.4	The junction layer	235
		5.4.5	Displacement in Ω	237
		5.4.6	The bottom layer	238
		5.4.7	Functions $v_i^{(m,j)}$	239
		5.4.8	Evaluation of the lock forces and moments at junction points	241
		5.4.9	Algebraic system for $\alpha^{(m)}$, $\beta^{(m)}$	242
		5.4.10	The recurrent procedure for the asymptotic expansion	243
	5.5		Estimate for the remainder of the asymptotic expansion	244
	5.6		Analysis of the leading term	245
	5.7		Physical interpretation	246
6	**Spectral analysis for 3D–1D multi-structures**			**248**
	6.1		An abstract scheme for the asymptotics of eigenvalues	249
	6.2		Spectral problem for the Laplacian	253
		6.2.1	The first eigenvalue	253
		6.2.2	The first eigenfunction	255
	6.3		Asymptotics of first eigenvalues of the Lamé operator	257
		6.3.1	Spectral problem	258
		6.3.2	Korn-type inequalities	258
		6.3.3	Spaces $\mathfrak{X}_0$ and $\mathbf{H}_0$	263
		6.3.4	Asymptotic formula for the eigenvalues	263
	6.4		Spectral problem for an inhomogeneous elastic multi-structure	267
		6.4.1	The spectral problem	267
		6.4.2	Asymptotic formulae for the eigenvalues	268

Bibliographical remarks — 274

Bibliography — 276

Index — 281

LIST OF SYMBOLS

Chapter 1

ε small positive parameter, 1
$\mathbf{X}, \mathbf{Y}$ scaled variables, 4
$\tilde{u}^{(N)}$ remainder of the asymptotic expansion, 7
g_ε small cavity, 10
Ω_ε 2D domain with a small cavity, 11
$m(U)$ average of U over ∂g_ε, 14
$M(U)$ average of U over Ω, 14
$I(\varepsilon)$ the energy increment, 20
$\delta(\mathbf{x})$ Dirac delta function, 22
Π_ε thin rectangle, 30
$\mathcal{W}_1, \mathcal{V}_1$ boundary layer fields, 36

Chapter 2

$\mathbf{a}^{(1)}, \ldots, \mathbf{a}^{(K)}$ junction points, 57
$N(\mathbf{x}, \mathbf{a})$ Neumann function, 58
Ω_ε 1D–3D multi-structure, 58
$\Pi_\varepsilon^{(j)}, S_\varepsilon^{(j)}$ thin cylinder and its base region, 58
$F, P, \Phi^{(j)}$ right-hand sides of a mixed boundary value problem, 59
G scaled junction region, 59
Π_-, Π_+ semi-infinite cylinders, 60
$\mathcal{N}(\mathbf{X}, \mathbf{Y})$ Neumann function for the half-space, 66
$\mathcal{W}(\mathbf{X})$ junction layer, 70
g scaled cross-section of a thin cylinder, 75
$H^1(\Omega)$ Sobolev space, 80

$H^{1/2}(\partial\Omega)$ the space of traces on $\partial\Omega$, 80
$A(\mathbf{x}, \varepsilon)$ solution of the problem with a constant right-hand side, 98
D_ε thin tube, 108

Chapter 3

$\mathbf{u}$ displacement vector, 115
$\varepsilon_{ij}, \sigma_{ij}$ components of strain and stress, 115
$\mathbf{L}(\frac{\partial}{\partial x})$ operator of the Lamé system, 116
$\boldsymbol{\sigma}^{(n)}$ vector of tractions, 116
ν Poisson ratio, 118
C_1, C_2, C_3, C_4 stiffness coefficients of an elastic rod, 121
$\mathcal{B}^{(j)}$ columns of the Boussinesq–Cerruti solution, 122
$\mathcal{M}^{(j)}$ columns of the Mindlin solution, 123
φ torsion potential, 127
χ_1, χ_2 bending potentials, 128
D_1, D_2, D_3, D_4 normalised stiffness coefficients, 134
$\mathcal{N}(\mathbf{x}, \mathbf{y})$ Green's matrix, 141
$\mathcal{E}(\mathbf{u}; D)$ elastic energy functional, 143

Chapter 4

$\mathbf{F}, \mathbf{P}, \boldsymbol{\Phi}^{(j)}, j = 1, \ldots, K$ the right-hand sides in the equilibrium system and the boundary conditions, 157

LIST OF SYMBOLS

$\mathbb{F}, \mathbb{M}$ principal force and moment vectors, 159

$\mathfrak{F}^{(j)}, \mathfrak{M}^{(j)}$ lock forces and moments, 159

$\mathcal{W}(\mathbf{X})$ junction layer, 166

$\mathcal{V}(\mathbf{Y})$ bottom layer, 170

$\mathfrak{U}(\zeta, z, \varepsilon)$ asymptotic field in a thin cylinder, 183

$\mathbf{T}(\mathbf{x}, \partial_x)$ operator of tractions, 189

Chapter 5

$\mu^{(1)}, \ldots, \mu^{(K)}$ unit vectors characterising orientations of one-dimensional segments, 213

$\boldsymbol{\alpha} + \boldsymbol{\beta} \times \mathbf{x}$ rigid-body displacement of the pile cap, 216

$\mathcal{L}^{(j)}, \mathcal{R}^{(j)}, \Lambda^{(j)}$ rotation matrices, 224

$B(\mathbf{a}, \rho)$ ball of radius ρ centred at $\mathbf{a}$, 226

$\mathcal{A}$ stiffness matrix, 231

Chapter 6

λ_j eigenvalues of the spectral problem in the multi-structure, 248

ρ_j eigenvalues of a matrix, 248

$\mathfrak{X}, \mathcal{H}$ Hilbert spaces, 249

$\mathbf{H}$ closed subspace of $\mathcal{H}$ dense in $\mathfrak{X}$, 250

$\mathbf{H}_0, \mathbf{H}_1$ subspaces the direct sum of which is $\mathbf{H}$, 250

$\boldsymbol{\Psi}^{(1)}, \boldsymbol{\Psi}^{(2)}, \boldsymbol{\Psi}^{(3)}$ special basis required in the analysis of the spectral problem for the elastic multi-structure, 259

$\mathfrak{X}_0$ linear hull of $\boldsymbol{\Psi}^{(1)}, \boldsymbol{\Psi}^{(2)}, \boldsymbol{\Psi}^{(3)}$, 263

$\psi_1, \psi_2, \psi_3, \phi_1, \phi_2, \phi_3$ special basis in the space of rigid-body displacements, 269

1
INTRODUCTION TO COMPOUND ASYMPTOTIC EXPANSIONS

This chapter contains an introduction to the technique of compound asymptotic expansions. We demonstrate the main features of this method by examples of several comparatively simple singularly perturbed boundary value problems. By 'singularly perturbed' we mean that the problem involves a small parameter ε, and it degenerates in some sense as $\varepsilon \to 0$: for example, the equation reduces its order or a part of the boundary reduces its dimension.

In these examples we represent solutions as series in powers of ε. The series, generally, do not converge. They are asymptotic in the sense that the more of their terms we keep the better approximation to the solution we obtain as $\varepsilon \to 0$. The series are multi-scaled, i.e. their coefficients depend on both original and stretched variables. In order to find these coefficients one needs to solve auxiliary ε-independent boundary value problems called model problems.

The chapter begins with two sections describing boundary value problems for second-order ordinary differential equations with a small parameter. Then we analyse solutions of boundary value problems for the Laplacian in domains with small cavities supplied with the Dirichlet or Neumann boundary data. Further, we pass to a mixed boundary value problem for the Laplacian posed in a thin rectangular region.

In the final section we present the asymptotic analysis of a boundary value problem posed for an elementary multi-structure associated with a junction of a system of thin rectangles. We always begin by constructing formal asymptotic expansions and then we justify the asymptotics by estimating the remainders.

1.1 Elementary examples of perturbation problems for ordinary differential equations

This section deals with two simple examples for ordinary differential equations involving a small parameter.

We begin with the following regularly perturbed boundary value problem

$$u''(x) - \varepsilon^2 u(x) = 1, \quad 0 \leq x \leq 1, \tag{1.1.1}$$
$$u(0) = 0, \quad u(1) = 1. \tag{1.1.2}$$

Here ε is assumed to be a small positive parameter. One may wish to obtain an approximation of u when $\varepsilon \to 0$.

The exact solution of (1.1.1), (1.1.2) is given by

$$u_{exact} = \frac{e^{\varepsilon x} + e^{-\varepsilon x} - 2}{2\varepsilon^2}$$

$$+ \frac{e^{\varepsilon x} - e^{-\varepsilon x}}{2\varepsilon^2(e^\varepsilon - e^{-\varepsilon})}(2 - e^\varepsilon - e^{-\varepsilon} + 2\varepsilon^2),$$

and its leading order asymptotic approximation (for small values of ε) is

$$u_0(x) = \frac{x^2}{2} + \frac{x}{2}. \tag{1.1.3}$$

The function u_0 satisfies the differential equation

$$u_0''(x) = 1, \quad 0 \leq x \leq 1,$$

and the boundary conditions (1.1.2).

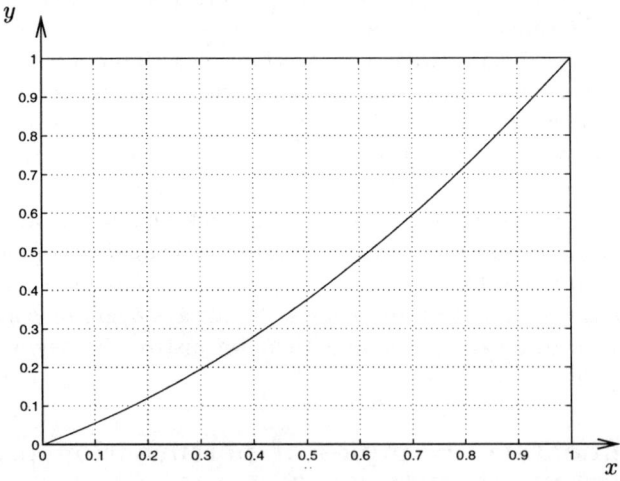

FIG. 1.1. *The graphs of both u_{exact} and u_0 for the case of the regularly perturbed problem ($\varepsilon = 0.1$).*

In Fig. 1.1 we have plotted the functions $y = u_{exact}(x)$ and $y = u_0(x)$ for $\varepsilon = 0.1$. The curves are practically indistinguishable, and the error of approximation is small (see the graph of $|u_{exact} - u_0|$ plotted in Fig. 1.2).

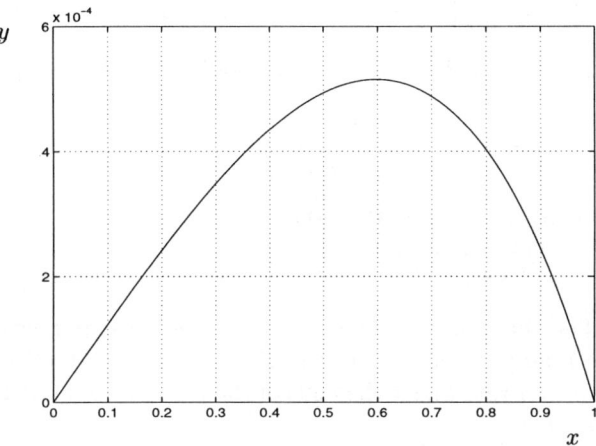

FIG. 1.2. Absolute error $|u_{exact} - u_0|$ of approximation for the case of the regularly perturbed problem ($\varepsilon = 0.1$).

Next, we give an example of a singularly perturbed boundary value problem. Let u satisfy the equation

$$\varepsilon^2 u''(x) - u(x) = 1, \quad 0 \leq x \leq 1, \tag{1.1.4}$$

and the boundary conditions at the ends of the interval

$$u(0) = 0, \quad u(1) = 1. \tag{1.1.5}$$

The solution u can be interpreted as the temperature of a thin rod connecting two large bodies which are maintained at constant temperature as shown in Fig. 1.3. The positive coefficient ε^2 denotes the normalised thermal conductivity, and the temperature of the surrounding medium is equal to -1.

Formally, if we set $\varepsilon = 0$ in (1.1.4) then the approximation u_0 is given by

$$u_0 = -1. \tag{1.1.6}$$

However, in contrast with the previous example u_0 does not satisfy the boundary conditions (1.1.5).

In this case, the exact solution of (1.1.4), (1.1.5) is

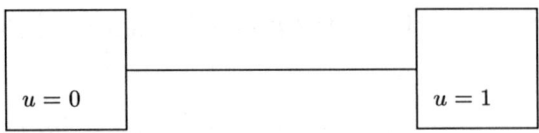

FIG. 1.3. A thin rod connecting two bodies of given temperature.

$$u_{exact} = -1 + \frac{e^{1/\varepsilon}(e^{1/\varepsilon} - 2)}{e^{2/\varepsilon} - 1} e^{-x/\varepsilon} + \frac{2e^{1/\varepsilon} - 1}{e^{2/\varepsilon} - 1} e^{x/\varepsilon}.$$

It has the asymptotic approximation

$$u_{asymp} = -1 + e^{-x/\varepsilon} + 2e^{-(1-x)/\varepsilon}, \tag{1.1.7}$$

which satisfies the equation (1.1.4) and leaves an exponentially small error in the boundary conditions (1.1.5). Two last correction terms in (1.1.7) are concentrated near the ends of the interval $(0, 1)$ and are called the *boundary layers*.

In Fig. 1.4 we plot the function $y = u_{exact}(x)$ for the case when $\varepsilon = 0.1$. It can be observed that the quantities u_{exact} and u_0 are quite close to each other in the middle region of the interval $(0, 1)$. However, $|u_{exact} - u_0|$ becomes large when we approach the end points $x = 0$ and $x = 1$. On the other hand, the functions u_{exact} and u_{asymp} are very close, so that it is hardly possible to distinguish between their graphs (when $\varepsilon = 0.1$ the difference $u_{exact} - u_{asymp}$ has the order 10^{-4}).

We note that the asymptotic approximation involving the boundary layer terms can be constructed for the solution of (1.1.4), (1.1.5) without knowledge of u_{exact}. That is, in order to compensate for the discrepancy left by u_0 in the boundary condition at $x = 0$, we make the change of the independent variable in (1.1.4), (1.1.5): $X = x/\varepsilon$, as if we have been looking at a neighbourhood of $x = 0$ through a microscope. The point $x = 1$ moves then to the right through a large distance $1/\varepsilon$, and we can forget about it for a moment. Now we obtain the model problem (independent of the small parameter):

$$\frac{d^2 U(X)}{dX^2} - U(X) = 0, \ X > 0, \tag{1.1.8}$$

$$U(0) = 1, \ U(X) \to 0 \ \text{as} \ X \to \infty, \tag{1.1.9}$$

which has the solution $U(X) = e^{-X}$. In a similar way, we introduce the scaled variable $Y = (1-x)/\varepsilon$ and compensate for the discrepancy of u_0 at $x = 1$ by solving the second model problem

$$\frac{d^2 V(Y)}{dY^2} - V(Y) = 0, \ Y > 0, \tag{1.1.10}$$

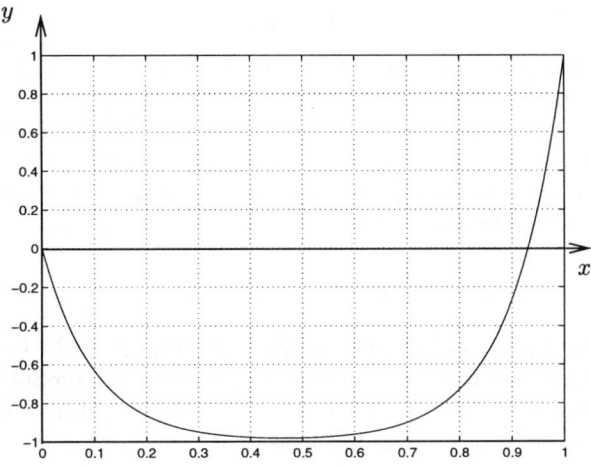

Fig. 1.4. The graphs of u_0, u_{exact} and u_{asymp} for the case of the singularly perturbed problem ($\varepsilon = 0.1$).

$$V(0) = 2, \quad V(Y) \to 0 \text{ as } Y \to \infty, \qquad (1.1.11)$$

which has the solution $V(Y) = 2\mathrm{e}^{-Y}$. The required asymptotic approximation is given by the formula

$$u_{asymp} = u_0 + U(X) + V(Y),$$

which is, in fact, the same as (1.1.7). This is a simple example of a multi-scaled compound asymptotic approximation invloving the boundary layer terms $U(X)$ and $V(Y)$. The above argument is an important ingredient of the method of compound asymptotic expansions discussed in the sequel.

1.2 A one-dimensional singularly perturbed problem

Here, we carry on with the analysis of the structure introduced in the previous section, but, in addition, we allow the coefficient of thermal conductivity of the rod to be a function of x, and the temperature of the surrounding medium to vary in x.

We consider the Dirichlet boundary value problem for a second-order differential equation that has a small coefficient near the derivative of the leading order:

$$\varepsilon^2 u''(x,\varepsilon) - r(x)u(x,\varepsilon) = p(x), \quad 0 \leq x \leq 1, \qquad (1.2.1)$$

$$u(0) = q_0, \quad u(1) = q_1, \qquad (1.2.2)$$

where r, p are smooth, and $r(x)$ is strictly positive on $[0,1]$; q_0, q_1 are given constants.

For every non-zero ε the problem (1.2.1), (1.2.2) is uniquely solvable; however, when $\varepsilon = 0$ the solution, in general, does not exist.

First, we solve the limit equation (1.2.1) (when $\varepsilon = 0$) disregarding the boundary conditions (1.2.2):

$$u_0(x) = -(r(x))^{-1} p(x). \qquad (1.2.3)$$

The function $u_0(x)$ does not necessarily satisfy (1.2.2). To fulfil the first boundary condition one has to compensate for the discrepancy produced by $u_0(x)$ at $x = 0$; thus, we freeze the coefficient r at $x = 0$ and solve the following boundary value problem on a semi-infinite interval:

$$\varepsilon^2 \frac{d^2}{dx^2} U_0(x/\varepsilon) - r(0) U_0(x/\varepsilon) = 0, \quad x > 0, \qquad (1.2.4)$$

$$U_0(0) = q_0 - u_0(0). \qquad (1.2.5)$$

Introducing the scaled variable $X = \varepsilon^{-1} x$ we rewrite (1.2.4) in the form

$$U_0''(X) - r(0) U_0(X) = 0, \quad X > 0. \qquad (1.2.6)$$

We observe that problem (1.2.6), (1.2.5) does not depend on the small parameter, and its solution, which decays at infinity, is given by

$$U_0(X) = (q_0 - u_0(0)) e^{-\sqrt{r(0)} X}. \qquad (1.2.7)$$

A similar consideration holds near the right end $x = 1$, and to satisfy the second boundary condition (by compensating for the discrepancy left by u_0) one has to solve the boundary value problem

$$V_0''(Y) - r(1) V_0(Y) = 0, \quad Y > 0, \qquad (1.2.8)$$

$$V_0(0) = q_1 - u_0(1), \qquad (1.2.9)$$

where $Y = \varepsilon^{-1}(1 - x)$. Its solution, which decays at infinity, is given by

$$V_0(Y) = (q_1 - u_0(1)) e^{-\sqrt{r(1)} Y}. \qquad (1.2.10)$$

As a result, the function u in (1.2.1), (1.2.2) can be represented in the form

$$u(x, \varepsilon) = u_0(x) + U_0(X) + V_0(Y) + \tilde{u}(x, \varepsilon). \qquad (1.2.11)$$

The remainder $\tilde{u}$ solves the boundary value problem

$$\varepsilon^2 \tilde{u}''(x,\varepsilon) - r(x)\tilde{u}(x,a) = -\varepsilon^2 u_0''(x) - (r(0) - r(x))U_0(x/\varepsilon)$$
$$-(r(1) - r(x))V_0((1-x)/\varepsilon), \quad (1.2.12)$$
$$\tilde{u}(0,\varepsilon) = -V_0(1/\varepsilon), \ \tilde{u}(1,\varepsilon) = -U_0(1/\varepsilon), \quad (1.2.13)$$

where the right-hand sides in (1.2.13) are exponentially small, and the modulus of the right-hand side of (1.2.12) is uniformly bounded by Const ε, with the constant coefficient independent of ε.

To derive an estimate for the solution we formulate the following auxiliary assertion:

LEMMA 1.1

The solution u of (1.2.1), (1.2.2) admits the estimate

$$|u(x,\varepsilon)| \leq \max\left\{|q_0|, |q_1|, \max_{0 \leq x \leq 1} \frac{|p(x)|}{r(x)}\right\}. \quad (1.2.14)$$

Proof

Let x_0 be the maximum point for u, and let $u(x_0) \geq 0$. If x_0 is the end point (i.e. $x_0 = 0$ or $x_0 = 1$) then $u(x_0)$ coincides with q_0 or q_1. When $x_0 \in (0,1)$ both terms of the left-hand side of (1.2.1), evaluated at x_0, are non-positive. Hence,

$$0 \leq u(x_0,\varepsilon) \leq \frac{|p(x_0)|}{r(x_0)}. \quad (1.2.15)$$

Then all positive values of u are estimated by the right-hand side of (1.2.14).

Now, let x_0 be the minimum point for u, and let $u(x_0) \leq 0$. Arguing as above, we arrive at the inequality

$$-u(x_0,\varepsilon) \leq \max\left\{|q_0|, |q_1|, \frac{|p(x_0)|}{r(x_0)}\right\}, \quad (1.2.16)$$

which completes the proof. □

Applying this lemma to problem (1.2.12), (1.2.13) we obtain that $|\tilde{u}(x,\varepsilon)| \leq$ Const ε, with the coefficient being independent of ε and

x. Thus, the first three summands in (1.2.11) give the *leading term* of the asymptotic approximation of u. The functions U_0 and V_0 decay exponentially at infinity, and they are classified as the *boundary layers* associated with the end points $x = 0$ and $x = 1$.

By analogy with (1.2.11) we seek an asymptotic expansion of u in the form

$$u(x,\varepsilon) \sim \sum_{j=0}^{\infty} \varepsilon^j \{u_j(x) + U_j(X) + V_j(Y)\}. \qquad (1.2.17)$$

Direct substitution of (1.2.17) into (1.2.1), (1.2.2) leads to the following system of recurrence relations:

$$u_{2j}(x) = (r(x))^{-1} u''_{2j-2}(x), \qquad (1.2.18)$$

$$u_{2j-1} = 0, \; j = 1, 2, \ldots, \qquad (1.2.19)$$

and

$$U''_k(X) - r(0)U_k(X) = \sum_{j=1}^{k} U_{k-j}(X) \frac{X^j}{j!} \frac{d^j r}{dx^j}(0), \; X > 0, \qquad (1.2.20)$$

$$U_k(0) = -u_k(0), \qquad (1.2.21)$$

$$V''_k(Y) - r(1)V_k(Y) = \sum_{j=1}^{k} V_{k-j}(Y) \frac{(-Y)^j}{j!} \frac{d^j r}{dx^j}(1), \; Y > 0, \qquad (1.2.22)$$

$$V_k(0) = -u_k(1), \qquad (1.2.23)$$

where the solutions of (1.2.20), (1.2.21) and (1.2.22), (1.2.23) are sought in the class of functions decaying at infinity. It is straightforward to prove that this decay is exponential.

The justification of the asymptotic expansion (1.2.17) is given by the following statement:

THEOREM 1.1

Series (1.2.17) is an asymptotic expansion of the solution $u(x,\varepsilon)$ to the boundary value problem (1.2.1), (1.2.2) in the sense that any partial sum

$$u^{(N)}(x,\varepsilon) = \sum_{j=0}^{N} \varepsilon^j \{u_j(x) + U_j(\varepsilon^{-1}x) + V_j(\varepsilon^{-1}(1-x))\} \qquad (1.2.24)$$

admits the estimate

$$|u(x,\varepsilon) - u^{(N)}(x,\varepsilon)| \leq \text{Const } \varepsilon^{N+1}, \quad x \in (0,1), \qquad (1.2.25)$$

where the constant coefficient is independent of ε and x.

Proof

Let
$$\tilde{u}^{(N)}(x,\varepsilon) = u(x,\varepsilon) - u^{(N)}(x,\varepsilon). \qquad (1.2.26)$$

Using (1.2.18) and (1.2.19) we obtain

$$\varepsilon^2 (\tilde{u}^{(N)})''(x,\varepsilon) - r(x)\tilde{u}^{(N)}(x,\varepsilon) = -\varepsilon^{N+1}(u''_{N-1}(x) + \varepsilon u''_N(x))$$

$$- \sum_{j=0}^{N} \varepsilon^j \{U''_j(X) - r(x)U_j(X)\}$$

$$- \sum_{j=0}^{N} \varepsilon^j \{V''_j(Y) - r(x)V_j(Y)\}. \qquad (1.2.27)$$

By Taylor's formula,

$$r(x) - \sum_{j=0}^{N} \frac{(\varepsilon X)^j}{j!} \frac{d^j r}{dx^j}(0) = O((\varepsilon X)^{N+1}),$$

and

$$r(x) - \sum_{j=0}^{N} \frac{(-\varepsilon Y)^j}{j!} \frac{d^j r}{dx^j}(1) = O((\varepsilon Y)^{N+1}).$$

Furthermore, there exists a positive β such that

$$U_j(X) = O(e^{-\beta X}), \quad X > 0,$$

and

$$V_j(Y) = O(e^{-\beta Y}), \quad Y > 0.$$

Therefore, by (1.2.20) and (1.2.22) the last two sums in (1.2.27) are $O(\varepsilon^{N+1})$.

By (1.2.21) and (1.2.23)

$$\tilde{u}^{(N)}(0,\varepsilon) = -\sum_{j=0}^{N} \varepsilon^j V_j(\varepsilon^{-1})$$

and

$$\tilde{u}^{(N)}(1,\varepsilon) = -\sum_{j=0}^{N} \varepsilon^j U_j(\varepsilon^{-1}).$$

Thus, the remainder $\tilde{u}^{(N)}$ satisfies

$$\varepsilon^2 (\tilde{u}^{(N)})''(x,\varepsilon) - r(x)\tilde{u}^{(N)}(x,\varepsilon) = O(\varepsilon^{N+1}),$$

$$\tilde{u}^{(N)}(0,\varepsilon) = O(e^{-\beta/\varepsilon}), \; \tilde{u}^{(N)}(1,\varepsilon) = O(e^{-\beta/\varepsilon}).$$

and estimate (1.2.25) follows from Lemma 1.1. □

Formula (1.2.17) is a simple example of a multi-scaled asymptotic expansion. The right-hand side of (1.2.17) includes terms of three types: the terms u_j are specified by the explicit recurrent formulae, whereas the boundary layer terms U_j and V_j depend on the scaled variables and satisfy boundary value problems on a semi-infinite axis. All terms of expansion (1.2.17) are specified via solutions of model problems independent of ε.

1.3 Neumann boundary value problem in a domain with small cavity

1.3.1 Formulation of the problem

Consider the heating of an infinite cylinder of cross-section Ω_ε that contains a thin rod of an infinite thermal resistance with a cross-section g_ε and assume that the axis of the cylinder and the axis of the thin rod are parallel to each other. We may specify in addition that the temperature u does not depend on the axial variable, so that the problem can be treated as two dimensional. To define the boundary conditions we suppose that the heat flux is prescribed on the exterior boundary.

Mathematically this problem can be formulated in the following way. We assume that a bounded domain $\Omega \subset \mathbb{R}^2$ has a smooth boundary and contains the origin. Let g_ε be given by

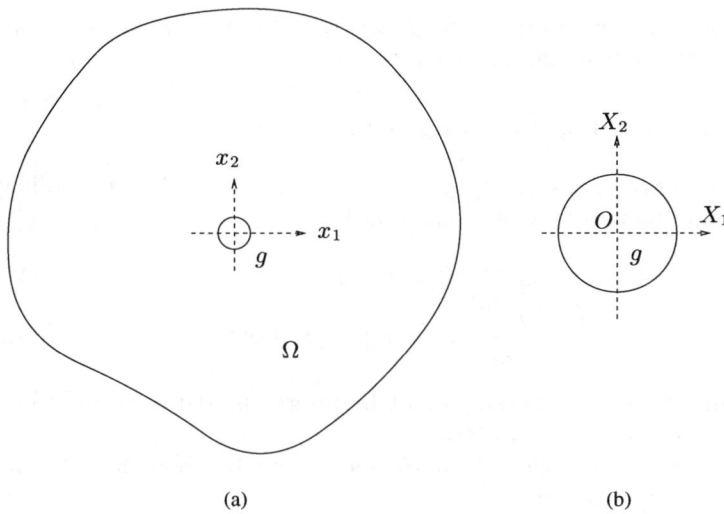

FIG. 1.5. Domain with small cavity.

$$g_\varepsilon = \{\mathbf{x} : \varepsilon^{-1}\mathbf{x} \in g\}, \tag{1.3.1}$$

where g is a unit disk with centre at the origin, and ε is a small positive parameter. We introduce the domain

$$\Omega_\varepsilon = \Omega \setminus \overline{g}_\varepsilon, \tag{1.3.2}$$

(see Fig. 1.5). Then the temperature u satisfies the boundary value problem

$$\Delta u(\mathbf{x},\varepsilon) = 0, \quad \mathbf{x} \in \Omega_\varepsilon, \tag{1.3.3}$$

$$\left.\frac{\partial u}{\partial r}\right|_{\partial g_\varepsilon} = 0, \tag{1.3.4}$$

$$\left.\frac{\partial u}{\partial n}\right|_{\partial \Omega} = p(\mathbf{x}). \tag{1.3.5}$$

Here $\partial/\partial n$ is the outward normal derivative on $\partial\Omega_\varepsilon$; (r,θ) denotes the polar coordinates centred at the origin.

The function $p(\mathbf{x})$ is assumed to be smooth, and it is also assumed to satisfy the orthogonality condition

$$\int_{\partial\Omega} p(\mathbf{x})ds = 0. \tag{1.3.6}$$

Then problem (1.3.3)–(1.3.5) has a solution specified up to an arbitrary

1.3.2 The leading order approximation

In the absence of the cavity g we would have to analyse the following Neumann boundary value problem in Ω:

$$\Delta v^{(0)}(\mathbf{x}) = 0, \quad \mathbf{x} \in \Omega, \qquad (1.3.7)$$

$$\frac{\partial v^{(0)}}{\partial n}(\mathbf{x}) = p(\mathbf{x}), \quad \mathbf{x} \in \partial\Omega, \qquad (1.3.8)$$

The function u, in general, cannot be approximated accurately by the field $v^{(0)}$ in the vicinity of ∂g_ε.

In general, the field $v^{(0)}$ does not satisfy the boundary condition (1.3.4), and it gives the discrepancy

$$\frac{\partial}{\partial r}(u - v_0) = -\frac{\partial}{\partial r}v^{(0)} \quad \text{on } \partial g_\varepsilon. \qquad (1.3.9)$$

Since ∂g_ε is a small circle, it is natural to approximate $v^{(0)}$ by the Taylor expansion, and to rewrite (1.3.9) in the form

$$\frac{\partial}{\partial r}(u(\mathbf{x},\varepsilon) - v^{(0)}(\mathbf{x}))$$

$$= -\cos\theta \, \frac{\partial v^{(0)}}{\partial x_1}(0) - \sin\theta \, \frac{\partial v^{(0)}}{\partial x_2}(0) + O(\varepsilon), \quad \mathbf{x} \in \partial g_\varepsilon.$$

In the vicinity of g_ε we approximate the quantity $u - v^{(0)}$ by the boundary layer associated with ∂g_ε. To obtain the corresponding boundary value problem one should 'forget' about the exterior part $\partial\Omega$ of the boundary and seek the approximation in the class of functions that decay at infinity. More precisely, the boundary layer is equal to $\varepsilon w^{(0)}(\mathbf{X})$, $\mathbf{X} = \mathbf{x}/\varepsilon$, where $w^{(0)}$ solves the following boundary value problem in $\mathbb{R}^2 \setminus g$:

$$\Delta w^{(0)}(\mathbf{X}) = 0, \quad \mathbf{X} \in \mathbb{R}^2 \setminus \overline{g}, \qquad (1.3.10)$$

$$\frac{\partial w^{(0)}}{\partial \rho}(\mathbf{X}) = -\cos\theta \frac{\partial v^{(0)}}{\partial x_1}(0) - \sin\theta \frac{\partial v^{(0)}}{\partial x_2}(0), \quad \|\mathbf{X}\| = 1, \qquad (1.3.11)$$

$$w^{(0)}(\mathbf{X}) \to 0, \quad \text{as } \|\mathbf{X}\| \to \infty, \qquad (1.3.12)$$

where $\rho = \|\mathbf{X}\| = r/\varepsilon$. Note that owing the choice of the scaled variables $\mathbf{X}$, the boundary value problem (1.3.10)–(1.3.12) is independent of the small parameter ε.

By separation of variables one can find the solution of (1.3.10)–(1.3.12):

$$w^{(0)}(\mathbf{X}) = \frac{1}{\|\mathbf{X}\|}\left(\cos\theta\frac{\partial v^{(0)}}{\partial x_1}(0) + \sin\theta\frac{\partial v^{(0)}}{\partial x_2}(0)\right), \quad (1.3.13)$$

where θ is the polar angle.

Thus, the field u can be written in the form

$$u(\mathbf{x},\varepsilon) = v^{(0)}(\mathbf{x}) + \varepsilon w^{(0)}(\mathbf{x}/\varepsilon) + \tilde{u}(\mathbf{x},\varepsilon), \quad (1.3.14)$$

where $\tilde{u}(\mathbf{x},\varepsilon)$ satisfies the boundary value problem

$$\Delta\tilde{u}(\mathbf{x},\varepsilon) = 0 \text{ in } \Omega_\varepsilon, \quad (1.3.15)$$

$$\frac{\partial\tilde{u}}{\partial r}(\mathbf{x},\varepsilon) = -\cos\theta\left(\frac{\partial v^{(0)}}{\partial x_1}(\mathbf{x}) - \frac{\partial v^{(0)}}{\partial x_1}(0)\right)$$
$$-\sin\theta\left(\frac{\partial v^{(0)}}{\partial x_2}(\mathbf{x}) - \frac{\partial v^{(0)}}{\partial x_2}(0)\right) \text{ on } \partial g_\varepsilon, \quad (1.3.16)$$

$$\frac{\partial\tilde{u}}{\partial n}(\mathbf{x},\varepsilon) = -\varepsilon\frac{\partial}{\partial n}w\left(\frac{\mathbf{x}}{\varepsilon}\right) \text{ on } \partial\Omega. \quad (1.3.17)$$

The right-hand sides in (1.3.16), (1.3.17) are smooth functions of orders $O(\varepsilon)$, $O(\varepsilon^2)$, respectively. Since $v^{(0)}(\mathbf{x})$ is harmonic in Ω,

$$\int_{\partial g_\varepsilon}\left\{\cos\theta\left(\frac{\partial v^{(0)}}{\partial x_1}(\mathbf{x}) - \frac{\partial v^{(0)}}{\partial x_1}(0)\right) + \sin\theta\left(\frac{\partial v^{(0)}}{\partial x_2}(\mathbf{x}) - \frac{\partial v^{(0)}}{\partial x_2}(0)\right)\right\}ds$$

$$= \int_{g_\varepsilon}\Delta v^{(0)}(\mathbf{x})d\mathbf{x} = 0. \quad (1.3.18)$$

Similarly, since $w(\mathbf{x}/\varepsilon)$ is harmonic outside g_ε and since it decays at infinity, we have

$$\int_{\partial\Omega}\frac{\partial}{\partial n}w\left(\frac{\mathbf{x}}{\varepsilon}\right)ds = -\int_{\mathbb{R}^2\setminus\Omega}\Delta w\left(\frac{\mathbf{x}}{\varepsilon}\right)d\mathbf{x} = 0. \quad (1.3.19)$$

By (1.3.18) and (1.3.19), problem (1.3.15)–(1.3.17) is solvable.

1.3.3 Remainder estimate

To estimate the energy norm of the remainder $\tilde{u}$ we need the following:

LEMMA 1.2

Let $U \in W_2^1(\Omega_\varepsilon)$ be a solution of the problem

$$\Delta U = 0 \quad \text{in } \Omega_\varepsilon, \tag{1.3.20}$$

$$\frac{\partial U}{\partial r} = q_1 \quad \text{on } \partial g_\varepsilon, \tag{1.3.21}$$

$$\frac{\partial U}{\partial n} = q_2 \quad \text{on } \partial\Omega, \tag{1.3.22}$$

where q_1, q_2 are smooth functions such that

$$\int_{\partial g_\varepsilon} q_1(\mathbf{x})ds = \int_{\partial\Omega} q_2(\mathbf{x})ds = 0. \tag{1.3.23}$$

Then the estimate

$$\left(\int_{\Omega_\varepsilon} |\nabla U|^2 dx\right)^{1/2} \leq C\{\varepsilon \max_{\partial g_\varepsilon} |q_1| + \max_{\partial\Omega} |q_2|\} \tag{1.3.24}$$

holds, where the constant C is independent of ε.

Proof

We begin with the inequality

$$\varepsilon^{-1} \int_{\partial g_\varepsilon} |U - m(U)|ds + \int_{\partial\Omega} |U - M(U)|ds$$

$$\leq C\left(\int_{\Omega_\varepsilon} |\nabla U|^2 dx\right)^{1/2}, \tag{1.3.25}$$

where the constant C is ε-independent, and

$$m(U) = \frac{1}{\text{mes}_1 \partial g_\varepsilon} \int_{\partial g_\varepsilon} U ds, \quad M(U) = \frac{1}{\text{mes}_1 \partial\Omega} \int_{\partial\Omega} U ds. \tag{1.3.26}$$

Let ω be a Lipschitz domain in $\mathbb{R}^2$, $\overline{\omega} \subset \Omega$. We assume that $\Omega \setminus \overline{\omega}$ is connected.

The inequality

$$\int_{\partial\Omega} v^2 ds \leq \text{Const} \int_{\Omega \setminus \omega} (|\nabla v|^2 + v^2) dx \tag{1.3.27}$$

is well known. It can be proved directly, and it is a very special case of Sobolev's embedding theorem.

We shall also use the standard Poincaré inequality

$$\|u - \bar{u}\|_{L_2(\Omega\setminus\omega)} \leq \text{Const } \|\nabla u\|_{L_2(\Omega\setminus\omega)} \qquad (1.3.28)$$

where $\bar{u}$ is the mean value of u on $\Omega \setminus \omega$. Clearly,

$$\int_{\partial\Omega} (u - M(u))^2 ds = \min_{c \in \mathbb{R}} \int_{\partial\Omega} (u - c)^2 ds,$$

and, hence,

$$\int_{\partial\Omega} |u - M(u)| ds \leq \text{Const } \left(\int_{\partial\Omega} (u - \bar{u})^2 ds\right)^{1/2}.$$

By setting $v = u - \bar{u}$ in (1.3.27) and using (1.3.28) we derive the required estimate (1.3.25) for the second integral.

This argument gives the estimate

$$\int_{\partial\omega} |u - \tilde{u}| ds \leq \text{Const } \left(\int_{\Omega\setminus\omega} \|\nabla u\|^2 dx\right)^{1/2}, \qquad (1.3.29)$$

where $\tilde{u}$ is the mean value of u on $\partial\omega$. The required inequality for the integral over ∂g_ε results from (1.3.29) by the scaling $\mathbf{X} = \varepsilon^{-1}\mathbf{x}$.

Multiplying (1.3.20) by U and integrating by parts we derive

$$\int_{\Omega_\varepsilon} |\nabla U|^2 dx = -\int_{\partial g_\varepsilon} q_1 U ds + \int_{\partial\Omega} q_2 U ds. \qquad (1.3.30)$$

By (1.3.23), the right-hand side of (1.3.30) is equal to

$$-\int_{\partial g_\varepsilon} q_1(U - m(U)) ds + \int_{\partial\Omega} q_2(U - M(U)) ds,$$

and its absolute value can be estimated by

$$\max_{\partial g_\varepsilon} |q_1| \int_{\partial g_\varepsilon} |U - m(U)| ds + \max_{\partial\Omega} |q_2| \int_{\partial\Omega} |U - M(U)| ds.$$

It follows from (1.3.30) and (1.3.25) that

$$\int_{\Omega_\varepsilon} |\nabla U|^2 dx \leq C(\varepsilon \max_{g_\varepsilon} |q_1| + \max_{\partial\Omega} |q_2|) \|\nabla U\|_{L_2(\Omega_\varepsilon)},$$

which implies (1.3.24). $\square$

Using Lemma 1.2 we derive the estimate for the remainder $\tilde{u}$ in (1.3.14):

$$\left(\int_{\Omega_\varepsilon}|\nabla\tilde{u}|^2 d\mathbf{x}\right)^{1/2} \leq \text{Const } \varepsilon^2,$$

where the constant coefficient is ε-independent.

Thus, the first two terms in (1.3.14) give the leading term of the asymptotic approximation of u. As in Section 1.2, scaled variables associated with the boundary layer occur in the expansion (1.3.14). We note that in the present case the boundary layer decays as $O(\|\mathbf{X}\|^{-1})$ at infinity, whereas in Section 1.2 the rate of decay of the boundary layer is exponential.

1.3.4 Complete asymptotic expansion

Formal asymptotics

As shown above, the coefficients of the asymptotic approximation of u represent solutions of two model problems independent of ε:

(1) the Neumann boundary value problems in a homogeneous region Ω (without any defect),

(2) the boundary layer problem posed in an infinite region with the cavity g.

The same model problems (possibly, with different right-hand sides) will be used to construct an asymptotic expansion in the form

$$u(\mathbf{x},\varepsilon) \sim \sum_{k=0}^{\infty} \varepsilon^k \left\{ v^{(k)}(\mathbf{x}) + \varepsilon w^{(k)}(\mathbf{x}/\varepsilon) \right\}, \qquad (1.3.31)$$

where $v^{(k)}$, $w^{(k)}$ are harmonic functions defined in Ω and $\mathbb{R}^2 \setminus g$, respectively.

The functions $v^{(k)}$ are defined as solutions of the Neumann boundary value problems in Ω, and they are specified up to an arbitrary additive constant. To provide uniqueness we impose the condition

$$\int_{\Omega} v^{(k)}(\mathbf{x}) d\mathbf{x} = 0. \qquad (1.3.32)$$

Since $v^{(k)}$ are harmonic in Ω they admit the following series expansion near the origin:

NEUMANN BOUNDARY VALUE PROBLEM 17

$$v^{(k)}(\mathbf{x}) = \sum_{j=1}^{2}\sum_{l=0}^{\infty} c_j^{(l,k)} p_j^{(l)}(\mathbf{x}), \tag{1.3.33}$$

where $p_1^{(l)}$, $p_2^{(l)}$ are homogeneous harmonic polynomials of order l

$$\begin{aligned} p_1^{(l)}(\mathbf{x}) &= \mathrm{Re}(x_1+ix_2)^l = \|\mathbf{x}\|^l \cos(l\theta), \\ p_2^{(l)}(\mathbf{x}) &= \mathrm{Im}(x_1+ix_2)^l = \|\mathbf{x}\|^l \sin(l\theta). \end{aligned} \tag{1.3.34}$$

The discrepancy in (1.3.4) produced by $\varepsilon^k v^{(k)}$ will be compensated by the boundary layer, and hence it is convenient to rewrite (1.3.33) in terms of the scaled variables

$$v^{(k)}(\mathbf{x}) = \sum_{j=1}^{2}\sum_{l=0}^{\infty} \varepsilon^l c_j^{(l,k)} p_j^{(l)}(\mathbf{X}). \tag{1.3.35}$$

The boundary layer term $\varepsilon^{k+1} w^{(k)}(\mathbf{X})$ compensates for the discrepancy of the order $O(\varepsilon^k)$ left by the functions $v^{(0)},\ldots,\varepsilon^k v^{(k)}$ in the boundary condition (1.3.4). Using expansion (1.3.33) we obtain the boundary value problem for $w^{(k)}$:

$$\Delta w^{(k)}(\mathbf{X}) = 0, \text{ in } \mathbb{R}^2 \setminus g, \tag{1.3.36}$$

$$\frac{\partial w^{(k)}}{\partial \rho}(\mathbf{X}) = -\frac{\partial}{\partial \rho}\left(\sum_{j=1}^{2}\sum_{l=1}^{k+1} c_j^{(l,k-l+1)} p_j^{(l)}(\mathbf{X})\right) \text{ on } \partial g, \tag{1.3.37}$$

$$w^{(k)}(\mathbf{X}) \to 0 \text{ as } \|\mathbf{X}\| \to \infty. \tag{1.3.38}$$

It is verified by direct substitution that the solution of (1.3.36)–(1.3.38) admits the explicit form

$$w^{(k)}(\mathbf{X}) = \sum_{l=1}^{k+1} \varepsilon^l \frac{1}{\|\mathbf{x}\|^l}\left\{c_1^{(l,k-l+1)}\cos(l\theta) + c_2^{(l,k-l+1)}\sin(l\theta)\right\}. \tag{1.3.39}$$

Since $\varepsilon^{k+1} w^{(k)}$ produces a discrepancy only in the Neumann boundary condition on $\partial\Omega$, we have re-expanded the function $w^{(k)}$ in terms of the variable $\mathbf{x}$. The function $\varepsilon^{k+1} w^{(k)}$ leaves a discrepancy in the boundary condition (1.3.5), and this error is to be compensated by the term $\varepsilon^{k+2} v^{(k+2)}$. The coefficients $v^{(k)}(\mathbf{x})$ from (1.3.31) are specified as solutions of the recurrent sequence of boundary value problems

$$\Delta v^{(k)}(\mathbf{x}) = 0 \text{ in } \Omega, \tag{1.3.40}$$

$$\frac{\partial v^{(k)}}{\partial n}(\mathbf{x}) = -\frac{\partial}{\partial n}\sum_{1\leq l\leq k/2}\frac{1}{\|\mathbf{x}\|^l}\left\{c_1^{(l,k-2l)}\cos(l\theta)\right.$$

$$+ c_2^{(l,k-2l)} \sin(l\theta) \Big\}, \quad \text{on } \partial\Omega, \ k \geq 1. \tag{1.3.41}$$

For $k = 1$ the right-hand side is equal to zero, and, by (1.3.32), $v^{(1)} = 0$.

Since the right-hand side in (1.3.41) is represented as the normal derivative of the harmonic function that vanishes at infinity, it gives zero after the integration along $\partial\Omega$. Therefore, problem (1.3.40), (1.3.41) is solvable.

Thus, the iterative procedure used to determine $v^{(k)}$, $w^{(k)}$ in (1.3.31) can be described as follows:

(1) We have specified the functions $v^{(0)}$, $w^{(0)}$, as described in Section 1.2, as well as the constant coefficients $c_j^{(l,0)}$ from (1.3.33).

(2) Assume that the functions v_j, w_j have been constructed for $j < k$. Hence, the coefficients $c_1^{(l,j)}$, $c_2^{(l,j)}$ are also known. Now, v_k and w_k can be found as solutions of the problems (1.3.36)–(1.3.38) and (1.3.40), (1.3.41).

Remainder estimate

The following statement justifies the asymptotic expansion (1.3.31).

THEOREM 1.2

Series (1.3.31) is an asymptotic expansion of the solution $u(\mathbf{x}, \varepsilon)$ to the boundary value problem (1.3.3)–(1.3.5) in the sense that any partial sum

$$u^{(N)}(\mathbf{x}, \varepsilon) = \sum_{k=0}^{N} \varepsilon^k \{v^{(k)}(\mathbf{x}) + \varepsilon w^{(k)}(\mathbf{x}/\varepsilon)\} \tag{1.3.42}$$

admits the estimate

$$\left(\int_{\Omega_\varepsilon} |\nabla \{u(\mathbf{x},\varepsilon) - u^{(N)}(\mathbf{x},\varepsilon)\}|^2 d\mathbf{x} \right)^{1/2} \leq \text{Const } \varepsilon^{N+2}. \tag{1.3.43}$$

Proof

Let $\tilde{u}^{(N)} = u - u^{(N)}$. This function is harmonic in Ω_ε. We find its normal derivatives on ∂g_ε and $\partial\Omega$. By (1.3.35) and (1.3.37), on ∂g_ε we have

$$\frac{\partial \tilde{u}^{(N)}}{\partial n}(\mathbf{x}) = \sum_{k=0}^{N} \varepsilon^k \left\{ \frac{\partial v^{(k)}}{\partial r}(\mathbf{x}) + \frac{\partial w^{(k)}}{\partial \rho}(\mathbf{X}) \right\}$$

$$= \sum_{j=1}^{2} \left\{ \sum_{l=1}^{\infty} \sum_{k=0}^{N} \varepsilon^{k+l-1} c_j^{(l,k)} \frac{\partial p_j^{(l)}}{\partial \rho}(\mathbf{X}) - \sum_{m=0}^{N} \varepsilon^m \sum_{l=1}^{m+1} c_j^{(l,m-l+1)} \frac{\partial p_j^{(l)}}{\partial \rho}(\mathbf{X}) \right\}.$$

Replacing $k + l - 1$ by m and changing the summation order we obtain

$$\frac{\partial \tilde{u}^{(N)}}{\partial n}(\mathbf{x}, \varepsilon) = \sum_{j=1}^{2} \sum_{m=N+1}^{\infty} \varepsilon^m \sum_{l=m+1-N}^{m+1} c_j^{(l,m-l+1)} \frac{\partial p_j^{(l)}}{\partial \rho}(\mathbf{X})$$

on ∂g_ε. It follows from (1.3.39) that

$$\frac{\partial \tilde{u}^{(N)}}{\partial n}(\mathbf{x}, \varepsilon) = p(\mathbf{x}) - \sum_{m=0}^{N} \varepsilon^m \frac{\partial v^{(m)}}{\partial n}(\mathbf{x})$$

$$- \frac{\partial}{\partial n} \sum_{k=0}^{N} \sum_{l=1}^{k+1} \frac{\varepsilon^{k+l+1}}{\|\mathbf{x}\|^l} \{c_1^{(l,k-l+1)} \cos(l\theta) + c_2^{(l,k-l+1)} \sin(l\theta)\}$$

on $\partial \Omega$. By (1.3.8) and due to the fact that $v^{(1)} = 0$, this equality can be rewritten as

$$\frac{\partial \tilde{u}^{(N)}}{\partial n}(\mathbf{x}, \varepsilon) = - \sum_{m=2}^{N} \varepsilon^m \frac{\partial v^{(m)}}{\partial n}(\mathbf{x})$$

$$- \frac{\partial}{\partial n} \sum_{k=0}^{N} \sum_{l=1}^{k+1} \frac{\varepsilon^{k+l+1}}{\|\mathbf{x}\|^l} \{c_1^{(l,k-l+1)} \cos(l\theta) + c_2^{(l,k-l+1)} \sin(l\theta)\}$$

on $\partial \Omega$. Changing the summation order, replacing $k + l + 1$ by m and using (1.3.41) we deduce that

$$\frac{\partial \tilde{u}^{(N)}}{\partial n}(\mathbf{x}) = - \frac{\partial}{\partial n} \sum_{m=N+1}^{2(N+1)} \varepsilon^m \sum_{\max\{1, m-N-1\} \leq l \leq m/2} \frac{1}{\|\mathbf{x}\|^l} \{c_1^{(l,m-2l)} \cos(l\theta)$$

$$+ c_2^{(l,m-2l)} \sin(l\theta)\}$$

on $\partial \Omega$. Hence, the remainder solves the Neumann problem

$$\Delta \tilde{u}^{(N)}(\mathbf{x}, \varepsilon) = 0, \quad \mathbf{x} \in \Omega_\varepsilon, \tag{1.3.44}$$

$$\frac{\partial \tilde{u}^{(N)}}{\partial n}(\mathbf{x}, \varepsilon) = \varepsilon^{N+1} q_1^{(N)}(\mathbf{x}, \varepsilon), \quad \mathbf{x} \in \partial g_\varepsilon, \tag{1.3.45}$$

$$\frac{\partial \tilde{u}^{(N)}}{\partial n}(\mathbf{x}, \varepsilon) = \varepsilon^{N+2} q_2^{(N)}(\mathbf{x}, \varepsilon), \quad \mathbf{x} \in \partial \Omega, \quad (1.3.46)$$

where $q_1^{(N)}$, $q_2^{(N)}$ are smooth functions of order $O(1)$ associated with expansions (1.3.33), (1.3.39), and

$$\int_{g_\varepsilon} q_1^{(N)} ds = \int_{\partial \Omega} q_2^{(N)} ds = 0.$$

Then applying Lemma 1.2 we obtain (1.3.43). $\square$

1.3.5 Asymptotic formula for the energy

Here we describe the influence of a small defect on the energy of the field in Ω.

The energy increment is defined by

$$I(\varepsilon) = \int_\Omega |\nabla v^{(0)}(\mathbf{x})|^2 d\mathbf{x} - \int_{\Omega_\varepsilon} |\nabla u(\mathbf{x}, \varepsilon)|^2 d\mathbf{x}. \quad (1.3.47)$$

To evaluate the leading order term in (1.3.47) it is sufficient to use formula (1.3.42) with $N = 2$:

$$u(\mathbf{x}, \varepsilon) = \sum_{k=0}^{2} \varepsilon^k \{v^{(k)}(\mathbf{x}) + \varepsilon w^{(k)}(\mathbf{x}/\varepsilon)\} + \tilde{u}^{(2)}.$$

By Theorem 1.2 the remainder term $\tilde{u}^{(2)}$ satisfies the estimate

$$\left(\int_{\Omega_\varepsilon} |\nabla \tilde{u}^{(2)}(\mathbf{x})|^2 d\mathbf{x} \right)^{1/2} \leq \text{Const } \varepsilon^4, \quad (1.3.48)$$

which implies (see (1.3.25)) that

$$\int_{\partial \Omega} |\tilde{u}^{(2)}(\mathbf{x}) - M(\tilde{u}^{(2)})| ds \leq \text{Const } \varepsilon^4, \quad (1.3.49)$$

where M is given by (1.3.26). Integrating by parts in (1.3.47) and using that $u(\mathbf{x}, \varepsilon)$ satisfies the homogeneous boundary condition on ∂g_ε, we deduce that

$$I(\varepsilon) = \int_{\partial \Omega} p(\mathbf{x})(v^{(0)}(\mathbf{x}) - u(\mathbf{x}, \varepsilon)) ds. \quad (1.3.50)$$

Since $v^{(1)} = 0$ the integral (1.3.50) has the form

$$I(\varepsilon) = -\int_{\partial\Omega} p(\mathbf{x})\{\varepsilon w^{(0)}(\mathbf{x}/\varepsilon) + \varepsilon^2 w^{(1)}(\mathbf{x}/\varepsilon) + \varepsilon^2 v^{(2)}(\mathbf{x})$$

$$+ \varepsilon^3 w^{(2)}(\mathbf{x}/\varepsilon) + \tilde{u}^{(2)}(\mathbf{x},\varepsilon) - M(\tilde{u}^{(2)})\}ds.$$

Here we also used the balance condition (1.3.6). By (1.3.39) and the fact that $v^{(1)} \equiv 0$, we deduce that $w^{(1)}(\mathbf{x}/\varepsilon) = O(\varepsilon^2)$ on $\partial\Omega$. Also, using (1.3.39) for $w^{(2)}$ together with the inequality (1.3.49) we obtain

$$I(\varepsilon) = -\int_{\partial\Omega} p(\mathbf{x})\left\{\varepsilon w^{(0)}(\mathbf{x}/\varepsilon) + \varepsilon^2 v^{(2)}(\mathbf{x})\right\}ds + O(\varepsilon^4). \quad (1.3.51)$$

Again, using (1.3.39) for $w^{(0)}$ we have

$$I(\varepsilon) = -\varepsilon^2 \int_{\partial\Omega} \frac{\partial v^{(0)}}{\partial n}(\mathbf{x}) T(\mathbf{x}) ds + O(\varepsilon^4),$$

where

$$T(\mathbf{x}) = v^{(2)}(\mathbf{x}) + \frac{1}{\|\mathbf{x}\|}\left(\frac{\partial v^{(0)}}{\partial x_1}(0)\cos\theta + \frac{\partial v^{(0)}}{\partial x_2}(0)\sin\theta\right)$$

is a function harmonic in Ω_ε, and by (1.3.41) for $k = 2$

$$\frac{\partial T}{\partial n}(\mathbf{x}) = 0 \text{ on } \partial\Omega.$$

Green's formula applied to the functions $v^{(0)}$ and T in Ω_ε yields

$$\int_{\partial\Omega} \frac{\partial v^{(0)}}{\partial n}(\mathbf{x}) T(\mathbf{x}) ds = \int_{\partial g_\varepsilon} \left(\frac{\partial v^{(0)}}{\partial r}(\mathbf{x}) T(\mathbf{x})\right.$$

$$\left. - v^{(0)}(\mathbf{x})\frac{\partial T}{\partial r}(\mathbf{x})\right) ds = 2\pi |\nabla v^{(0)}(0)|^2.$$

Finally, the change of energy is given by

$$I(\varepsilon) = -2\pi\varepsilon^2 |\nabla v^{(0)}(0)|^2 + O(\varepsilon^4). \quad (1.3.52)$$

It follows that the presence of a small cavity supplied with the homogeneous Neumann boundary data reduces the energy of u.

1.4 Dirichlet boundary value problem in a domain with small inclusion

So far we have been dealing with asymptotic expansions in powers of ε whereas quite frequently the asymptotic approximations involve powers of $\log \varepsilon$ as well. Consider the following boundary value problem in Ω_ε:

$$\Delta u(\mathbf{x}, \varepsilon) = 0, \quad \mathbf{x} \in \Omega_\varepsilon, \tag{1.4.1}$$
$$u(\mathbf{x}, \varepsilon) = \varphi(\mathbf{x}), \quad \mathbf{x} \in \partial\Omega, \tag{1.4.2}$$
$$u(\mathbf{x}, \varepsilon) = 0, \quad \mathbf{x} \in \partial g_\varepsilon, \tag{1.4.3}$$

where Ω_ε, g_ε are the same as in Section 1.2; φ is smooth. In contrast with the previous section we assume that the thin cylindrical inclusion is kept at zero temperature, and a certain distribution of temperature is given on the exterior surface. Note that the boundary value problem (1.4.1)–(1.4.3) is solvable for every small, positive ε. The formulation above can be regarded as a singular perturbation of the following boundary value problem corresponding to the region Ω:

$$\Delta v^{(0)}(\mathbf{x}) = 0, \quad \mathbf{x} \in \Omega, \tag{1.4.4}$$
$$v^{(0)}(\mathbf{x}) = \varphi(\mathbf{x}), \quad \mathbf{x} \in \partial\Omega. \tag{1.4.5}$$

One can try to compensate for the error produced by v in the boundary condition (1.4.3) and follow the algorithm given in the previous section in order to construct the boundary layer. This attempt fails since the Dirichlet boundary value problem in $\mathbb{R}^2 \setminus \bar{g}$ does not necessarily have a solution that decays at infinity.

1.4.1 The leading order approximation

When $\varepsilon \to 0$ the domain Ω_ε tends to $\Omega \setminus \{0\}$. Thus, it is reasonable to anticipate that the solution of the problem

$$-\Delta v(\mathbf{x}) = \beta \delta(\mathbf{x}), \quad \mathbf{x} \in \Omega,$$
$$v(\mathbf{x}) = \varphi(\mathbf{x}), \quad \mathbf{x} \in \partial\Omega,$$

where β is a constant coefficient, gives a more accurate approximation of the field $u(\mathbf{x}, \varepsilon)$ in comparison with $v^{(0)}$. We represent the function v in the form

$$v(\mathbf{x}) = v^{(0)}(\mathbf{x}) + \beta G(\mathbf{x}),$$

DIRICHLET BOUNDARY VALUE PROBLEM

where

$$-\Delta G(\mathbf{x}) = \delta(\mathbf{x}), \quad \mathbf{x} \in \Omega, \qquad (1.4.6)$$
$$G(\mathbf{x}) = 0, \quad \mathbf{x} \in \partial\Omega, \qquad (1.4.7)$$

and the field u in the exterior of a neighbourhood of g_ε is approximated by

$$u \simeq v^{(0)}(\mathbf{x}) + \beta G(\mathbf{x}). \qquad (1.4.8)$$

Since $(2\pi)^{-1}\log\|\mathbf{x}\|^{-1}$ is Green's function in an infinite plane, in the vicinity of the origin the quantity G admits the representation

$$G(\mathbf{x}) = -\{(2\pi)^{-1}\log\|\mathbf{x}\| + \mathcal{G}(\mathbf{x})\}, \quad \text{as } \|\mathbf{x}\| \to 0, \qquad (1.4.9)$$

with $\mathcal{G}$ being harmonic in $\bar{\Omega}$.

It should be noted that approximation (1.4.8) does not satisfy the boundary conditions, and in order to compensate for the discrepancy we introduce the boundary layer $V(\mathbf{X},\varepsilon)$ as a bounded solution of the boundary value problem:

$$\Delta_X V^{(0)}(\mathbf{X},\varepsilon) = 0, \quad \mathbf{X} \in \mathbb{R}^2 \setminus g, \qquad (1.4.10)$$
$$V^{(0)}(\mathbf{X},\varepsilon) = -v^{(0)}(0)$$
$$+ \beta\left\{\frac{1}{2\pi}\log\varepsilon + \mathcal{G}(0)\right\}, \quad \mathbf{X} \in \partial g. \qquad (1.4.11)$$

The solution does not decay at infinity unless

$$\beta = \left\{\frac{1}{2\pi}\log\varepsilon + \mathcal{G}(0)\right\}^{-1} v^{(0)}(0). \qquad (1.4.12)$$

In the latter case $V^{(0)}$ is identically zero. Consequently, the solution u of (1.4.1)–(1.4.3) can be represented by

$$u(\mathbf{x},\varepsilon) = v^{(0)}(\mathbf{x}) + \alpha(\varepsilon)v^{(0)}(0)G(\mathbf{x}) + \tilde{u}^{(0)}(\mathbf{x},\varepsilon), \qquad (1.4.13)$$

where

$$\alpha(\varepsilon) = \{(2\pi)^{-1}\log\varepsilon + \mathcal{G}(0)\}^{-1}, \qquad (1.4.14)$$

and the function $\tilde{u}^{(0)}(\mathbf{x},\varepsilon)$ solves the problem

$$\Delta\tilde{u}^{(0)}(\mathbf{x},\varepsilon) = 0 \quad \text{in } \Omega_\varepsilon, \qquad (1.4.15)$$
$$\tilde{u}^{(0)}(\mathbf{x},\varepsilon) = 0 \quad \text{on } \partial\Omega, \qquad (1.4.16)$$
$$\tilde{u}^{(0)}(\mathbf{x},\varepsilon) = -(v^{(0)}(\mathbf{x}) - v^{(0)}(0))$$

$$+ \alpha(\varepsilon)v^{(0)}(0)(\mathcal{G}(\mathbf{x}) - \mathcal{G}(0)) \quad \text{on } \partial g_\varepsilon. \tag{1.4.17}$$

Since the right-hand side in (1.4.17) is of order $O(\varepsilon)$, it follows from the maximum principle that

$$|\tilde{u}^{(0)}(\mathbf{x}, \varepsilon)| < \text{Const } \varepsilon,$$

where the constant is independent of ε.

1.4.2 The next approximation

To obtain the second-order approximation, we keep linear terms of the expansion of the right-hand side in (1.4.17), and construct the boundary layer $V^{(1)}$, which removes the discrepancy:

$$\Delta V^{(1)}(\mathbf{X}, \alpha(\varepsilon)) = 0, \quad \mathbf{X} \in \mathbb{R}^2 \setminus \bar{g}, \tag{1.4.18}$$

$$V^{(1)}(\mathbf{X}, \alpha(\varepsilon)) = \mathbf{X} \cdot \nabla v^{(0)}(0)$$
$$- \alpha(\varepsilon)v^{(0)}(0)\mathbf{X} \cdot \nabla \mathcal{G}(0), \quad \mathbf{X} \in \partial g. \tag{1.4.19}$$

Since the right-hand side in (1.4.19), being integrated along g, gives zero, the problem (1.4.18), (1.4.19) has a solution that decays at infinity, and it can be written in the form

$$V^{(1)}(\mathbf{X}, \alpha(\varepsilon)) = \frac{1}{\|\mathbf{X}\|} \left\{ \cos\theta \left(\frac{\partial v^{(0)}}{\partial x_1}(0) - \alpha(\varepsilon)v^{(0)}(0)\frac{\partial \mathcal{G}}{\partial x_1}(0) \right) \right.$$
$$\left. + \sin\theta \left(\frac{\partial v^{(0)}}{\partial x_2}(0) - \alpha(\varepsilon)v^{(0)}(0)\frac{\partial \mathcal{G}}{\partial x_2}(0) \right) \right\}. \tag{1.4.20}$$

THEOREM 1.3

The solution u of (1.4.1)–(1.4.3) admits the representation

$$u(\mathbf{x}, \varepsilon) = v^{(0)}(\mathbf{x}) + \alpha(\varepsilon)v^{(0)}(0)G(\mathbf{x})$$
$$+ \varepsilon V^{(1)}(\mathbf{X}, \alpha(\varepsilon)) + \tilde{u}^{(1)}(\mathbf{x}, \varepsilon). \tag{1.4.21}$$

The remainder term $\tilde{u}^{(1)}$ satisfies

$$|\tilde{u}^{(1)}(\mathbf{x}, \varepsilon)| < \text{Const } \varepsilon^2, \quad \mathbf{x} \in \Omega_\varepsilon, \tag{1.4.22}$$

where the coefficient is independent of ε.

Proof

The function $\tilde{u}^{(1)}(\mathbf{x}, \varepsilon)$ is harmonic in Ω_ε, and $\tilde{u}^{(1)}(\mathbf{x}, \varepsilon) = O(\varepsilon^2)$ on $\partial\Omega_\varepsilon$. Then, inequality (1.4.22) follows from the maximum principle. $\square$

1.4.3 The complete asymptotic expansion

General term of the series

We seek an asymptotic expansion of $u(\mathbf{x}, \varepsilon)$ in the form

$$u(\mathbf{x}, \varepsilon) \sim \sum_{k=0}^{\infty} \varepsilon^k \{v^{(k)}(\mathbf{x}, \alpha) + q^{(k)}(\alpha)G(\mathbf{x}) + V^{(k)}(\mathbf{X}, \alpha)\}, \quad (1.4.23)$$

where $\alpha = \alpha(\varepsilon)$ is given by (1.4.14), $v^{(k)}(\mathbf{x}, \alpha)$ is a polynomial in α of order $< k$ whose coefficients are harmonic functions in $\bar{\Omega}$, $q^{(k)}(\alpha)$ is a polynomial of order $\leq k$, and $V^{(k)}(\mathbf{X}, \alpha)$ is a polynomial of order $\leq k$ whose coefficients are harmonic in $\mathbb{R}^2 \setminus g$ and vanish at infinity.

In the previous two sections we constructed the coefficients in ε^0 and ε^1 of the expansion (1.4.23). Using the notation adopted in (1.4.23) we can write $v^{(0)}(\mathbf{x}, \alpha) = v^{(0)}(\mathbf{x})$, $q^{(0)}(\alpha) = v^{(0)}(0)\alpha$, $V^{(0)} = 0$, and $v^{(1)} = 0, q^{(1)} = 0$; the function $V^{(1)}$ is given by (1.4.20).

To obtain the boundary conditions for $V^{(k)}$ on ∂g, we expand $v^{(j)}$ and the smooth part $\mathcal{G}$ of Green's function in Taylor's series in the vicinity of the origin. In the scaled coordinates $\mathbf{X}$ these expansions have the form

$$v^{(j)}(\mathbf{x}, \alpha) = \sum_{\beta \geq 0} \varepsilon^{|\beta|} \frac{\mathbf{X}^\beta}{\beta!} \frac{\partial^\beta v^{(j)}}{\partial \mathbf{x}^\beta}(0, \alpha), \quad (1.4.24)$$

and

$$\mathcal{G}(\mathbf{x}) = \sum_{\beta \geq 0} \varepsilon^{|\beta|} \frac{\mathbf{X}^\beta}{\beta!} \frac{\partial^\beta \mathcal{G}}{\partial \mathbf{x}^\beta}(0), \quad (1.4.25)$$

where $\beta = (\beta_1, \beta_2)$ is a multi-index. Since $v^{(j)}$ and $\mathcal{G}$ are harmonic, they are real analytic and therefore the series (1.4.24) and (1.4.25) converge in a ball $\|\varepsilon\mathbf{X}\| \leq$ Const. Moreover, the sum of all terms of order l is a harmonic polynomial and hence

$$\sum_{|\beta|=l} \frac{\mathbf{X}^\beta}{\beta!} \frac{\partial^\beta v^{(j)}}{\partial \mathbf{x}^\beta}(0, \alpha) = \|\mathbf{X}\|^l \{p_j^{(l,1)}(\alpha) \cos(l\theta)$$

$$+ p_j^{(l,2)}(\alpha)\sin(l\theta)\}, \qquad (1.4.26)$$

$$\sum_{|\beta|=l} \frac{\mathbf{X}^\beta}{\beta!} \frac{\partial^\beta \mathcal{G}}{\partial \mathbf{x}^\beta}(0) = \|\mathbf{X}\|^l \{g^{(l,1)}\cos(l\theta) + g^{(l,2)}\sin(l\theta)\}, \quad (1.4.27)$$

where $p_j^{(l,k)}(\alpha)$ are polynomials in α of order $< j$, and $g^{(l,k)}$ are constants.

Boundary condition for $V^{(k)}$

The term $V^{(k)}$ compensates for the discrepancy of order $O(\varepsilon^k)$ in the boundary condition on ∂g produced by the expansions (1.4.24) and (1.4.25). Thus,

$$V^{(k)}(\mathbf{X}, \alpha) = -\sum_{s=0}^{k} \|\mathbf{X}\|^{k-s}\{p_s^{(k-s,1)}(\alpha)\cos(k-s)\theta$$

$$+ p_s^{(k-s,2)}(\alpha)\sin(k-s)\theta\} + q^{(k)}(\alpha)\alpha^{-1}$$

$$-\sum_{s=0}^{k-1} q^{(s)}(\alpha)\|\mathbf{X}\|^{k-s}\{g^{(k-s,1)}\cos(k-s)\theta + g^{(k-s,2)}\sin(k-s)\theta\}, \quad (1.4.28)$$

for $\|\mathbf{X}\| = 1$. The function $V^{(k)}(\mathbf{X}, \alpha)$, which is harmonic in $\mathbb{R}^2 \setminus g$, satisfies this boundary condition and is given by the right-hand of side (1.4.28) where $\|\mathbf{X}\|^{k-s}$ is replaced by $\|\mathbf{X}\|^{s-k}$. This field decays at infinity if and only if

$$q^{(k)}(\alpha) = \alpha p_k^{(0,1)}(\alpha). \qquad (1.4.29)$$

Boundary condition for $v^{(k)}$

To obtain the boundary condition for $v^{(k)}(\mathbf{x}, \alpha)$ on $\partial \Omega$ we should expand $V^{(j)}(\mathbf{X}, \alpha)$ as $\|\mathbf{X}\| \to \infty$. In our case this expansion has a finite number of terms, and in the coordinates $\mathbf{x}$ it is given by

$$V^{(j)}(\mathbf{X}, \alpha) = -\sum_{s=0}^{j-1} \varepsilon^{j-s}\|\mathbf{x}\|^{s-j}\Big\{(p_s^{(j-s,1)}(\alpha) + g^{(j-s,1)})\cos(j-s)\theta$$

$$+ (p_s^{(j-s,2)}(\alpha) + g^{(j-s,2)})\sin(j-s)\theta\Big\}. \qquad (1.4.30)$$

The term $v^{(k)}(\mathbf{x}, \alpha)$ compensates for the discrepancy of order $O(\varepsilon^k)$ in the boundary condition on $\partial \Omega$ produced by expansion (1.4.30). Thus,

$$v^{(k)}(\mathbf{x},\alpha) = \sum_{1 \leq s \leq k/2} \|\mathbf{x}\|^{-s}\{(p_{k-2s}^{(s,1)}(\alpha) + g^{(s,1)})\cos(s\theta)$$

$$+ (p_{k-2s}^{(s,2)}(\alpha) + g^{(s,2)})\sin(s\theta)\}, \qquad (1.4.31)$$

for $\mathbf{x} \in \partial\Omega$.

Construction of the asymptotic series

Now, we are able to construct all terms of the asymptotic expansion (1.4.23) using the induction in k. First, when $k = 0$ and $k = 1$ the terms $v^{(k)}$, $q^{(k)}$ and $V^{(k)}$ are constructed as in earlier subsections. Next, assume that we have all terms $v^{(k)}$, $q^{(k)}$ and $V^{(k)}$ for $k < j$. Then $v^{(j)}$ is defined as a harmonic function in Ω which satisfies the Dirichlet boundary value problem (1.4.31) where the index k is replaced by j. Since all the terms of the right-hand side are polynomials in α of order smaller than j with coefficients smooth on $\partial\Omega$, the function $v^{(j)}$ itself is a polynomial in α of order smaller than j with coefficients, which are smooth in $\bar{\Omega}$. The polynomial $q^{(j)}$ is defined by the formula (1.4.29) where the index k is replaced by j. Finally, the field $V^{(j)}$ is defined as a harmonic function in $\mathbb{R}^2 \setminus g$, which satisfies the boundary condition (1.4.28) (with k being replaced by j) and vanishes at infinity. Thus, we have described the recurrent procedure for all terms of the asymptotic expansion (1.4.23).

Remainder estimate

Let $u^{(N)}$ denote a finite sum

$$u^{(N)}(\mathbf{x},\varepsilon) = \sum_{k=0}^{N} \varepsilon^k \{v^{(k)}(\mathbf{x},\alpha(\varepsilon)) + q^{(k)}(\alpha(\varepsilon))G(\mathbf{x})$$

$$+ V^{(k)}(\mathbf{X},\alpha(\varepsilon))\}, \qquad (1.4.32)$$

and $\tilde{u}^{(N)} = u - u^{(N)}$. Here, $u^{(N)}$ is harmonic in Ω_ε. By (1.4.9), (1.4.24) and (1.4.25)

$$\tilde{u}^{(N)}\big|_{\partial g_\varepsilon} = -\sum_{k=0}^{N} \varepsilon^k \Bigg\{ \sum_{\beta \geq 0} \varepsilon^{|\beta|} \frac{\mathbf{X}^\beta}{\beta!} \frac{\partial^\beta v^{(j)}}{\partial x^\beta}(0,\alpha)$$

$$- q^{(k)}(\alpha)\Big(\frac{1}{2\pi}\log\varepsilon + \sum_{\beta \geq 0} \varepsilon^{|\beta|}\frac{\mathbf{X}^\beta}{\beta!}\frac{\partial^\beta \mathcal{G}}{\partial \mathbf{x}^\beta}(0)\Big) + V^{(k)}(\mathbf{X},\alpha)\Bigg\}.$$

Using (1.4.26)–(1.4.29) we arrive at

$$\tilde{u}^{(N)}\Big|_{g_\varepsilon} = -\sum_{k=0}^{N}\sum_{l=0}^{\infty}\varepsilon^{k+l}\{p_k^{(l,1)}(\alpha)\cos(l\theta)+p_k^{(l,2)}(\alpha)\sin(l\theta)\}$$

$$-q^{(k)}(\alpha)\sum_{k=0}^{N}\sum_{l=1}^{\infty}\varepsilon^{k+l}\{g^{(l,1)}\cos(l\theta)+g^{(l,2)}\sin(l\theta)\}$$

$$+\sum_{m=0}^{N}\varepsilon^m\Bigg(\sum_{k=0}^{m}\{p_k^{(m-k,1)}(\alpha)\cos((m-k)\theta)+p_k^{(m-k,2)}(\alpha)\sin((m-k)\theta)\}$$

$$+\sum_{k=0}^{m-1}q^{(k)}(\alpha)\{g^{(m-k,1)}\cos((m-k)\theta)+g^{(m-k,2)}\sin((m-k)\theta)\}\Bigg).$$

Replacing $k+l$ by m and changing the summation order in the first two sums in the right-hand side we obtain

$$\tilde{u}^{(N)}\Big|_{g_\varepsilon} = -\sum_{m=N+1}^{\infty}\varepsilon^m\sum_{k=0}^{N}\Bigg(\{p_k^{(m-k,1)}(\alpha)\cos((m-k)\theta)$$

$$+p_k^{(m-k,2)}(\alpha)\sin((m-k)\theta)\}$$

$$+q^{(k)}(\alpha)\{g^{(m-k,1)}\cos((m-k)\theta)+g^{(m-k,2)}\sin((m-k)\theta)\}\Bigg).$$

Furthermore, using the equations

$$G\big|_{\partial\Omega}=0,\quad V^{(0)}=0,$$

and

$$v^{(0)}(\mathbf{x},\varepsilon)+q^{(0)}(\alpha)G(\mathbf{x})=\varphi(\mathbf{x})\quad\text{on }\partial\Omega$$

we conclude that

$$\tilde{u}^{(N)}(\mathbf{x},\varepsilon)=-\sum_{k=1}^{N}\varepsilon^k\{v^{(k)}(\mathbf{x},\alpha)+V^{(k)}(\mathbf{X},\alpha)\}$$

on $\partial\Omega$. By (1.4.30) we rewrite the last equation as

$$\tilde{u}^{(N)}(\mathbf{x})=-\sum_{k=1}^{N}\varepsilon^k v^{(k)}(\mathbf{x},\alpha)$$

$$+ \sum_{j=1}^{N} \sum_{m=0}^{j-1} \varepsilon^{2j-m} \|\mathbf{x}\|^{m-j} \bigg(\{p_m^{(j-m,1)}(\alpha) + g^{(j-m,1)}\} \cos((j-m)\theta)$$

$$+ \{p_m^{(j-m,2)}(\alpha) + g^{(j-m,2)}\} \sin((j-m)\theta) \bigg) \quad \text{on } \partial\Omega.$$

Replacing $2j - m$ by k and $j - m$ by s simultaneously, and changing the summation order, we arrive at

$$\tilde{u}^{(N)}(\mathbf{x}, \varepsilon) = -\sum_{k=1}^{N} \varepsilon^k v^{(k)}(\mathbf{x}, \alpha)$$

$$+ \bigg(\sum_{k=1}^{N} \sum_{1 \leq s \leq k/2} + \sum_{k=N+1}^{2N} \sum_{k-N \leq s \leq k/2} \bigg) \varepsilon^k \|\mathbf{x}\|^{-s} \{(p_{k-2s}^{(s,1)}(\alpha) + g^{(s,1)}) \cos(s\theta)$$

$$+ (p_{k-2s}^{(s,2)}(\alpha) + g^{(s,2)}) \sin(s\theta)\}.$$

By (1.4.31),

$$\tilde{u}^{(N)}(\mathbf{x}, \varepsilon) = \sum_{k=N+1}^{2N} \sum_{k-N \leq s \leq k/2} \varepsilon^k \|\mathbf{x}\|^{-s} \{(p_{k-2s}^{(s,1)}(\alpha) + g^{(s,1)}) \cos(s\theta)$$

$$+ (p_{k-2s}^{(s,2)}(\alpha) + g^{(s,2)}) \sin(s\theta)\} \quad \text{on } \partial\Omega.$$

Hence, $\tilde{u}^{(N)} = O(\varepsilon^{N+1})$ on $\partial\Omega_\varepsilon$, and $\tilde{u}^{(N)}$ is harmonic in Ω_ε. Using the maximum principle we deduce that

$$|\tilde{u}^{(N)}| < \text{Const } \varepsilon^{N+1}, \quad \mathbf{x} \in \bar{\Omega}_\varepsilon. \qquad (1.4.33)$$

Thus, we arrive at the following statement:

THEOREM 1.4

The solution u of the boundary value problem (1.4.1)–(1.4.3) admits the representation

$$u = u^{(N)} + \tilde{u}^{(N)}, \quad N = 0, 1, \ldots,$$

where $u^{(N)}$ is given in the form (1.4.32) and the remainder $\tilde{u}^{(N)}$ satisifes (1.4.33).

1.5 Mixed boundary value problem for the Laplacian in a thin rectangle

In this section we discuss an example involving a mixed boundary value problem for the Laplacian posed in a thin rectangle:

$$\Pi_\varepsilon = \{(x_1, x_2) : 0 < x_1 < l, |x_2| < \varepsilon/2\},$$

where l is a given constant, and ε is a small parameter. We prescribe the Dirichlet and Neumann conditions on the vertical and horizontal parts of $\partial\Pi_\varepsilon$, respectively. Since the limit region (skeleton) is a one-dimensional segment, the boundary value problem discussed below should be regarded as a singularly perturbed problem. When constructing an asymptotic power series for the solution we shall deal with *three model problems* independent of the small parameter ε. These problems are:

(1) the Neumann boundary value problem for the second-order ordinary differential equation on the scaled cross-section of the rectangle;
(2) the Dirichlet boundary value problem for the second-order ordinary differential equation on $(0, l)$;
(3) the problem of the boundary layer type (in a semi-infinite rectangular strip) associated with the ends of the thin region.

1.5.1 Formulation of the boundary value problem

Let the function $u(\mathbf{x}, \varepsilon)$ satisfy the following boundary value problem:

$$\Delta u(\mathbf{x}, \varepsilon) = 0, \quad \mathbf{x} \in \Pi_\varepsilon, \qquad (1.5.1)$$

$$\left.\frac{\partial u}{\partial x_2}\right|_{x_2=\pm\varepsilon/2} = \pm\varepsilon p_\pm(x_1), \quad 0 < x_1 < l, \qquad (1.5.2)$$

$$u(0, x_2) = q_1, \quad u(l, x_2) = q_2, \quad |x_2| < \varepsilon/2, \qquad (1.5.3)$$

where $p_\pm$ are given smooth functions, and q_1, q_2 are constant quantities. This mixed boundary value problem is uniquely solvable.

In terms of the variables $(x_1, t) = (x_1, x_2/\varepsilon)$, problem (1.5.1)–(1.5.3) can be rewritten as

$$\left\{\varepsilon^2 \frac{\partial^2}{\partial x_1^2} + \frac{\partial^2}{\partial t^2}\right\} u(\mathbf{x}, \varepsilon) = 0, \quad 0 < x_1 < l, \ |t| < 1/2, \qquad (1.5.4)$$

$$\left.\frac{\partial u}{\partial t}\right|_{t=\pm 1/2} = \pm\varepsilon^2 p_\pm(x_1), \quad 0 < x_1 < l, \qquad (1.5.5)$$

MIXED BOUNDARY VALUE PROBLEM

$$u|_{x_1=0} = q_1, \quad u|_{x_1=l} = q_2, \quad |t| < 1/2. \tag{1.5.6}$$

1.5.2 Two-term approximation

Setting $\varepsilon = 0$ formally in (1.5.4) and (1.5.5) one could try to approximate the solution by a function $w_0(x_1)$ subject to

$$w_0(0) = q_1, \quad w_0(l) = q_2. \tag{1.5.7}$$

Since the discrepancies in (1.5.4) and (1.5.5) will be of order ε^2, it is natural to compensate them by adding a function $\varepsilon^2 V_0(x_1, t)$. Substituting

$$w_0(x_1) + \varepsilon^2 V(x_1, t)$$

into (1.5.4) and (1.5.5) we obtain the equation

$$\frac{\partial^2 V_0}{\partial t^2} = -\frac{d^2 w_0}{dx_1^2}, \quad |t| < 1/2, \tag{1.5.8}$$

and the boundary condition

$$\left.\frac{\partial V_0}{\partial t}\right|_{t=\pm 1/2} = \pm p_\pm, \quad 0 < x_1 < l. \tag{1.5.9}$$

The solvability criterion for (1.5.8), (1.5.9) is

$$\frac{d^2 w_0}{dx_1^2} = -(p_+(x_1) + p_-(x_1)), \quad x_1 \in (0, l). \tag{1.5.10}$$

Equations (1.5.10), (1.5.7) give the boundary value problem for w_0, which is uniquely solvable.

The function V_0 is specified as a general solution of (1.5.8), (1.5.9) and can be represented in the form

$$V_0(x_1, t) = W_0(x_1, t) + w_1(x_1), \tag{1.5.11}$$

where

$$\int_{-1/2}^{1/2} W_0(x_1, t) dt = 0 \tag{1.5.12}$$

and $w_1(x_1)$ is a function to be found in later steps of the asymptotic algorithm. The function W_0 solves the same problem as V_0 and is determined uniquely.

Now the field u is represented in the form

$$u(\mathbf{x},\varepsilon) = w_0(x_1) + \varepsilon^2 W_0(x_1,t) + \tilde{u}^{(0)}(\mathbf{x},\varepsilon), \qquad (1.5.13)$$

where $\tilde{u}^{(0)}$ solves the boundary value problem

$$-\Delta \tilde{u}^{(0)} = \varepsilon^2 \frac{\partial^2}{\partial x_1^2} W_0(x_1, x_2/\varepsilon), \quad 0 < x_1 < l, \ |x_2| < \varepsilon/2,$$

$$\left.\frac{\partial \tilde{u}^{(0)}}{\partial x_2}\right|_{x_2=\pm\varepsilon/2} = 0, \quad 0 < x_1 < l,$$

$$\tilde{u}^{(0)}(0,x_2,\varepsilon) = -\varepsilon^2 W_0(0,x_2/\varepsilon),$$

$$\tilde{u}^{(0)}(l,x_2,\varepsilon) = -\varepsilon^2 W_0(l,x_2/\varepsilon), \quad |x_2| < \varepsilon/2.$$

From the next lemma, the remainder $\tilde{u}^{(0)}$ admits the estimate

$$\varepsilon^{-2} \int_{\Pi_\varepsilon} (\tilde{u}^{(0)})^2 d\mathbf{x} + \int_{\Pi_\varepsilon} |\nabla \tilde{u}^{(0)}|^2 d\mathbf{x} \leq \text{Const } \varepsilon^4, \qquad (1.5.14)$$

where the constant is independent of ε.

LEMMA 1.3

Let $v \in W_2^1(\Pi_\varepsilon)$ be a solution to the problem

$$-\Delta v = \mathcal{F} \quad \text{on } \Pi_\varepsilon, \qquad (1.5.15)$$

$$\frac{\partial v}{\partial x_2}(x_1, \pm\varepsilon/2) = 0, \ 0 < x_1 < l, \qquad (1.5.16)$$

$$v(0,x_2) = h_0(x_2), \quad v(l,x_2) = h_1(x_2), \quad |x_2| < \varepsilon/2, \qquad (1.5.17)$$

where $\mathcal{F} \in L_2(\Pi_\varepsilon)$ and $h_k \in W_2^1(-\varepsilon/2, \varepsilon/2)$, $k = 0,1$. Furthermore, let

$$\int_{\varepsilon/2}^{\varepsilon/2} \mathcal{F}(x_1,\tau)d\tau = 0 \qquad (1.5.18)$$

for almost all $x_1 \in (0,l)$ and

$$\int_{-\varepsilon/2}^{\varepsilon/2} h_k(\tau)d\tau = 0, \quad k = 0,1. \qquad (1.5.19)$$

Then

$$\int_{\Pi_\varepsilon} (|\nabla v|^2 + \varepsilon^{-2} v^2) d\mathbf{x} \leq c_l \Big(\int_{\Pi_\varepsilon} \mathcal{F}^2 d\mathbf{x} + \varepsilon \sum_{k=0}^{1} \int_{-\varepsilon/2}^{\varepsilon/2} |h'_k(x_2)|^2 dx_2 \Big),$$

where c_l depends only on l.

Proof

Let
$$m(x_1) = \int_{-\varepsilon/2}^{\varepsilon/2} v(x_1, \tau) d\tau.$$

By (1.5.15)–(1.5.19)
$$m''(x_1) = 0$$

and
$$m(0) = m(l) = 0.$$

Hence,
$$\int_{-\varepsilon/2}^{\varepsilon/2} v(x_1, \tau) d\tau = 0$$

for almost all $x_1 \in (0, l)$.

Since $\pi^2 \varepsilon^{-2}$ is the first positive eigenvalue of the operator $-d^2/d\tau^2$ on $(-\varepsilon/2, \varepsilon/2)$ with zero Neumann boundary conditions at $\pm \varepsilon/2$, we have
$$\int_{\Pi_\varepsilon} v^2 d\mathbf{x} \leq \frac{\varepsilon^2}{\pi^2} \int_{\Pi_\varepsilon} |\nabla v|^2 d\mathbf{x}. \qquad (1.5.20)$$

Without loss of generality we assume that h_0 and h_1 are smooth on $[-\varepsilon/2, \varepsilon/2]$ and $h'_k(\pm \varepsilon/2) = 0$. (This restriction can be removed by approximation.)

Let $\eta = \eta(t)$ be a smooth function on $(0, \infty)$ equal to 1 for $t < 1$ and 0 for $t > 2$. We introduce the function
$$H_\varepsilon(\mathbf{x}) = \eta(x_1/\varepsilon) h_0(x_2) + \eta((l - x_1)/\varepsilon) h_1(x_2). \qquad (1.5.21)$$

Then $\tilde{v} = v - H_\varepsilon$ is subject to zero boundary conditions (1.5.16), (1.5.17) and satisfies
$$\int_{\Pi_\varepsilon} |\nabla \tilde{v}|^2 d\mathbf{x} = \int_{\Pi_\varepsilon} \mathcal{F} \tilde{v} d\mathbf{x} + \int_{\Pi_\varepsilon} \nabla H_\varepsilon \cdot \nabla \tilde{v} d\mathbf{x}.$$

Hence
$$\int_{\Pi_\varepsilon} |\nabla \tilde{v}|^2 d\mathbf{x} \leq 2 \|\mathcal{F}\|_{L_2(\Pi_\varepsilon)} \|\tilde{v}\|_{L_2(\Pi_\varepsilon)} + \int_{\Pi_\varepsilon} |\nabla H_\varepsilon|^2 d\mathbf{x}. \qquad (1.5.22)$$

By using $h_k \perp 1$ on $(-\varepsilon/2, \varepsilon/2)$ we verify the estimate

$$\int_{\Pi_\varepsilon} |\nabla H_\varepsilon|^2 d\mathbf{x} \leq \text{Const } \varepsilon \sum_{k=0}^{1} \int_{-\varepsilon/2}^{\varepsilon/2} |h'_k(x_2)|^2 dx_2. \quad (1.5.23)$$

This together with (1.5.22) and with (1.5.20) for $v = \tilde{v}$ leads to

$$\int_{\Pi_\varepsilon} |\nabla \tilde{v}|^2 d\mathbf{x} \leq \text{Const}(\varepsilon^2 \int_{\Pi_\varepsilon} \mathcal{F}^2 d\mathbf{x} + \varepsilon \sum_{k=0}^{1} \int_{-\varepsilon/2}^{\varepsilon/2} |h'_k(x_2)|^2 dx_2). \quad (1.5.24)$$

Combining the two last estimates and (1.5.20) we complete the proof. $\square$

1.5.3 The next approximation

Now, we construct the next group of terms of the asymptotic expansion. Looking for a two-term approximation of the remainder $\tilde{u}_0(\mathbf{x}, \varepsilon)$ we approximate the solution of (1.5.1)–(1.5.3) as

$$u(\mathbf{x}, \varepsilon) \simeq w_0(x_1) + \varepsilon^2 W_0(x_1, t) + \varepsilon^2 w_1(x_1) + \varepsilon^4 V_1(x_1, t). \quad (1.5.25)$$

Substituting this sum into (1.5.4), (1.5.5) we obtain the equations for V_1:

$$\frac{\partial^2 V_1}{\partial t^2}(x_1, t) = -\frac{d^2 w_1}{dx_1^2}(x_1) - \frac{\partial^2 W_0}{\partial x_1^2}(x_1, t), \quad |t| < 1/2, \quad (1.5.26)$$

and

$$\frac{\partial V_1}{\partial t}(x_1, \pm 1/2) = 0. \quad (1.5.27)$$

The boundary value problem (1.5.26), (1.5.27) is solvable if and only if

$$\frac{\partial^2 w_1}{\partial x_1^2} = -\int_{-1/2}^{1/2} \frac{\partial^2 W_0}{\partial x_1^2} dt = 0, \quad x_1 \in (0, l). \quad (1.5.28)$$

When $x_1 = 0$ and $x_1 = l$, the quantity (1.5.25) is equal to

$$q_1 + \varepsilon^2 (W_0(0, t) + w_1(0)) + O(\varepsilon^4),$$

and

$$q_2 + \varepsilon^2 (W_0(l, t) + w_1(l)) + O(\varepsilon^4),$$

respectively.

Thus, expansion (1.5.25) generates a discrepancy $O(\varepsilon^2)$ in the boundary conditions (1.5.6). Since W_0 depends on the transverse scaled variable t, it is impossible, in general, to remove the error by the choice of boundary conditions for w_1. Hence, we have to compensate for this discrepancy by adding the boundary layer terms $\varepsilon^2 \mathcal{V}_1$ and $\varepsilon^2 \mathcal{W}_1$ near the ends of Π_ε. Using the scaled coordinates

$$(\zeta_1, \zeta_2) = \varepsilon^{-1}(x_1, x_2), \quad (\xi_1, \xi_2) = \varepsilon^{-1}(x_1 - l, x_2),$$

near the ends $x_1 = 0$ and $x_1 = l$ we define $\mathcal{W}_1$ and $\mathcal{V}_1$ as solutions of the boundary value problems

$$\Delta_\zeta \mathcal{W}_1(\zeta) = 0 \text{ in } \{\zeta : |\zeta_2| < 1/2, \zeta_1 > 0\}, \quad (1.5.29)$$

$$\left.\frac{\partial \mathcal{W}_1}{\partial \zeta_2}\right|_{\zeta_2 = \pm 1/2} = 0, \ \zeta_1 > 0, \quad (1.5.30)$$

$$\mathcal{W}_1|_{\zeta_1 = 0} = -w_1(0) - W_0(0, \zeta_2), \quad (1.5.31)$$

$$\mathcal{W}_1(\zeta) \to 0 \text{ as } \zeta_1 \to +\infty, \quad (1.5.32)$$

and

$$\Delta_\xi \mathcal{V}_1(\xi) = 0 \text{ in } \{\xi : |\xi_2| < 1/2, \xi_1 < 0\}, \quad (1.5.33)$$

$$\left.\frac{\partial \mathcal{V}_1}{\partial \xi_2}\right|_{\xi_2 = \pm 1/2} = 0, \ \xi_1 < 0, \quad (1.5.34)$$

$$\mathcal{V}_1|_{\xi_1 = 0} = -w_1(l) - W_0(l, \xi_2), \quad (1.5.35)$$

$$\mathcal{V}_1(\xi) \to 0 \text{ as } \xi_1 \to -\infty. \quad (1.5.36)$$

The boundary value problems (1.5.29)–(1.5.32) and (1.5.33)–(1.5.36) have the solutions $\mathcal{W}_1$ and $\mathcal{V}_1$ that decay at infinity if and only if

$$w_1(0) = -\int_{-1/2}^{1/2} W_0(0, t)dt = 0, \quad (1.5.37)$$

and

$$w_1(l) = -\int_{-1/2}^{1/2} W_0(l, t)dt = 0. \quad (1.5.38)$$

By (1.5.28) we obtain that $w_1 = 0$. It follows from (1.5.12) and (1.5.37), (1.5.38) that V_1 is specified as a solution of (1.5.26), (1.5.27), and can be written as

$$V_1(x_1, t) = W_1(x_1, t) + w_2(x_1),$$

where
$$\int_{-1/2}^{1/2} W_1(x_1, t)dt = 0,$$

and w_2 is to be determined at the next step. Thus, we have obtained the asymptotic approximation for $u(x, \varepsilon)$:

$$u(\mathbf{x}, \varepsilon) = w_0(x_1) + \varepsilon^2 W_0(x_1, t)$$
$$+ \varepsilon^2 \{\varepsilon^2 W_1(x_1, t) + \mathcal{V}_1(\boldsymbol{\xi}) + \mathcal{W}_1(\boldsymbol{\zeta})\} + \tilde{u}^{(1)}(\mathbf{x}, \varepsilon). \quad (1.5.39)$$

The remainder $\tilde{u}^{(1)}(\mathbf{x}, \varepsilon)$ leaves discrepancies in (1.5.1) and (1.5.3) of order ε^4, and satisfies the homogeneous Neumann boundary condition on the horizontal parts of $\partial \Pi_\varepsilon$.

1.5.4 Higher-order approximation

Suppose that the approximation

$$u^{(N)}(\mathbf{x}, \varepsilon) = w_0(x_1) + \varepsilon^2 W_0(x_1, t)$$
$$+ \sum_{k=1}^{N-1} \varepsilon^{2k} \{\varepsilon^2 W_k(x_1, t) + \mathcal{V}_k(\boldsymbol{\xi}) + \mathcal{W}_k(\boldsymbol{\zeta})\} \quad (1.5.40)$$

has been constructed, where W_k are solutions of a Neumann problem on the interval $|t| \leq 1/2$, orthogonal to 1 in $L_2(-1/2, 1/2)$, $\mathcal{V}_k$ and $\mathcal{W}_k$ are boundary layer terms corresponding to the right and left sides of the rectangle, and exponentially decaying at infinity.

We assume that (1.5.40) gives rise to the discrepancies $O(\varepsilon^{2N})$ in equation (1.5.4) and in the Dirichlet conditions (1.5.6) and satisfies the Neumann condition (1.5.5).

To find the next-order approximation, we add

$$\varepsilon^{2N} \{\varepsilon^2 \mathcal{V}_N(x_1, t) + \mathcal{V}_N(\boldsymbol{\xi}) + \mathcal{W}_N(\boldsymbol{\zeta})\} \quad (1.5.41)$$

to the sum (1.5.40). Substituting the new approximation into the problem (1.5.4)–(1.5.6) we obtain

$$\frac{\partial^2 \mathcal{V}_N}{\partial t^2}(x_1, t) = -\frac{\partial^2 \mathcal{W}_{N-1}}{\partial x_1^2}(x_1, t), \quad |t| < 1/2, \quad (1.5.42)$$

$$\frac{\partial \mathcal{V}_N}{\partial t}(x_1, \pm 1/2) = 0. \quad (1.5.43)$$

Moreover, the functions $\mathcal{V}_N$ and $\mathcal{W}_N$ prove to be harmonic in the corresponding semi-strips and subject to the homogeneous Neumann condition on the horizontal parts of the boundaries.

MIXED BOUNDARY VALUE PROBLEM 37

Problem (1.5.42), (1.5.43) is solvable owing to the fact that W_{N-1} is orthogonal to 1; we use the notation W_N for its solution subject to the orthogonality condition

$$\int_{-1/2}^{1/2} W_N(x_1,t)dt = 0, \quad 0 \le x_1 \le l. \qquad (1.5.44)$$

Inserting the sum of (1.5.40) and (1.5.41) into the boundary conditions (1.5.6) we find

$$\mathcal{V}_N(0,\xi_2) = -W_{N-1}(l,\xi_2)$$

and

$$\mathcal{W}_N(0,\zeta_2) = -W_{N-1}(0,\zeta_2).$$

Since the right-hand sides are orthogonal to 1 in $L_2(-1/2, 1/2)$, the functions $\mathcal{V}_N$ and $\mathcal{W}_N$ exponentially decaying at infinity can be uniquely determined. Hence, the next-order approximation has been constructed. When justifying this asymptotic procedure we need the orthogonality relations

$$\int_{-1/2}^{1/2} \mathcal{V}_k(\xi)d\xi_2 = 0, \quad \int_{-1/2}^{1/2} \mathcal{W}_k(\zeta)d\zeta_2 = 0 \qquad (1.5.45)$$

to hold for all $\xi_1 < 0$ and $\zeta_1 > 0$, respectively. They result from the identities

$$\frac{d}{d\xi_1} \int_{-1/2}^{1/2} \mathcal{V}_k(\xi)d\xi_2 = 0,$$

and

$$\frac{d}{d\zeta_1} \int_{-1/2}^{1/2} \mathcal{W}_k(\xi)d\zeta_2 = 0,$$

which follow from Green's formula.

Let us represent the solution of problem (1.5.1)–(1.5.3) in the form

$$u(\mathbf{x},\varepsilon) = w_0(x_1) + \varepsilon^2 W_0(x_1,t)$$

$$+ \sum_{k=1}^{N} \varepsilon^{2k}\{\varepsilon^2 W_k(x_1,t) + \mathcal{V}_k(\xi) + \mathcal{W}_k(\zeta)\} + \tilde{u}^{(N)}(\mathbf{x},\varepsilon).$$

The remainder $\tilde{u}^{(N)}$ satisfies

$$-\Delta \tilde{u}^{(N)} = \mathcal{F}^{(N)} \quad \text{on} \quad \Pi_\varepsilon, \qquad (1.5.46)$$

$$\tilde{u}^{(N)}(0, x_2, \varepsilon) = h_0^{(N)}, \quad \tilde{u}^{(N)}(l, x_2, \varepsilon) = h_1^{(N)}, \qquad (1.5.47)$$

and its normal derivative on the upper and lower sides of Π_ε is zero. Here

$$\mathcal{F}^{(N)}(\mathbf{x}, \varepsilon) = \varepsilon^{2N+2} \frac{\partial^2}{\partial x_1^2} W_N\left(x_1, \frac{x_2}{\varepsilon}\right),$$

$$h_0^{(N)}(x_2, \varepsilon) = -\varepsilon^{2N+2} W_N\left(0, \frac{x_2}{\varepsilon}\right) - \sum_{k=1}^{N} \varepsilon^{2k} \mathcal{V}_k\left(-\frac{l}{\varepsilon}, \frac{x_2}{\varepsilon}\right),$$

$$h_1^{(N)}(x_2, \varepsilon) = -\varepsilon^{2N+2} W_N\left(l, \frac{x_2}{\varepsilon}\right) - \sum_{k=1}^{N} \varepsilon^{2k} \mathcal{W}_k\left(\frac{l}{\varepsilon}, \frac{x_2}{\varepsilon}\right).$$

By (1.5.44) and (1.5.45), the right-hand sides in (1.5.46) and (1.5.47) are subject to the orthogonality conditions (1.5.18) and (1.5.19) and

$$\mathcal{F}^{(N)} = O(\varepsilon^{2N+2}) \text{ on } \overline{\Pi}_\varepsilon,$$

$$\frac{dh_k^{(N)}}{dx_2} = O(\varepsilon^{2N+1}) \text{ on } [-\varepsilon/2, \varepsilon/2], \, k = 0, 1.$$

Therefore, Lemma 1.3 implies

$$\int_{\Pi_\varepsilon} (|\nabla \tilde{u}^{(N)}|^2 + \varepsilon^{-2}(\tilde{u}^{(N)})^2) d\mathbf{x} = O(\varepsilon^{4(N+1)}),$$

which justifies the validity of the asymptotic expansion

$$u(\mathbf{x}, \varepsilon) \sim w_0(x_1) + \varepsilon^2 W_0(x_1, t)$$

$$+ \sum_{k=1}^{\infty} \varepsilon^{2k} \{W_k(x_1, t) + \mathcal{V}_k(\boldsymbol{\xi}) + \mathcal{W}_k(\boldsymbol{\zeta})\} \qquad (1.5.48)$$

for the solution of (1.5.1)–(1.5.3).

1.6 Problem of junction between thin bodies

Consider a domain $\Omega(\varepsilon)$ shown in Fig. 1.6 and consisting of four thin rectangles $\Pi_\varepsilon^{(i)}$ connected at a *junction region*. The normalised thickness of thin rectangles is characterised by a small, positive, non-dimensional parameter ε. We formulate the steady-state heat transfer problem and assume that internal sources of heat are absent, the zero heat flux is

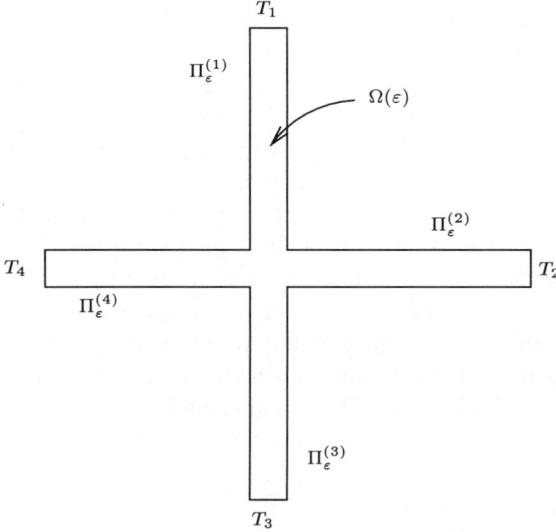

FIG. 1.6. A junction including four thin rectangles.

prescribed on the lateral surface of thin rectangles $\partial\Omega_{\mathcal{N}}(\varepsilon)$ and the temperature is given at free ends $\partial\Omega_{\mathcal{D}}(\varepsilon) = \cup_{i=1}^{4}\partial\Omega_{\mathcal{D}}^{(i)}(\varepsilon)$ of thin domains $\Pi_{\varepsilon}^{(i)}$, $i = 1, 2, 3, 4$. The subscript indices $\mathcal{N}$ and $\mathcal{D}$ stand for the Neumann and Dirichlet boundary conditions.

The temperature $u(x_1, x_2, \varepsilon)$ satisfies the mixed boundary value problem for the Laplacian

$$\Delta u = 0 \text{ in } \Omega(\varepsilon), \tag{1.6.1}$$

$$\frac{\partial u}{\partial n} = 0 \text{ on } \partial\Omega_{\mathcal{N}}(\varepsilon), \tag{1.6.2}$$

$$u = T_i \text{ on } \partial\Omega_{\mathcal{D}}^{(i)}(\varepsilon). \tag{1.6.3}$$

Here, the quantities T_i are assumed to be constant. The problem interpreted in the variational form is uniquely solvable in the class of functions with the finite Dirichlet integral evaluated over $\Omega(\varepsilon)$.

Similar to the previous sections, the solution u can be represented as a multi-scaled asymptotic series. To find the coefficients of the asymptotic expansion one has to solve model problems in domains independent of ε. In our particular case of the right-hand sides these model problems are:

(1) The Neumann boundary value problem in an infinite region Ξ with four extensions to infinity (the junction layer problem). This domain is shown in Fig. 1.7.

(2) Model problems for a second-order ordinary differential equation on a segment.
(3) A system of linear algebraic equations (formulated for the limit structure) with respect to the temperature at the junction point and the values of the heat flux in thin rods. For a general case, in (1.6.1)–(1.6.3) we also have to specify the Neumann data on the lateral surface of thin rectangles, and the Dirichlet data at the end regions may also depend on the scaled variables. In this situation, two additional model problems will be required:
  - a model problem on a scaled cross-section of a thin rectangle;
  - a mixed boundary value problem in a semi-infinite strip, where the Neumann data are given on the lateral surface and the Dirichlet condition is specified at the end region.

1.6.1 Model problems

Junction solution

Let Ξ be an unbounded region with four extensions at infinity (like the one shown in Fig. 1.7).

Consider the Neumann problem

$$-\Delta_X W(\mathbf{X}) = \mathcal{F}(\mathbf{X}), \quad \mathbf{X} \in \Xi, \quad (1.6.4)$$

$$\frac{\partial W}{\partial n}(\mathbf{X}) = 0, \quad \mathbf{X} \in \partial\Xi, \quad (1.6.5)$$

where $\mathcal{F}$ is a function from $L_2(\Xi)$ with compact support in Ξ.

Also, assume that

$$\int_\Xi \mathcal{F}(\mathbf{X})d\mathbf{X} = 0. \quad (1.6.6)$$

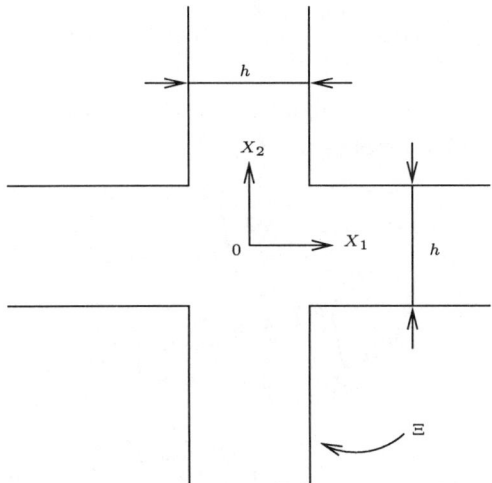

FIG. 1.7. The junction layer region.

Space $E(\Xi)$

We seek a solution of (1.6.4)–(1.6.6) which belongs to the space $E(\Xi)$ of functions W with finite norm

$$\|W\|_{E(\Xi)} = \left(\int_\Xi |\nabla W|^2 d\mathbf{X} + \int_{\Xi \cap D_h} W^2 d\mathbf{X} \right)^{1/2},$$

where $D_h = \{\mathbf{X} : |\mathbf{X}| < h\}$.

LEMMA 1.4

For any function W from $E(\Xi)$ the following estimate holds:

$$\sup_{|X_1| > \frac{h}{2}} |X_1|^{-1} \int_{-h/2}^{h/2} W^2(X_1, X_2) dX_2 \qquad (1.6.7)$$
$$+ \sup_{|X_2| > \frac{h}{2}} |X_2|^{-1} \int_{-h/2}^{h/2} W^2(X_1, X_2) dX_1 \leq \text{Const } \|W\|_{E(\Xi)}^2.$$

Proof

Let $|\tau| < h/2$ and $X_1 > h/2$. By the Newton–Leibniz formula

$$W(X_1, X_2) = \int_\tau^{X_1} \frac{\partial}{\partial s} W(s, X_2) ds + W(\tau, X_2).$$

Taking the square of the left-hand and the right-hand sides and using the Cauchy inequality we obtain

$$\int_{-h/2}^{h/2} W^2(X_1, X_2) dX_2 \leq \text{Const} \left(X_1^2 \int_{-h/2}^{X_1} \int_{-h/2}^{h/2} \left| \frac{\partial}{\partial s} W(s, X_2) \right|^2 ds dX_2 \right.$$

$$\left. + \inf_{|\tau| < \frac{h}{2}} \int_{-h/2}^{h/2} W^2(\tau, X_2) dX_2 \right).$$

Since

$$\inf_{|\tau| < \frac{h}{2}} \int_{-h/2}^{h/2} W^2(\tau, X_2) dX_2 \leq \frac{1}{h} \int_{-h/2}^{h/2} \int_{-h/2}^{h/2} W^2(X_1, X_2) dX_1 dX_2$$

we obtain

$$X_1^{-1} \int_{-h/2}^{h/2} W^2(X_1, X_2) dX_2 \leq \text{Const} \, \|W\|_{E(\Xi)}^2.$$

The remaining estimates in (1.6.7) are proved in a similar way. □

Let $E_0(\Xi)$ denote the subspace of functions $W \in E(\Xi)$ such that

$$\int_{\Xi \cap D_h} W d\mathbf{X} = 0.$$

For such functions one can apply the Poincaré inequality and obtain the estimate

$$\int_{\Xi \cap D_h} W^2 d\mathbf{X} \leq \text{Const} \int_{\Xi \cap D_h} |\nabla W|^2 d\mathbf{X}.$$

Hence,

$$\|W\|_{E(\Xi)} \leq \text{Const} \left(\int_\Xi |\nabla W|^2 d\mathbf{X} \right)^{1/2}$$

for any $W \in E_0(\Xi)$. It is easy to see that any constant belongs to $E(\Xi)$. The space $E_0(\Xi)$ is the orthogonal complement to the subspace of constants in $E(\Xi)$.

Solvability and asymptotics at infinity

By a solution of the problem (1.6.4)–(1.6.6) we understand a function $W \in E(\Xi)$ such that

$$\int_\Xi \nabla W \cdot \nabla V \, d\mathbf{X} = \int_\Xi \mathcal{F} V \, d\mathbf{X} \qquad (1.6.8)$$

for all functions $V \in E(\Xi)$.

We use the notation $\Pi^{(j)}$, $j = 1, 2, 3, 4$, for the unbounded extensions at infinity for the domain Ξ (see Fig. 1.7), i.e.

$$\Pi^{(1)} = \{\mathbf{X} : |X_1| < h/2, X_2 > h/2\},$$

$$\Pi^{(2)} = \{\mathbf{X} : |X_2| < h/2, X_1 > h/2\},$$

$$\Pi^{(3)} = \{\mathbf{X} : |X_1| < h/2, X_2 < -h/2\},$$

$$\Pi^{(4)} = \{\mathbf{X} : |X_2| < h/2, X_1 < -h/2\}.$$

Also, we need the local Cartesian coordinates $\mathbf{X}^{(j)} = (X_1^{(j)}, X_2^{(j)})$ for the semi-infinite strip $\Pi^{(j)}$ introduced in such a way that the $OX_1^{(j)}$ axis is directed along $\Pi^{(j)}$ and $X_1^{(j)} = 0$ at the base of $\Pi^{(j)}$; within $\Pi^{(j)}$ we have $|X_2^{(j)}| < h/2$ and $X_2^{(j)} = 0$ on the axis of $\Pi^{(j)}$. In the new coordinates

$$\Pi^{(j)} = \{\mathbf{X}^{(j)} : X_1^{(j)} > 0, |X_2^{(j)}| < h/2\}.$$

THEOREM 1.5

Let $\mathcal{F} \in L_2(\Xi)$ have a bounded support, and assume that (1.6.6) holds. There exists a solution W of the problem (1.6.4)–(1.6.6) from the space $E_0(\Xi)$. This solution is unique. It has the following asymptotic representation at infinity:

$$W(\mathbf{X}) = q^{(j)} + O\bigl(e^{-\pi h^{-1} X_1^{(j)}}\bigr), \quad \text{as } X_1^{(j)} \longrightarrow \infty, \qquad (1.6.9)$$

where $q^{(j)}$, $j = 1, 2, 3, 4$, are constants.

Proof

Solvability. Consider the functional

$$V \longrightarrow \int_\Xi \mathcal{F} V \, d\mathbf{X}.$$

By Lemma 1.4 it is continuous in $E_0(\Xi)$, and, due to the Riesz theorem, there exists a unique function $W \in E_0(\Xi)$ such that (1.6.8) holds for any $V \in E_0(\Xi)$. That equality is also satisfied for any function $V \in E(\Xi)$ due to (1.6.6). Thus, W is a solution of problem (1.6.4)–(1.6.6).

Asymptotics at infinity. Let $W \in E(\Xi)$ satisfy the boundary value problem (1.6.4)–(1.6.6). Since the right-hand side $\mathcal{F}$ has a compact support, the function W satisfies the homogeneous boundary value problem

$$\Delta_{\mathbf{X}^{(j)}} W = 0 \quad \text{in } \Pi^{(j)},$$
$$\frac{\partial W}{\partial n} = 0 \quad \text{on } \partial \Pi^{(j)},$$

when $X_1 > R$ for sufficiently large R.

Using the Fourier method and the inclusion $W \in E(\Xi)$ we obtain (1.6.9). □

Junction layer

Let $\chi(\tau)$ denote a smooth function which vanishes for $\tau < h/2$ and equals one for $\tau > h$.

Let $W \in E_0(\Xi)$ satisfy problem (1.6.4), (1.6.5), and let $q^{(j)}$ be the constants in (1.6.9). Then the function

$$\mathcal{W} = W - \sum_{j=1}^{4} q^{(j)} \chi(X_1^{(j)}) \tag{1.6.10}$$

vanishes exponentially at infinity. The function $\mathcal{W}$ satisfies the boundary value problem

$$-\Delta \mathcal{W}(\mathbf{X}) = \mathcal{F}(\mathbf{X}) + \sum_{j=1}^{4} q^{(j)} \chi''(X_1^{(j)}) \quad \text{in } \Xi, \tag{1.6.11}$$

$$\frac{\partial \mathcal{W}}{\partial n} = 0 \quad \text{on } \partial \Xi. \tag{1.6.12}$$

Since the sum in (1.6.10) vanishes on $D_h = \{\mathbf{X} : |\mathbf{X}| < h\}$, the function $\mathcal{W}$ belongs to $E_0(\Xi)$.

PROBLEM OF JUNCTION BETWEEN THIN BODIES   45

Equations (1.6.11), (1.6.12) describe the problem with respect to an unknown function $\mathcal{W}$, exponentially vanishing at infinity, and four constants $q^{(j)}$, $j = 1, 2, 3, 4$. The function $\mathcal{W}$ will be called the *junction layer*.

THEOREM 1.6

Let $\mathcal{F} \in L_2(\Xi)$ have a bounded support, and assume that (1.6.6) holds. Then there exists a unique set

$$\{\mathcal{W}, q^{(1)}, q^{(2)}, q^{(3)}, q^{(4)}\} \in E_0(\Xi) \times \mathbb{R}^4 \qquad (1.6.13)$$

such that equations (1.6.11), (1.6.12) are satisfied and

$$\mathcal{W}(\mathbf{X}) = O(e^{-\pi h^{-1} X_1^{(j)}}) \quad \text{as} \quad X_1^{(j)} \to \infty.$$

Proof

The existence of the solution follows from Theorem 1.5 and the structure of the right-hand sides in (1.6.11), (1.6.12).

To prove the uniqueness, we consider the case when $\mathcal{F} = 0$. Let $q^{(1)}, q^{(2)}, q^{(3)}, q^{(4)}$ be arbitrary constants. By Theorem 1.5, there exists the unique solution $W \in E_0(\Xi)$ of (1.6.11), (1.6.12), and

$$W = Q^{(j)} + O(e^{-\pi h^{-1} X_1^{(j)}}) \quad \text{as} \quad X_1^{(j)} \to \infty.$$

Thus, we have specified a linear map

$$(q^{(1)}, q^{(2)}, q^{(3)}, q^{(4)}) \to (Q^{(1)}, Q^{(2)}, Q^{(3)}, Q^{(4)}). \qquad (1.6.14)$$

We verify that the kernel of this map is trivial. To show this we check that the map is surjective. This follows from the fact that for any set $(Q^{(1)}, Q^{(2)}, Q^{(3)}, Q^{(4)})$ there exists the solution

$$W = -\sum_{j=1}^{4} Q^{(j)} \chi(X_1^{(j)}) \in E_0(\Xi)$$

of the problem (1.6.11), (1.6.12) where $\mathcal{F} = 0$ and $q^{(j)} = Q^{(j)}$.

Let (1.6.13) satisfy (1.6.11), (1.6.12) with $\mathcal{F} = 0$. Then $Q^{(j)} = 0$, $j = 1, 2, 3, 4$, and, hence, $q^{(j)} = 0$, $j = 1, 2, 3, 4$. By Theorem 1.5, $\mathcal{W} = 0$.

□

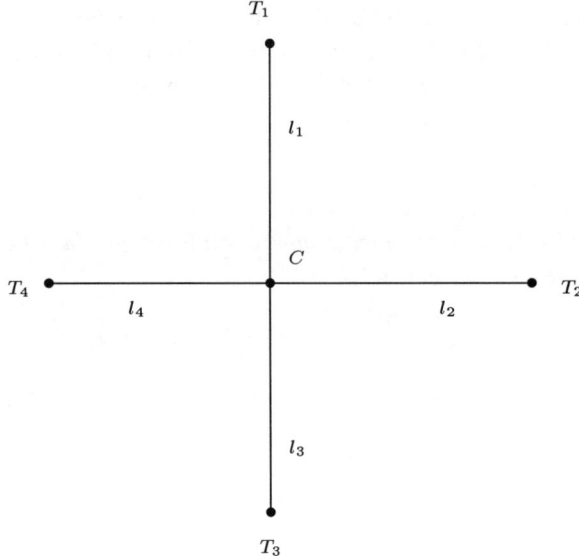

FIG. 1.8. The system of one-dimensional rods: the skeleton of multi-structure.

Algebraic system

To construct the asymptotic expansion of the solution of (1.6.4)–(1.6.6) we need to solve a certain system of linear algebraic equations. Here we give a simplified model which gives rise to this system.

As $\varepsilon \to 0$, the set $\Omega(\varepsilon)$ shrinks to a union of four one-dimensional segments (see Fig. 1.8).

Let C be an unknown temperature at the junction point, and let $B^{(i)}$ be an unknown normalised heat flux at the junction point along the i-th segment. The given temperature values at the ends of the segments are denoted by T_1, T_2, T_3, T_4. Assuming a linear distribution of temperature along the segments we obtain

$$B^{(i)} l_i + C = T_i, \quad i = 1, 2, 3, 4. \tag{1.6.15}$$

The balance relation for the heat fluxes at the junction point has the form

$$\sum_{i=1}^{4} B^{(i)} = 0. \tag{1.6.16}$$

Solving the system (1.6.15)–(1.6.16) we obtain

$$C = \left(\sum_{i=1}^{4} l_i^{-1}\right)^{-1} \sum_{i=1}^{4} l_i^{-1} T_i \quad (1.6.17)$$

$$B^{(i)} = l_i^{-1}(T_i - C). \quad (1.6.18)$$

1.6.2 The leading order approximation

The local system of coordinates $Ox_1^{(i)} x_2^{(i)}$ is introduced for each thin rectangle $\Pi_\varepsilon^{(i)}$ (see Fig. 1.9). Then

$$\Pi_\varepsilon^{(i)} = \{(x_1^{(i)}, x_2^{(i)}) : 0 < x_1^{(i)} < l_i, |x_2^{(i)}| < \varepsilon h/2\}.$$

Clearly, $(X_1^{(i)}, X_2^{(i)}) = \varepsilon^{-1}(x_1^{(i)}, x_2^{(i)})$, where $X_1^{(i)}, X_2^{(i)}$ were introduced in the previous section.

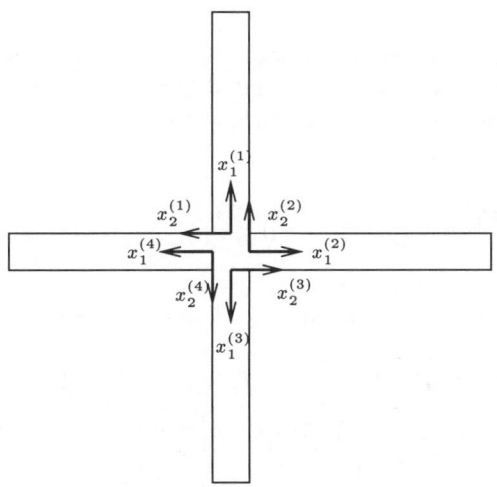

FIG. 1.9. Local coordinate systems.

Also, let $\mathbf{X} = (X_1, X_2) = \varepsilon^{-1}(x_1, x_2)$.

As before, $\chi(\tau)$ denotes a smooth function, which vanishes for $\tau < h/2$ and which is equal to one for $\tau > h$.

We seek the leading term $\mathcal{T}_0$ in the form

$$\mathcal{T}_0(\mathbf{x}, \varepsilon) = C_0 + \sum_{i=1}^{4} w_0^{(i)}(x_1^{(i)}) \chi(x_1^{(i)}/\varepsilon) + \mathcal{W}_0(\mathbf{X}), \quad (1.6.19)$$

where $\mathcal{W}_0$ is the junction layer. Substituting (1.6.19) in (1.6.1)–(1.6.3) we obtain

$$-\Delta T_0 = -\varepsilon^{-2}\Delta_{\mathbf{X}}\mathcal{W}_0(\mathbf{X}) - \sum_{i=1}^{4}\left\{\varepsilon^{-2}\chi''(X_1^{(i)})w_0^{(i)}(x_1^{(i)})\right. \quad (1.6.20)$$

$$\left. + 2\varepsilon^{-1}\chi'(X_1^{(i)})\frac{dw_0^{(i)}}{dx_1^{(i)}}(x_1^{(i)}) + \chi(X_1^{(i)})\frac{d^2w_0^{(i)}}{dx_1^{(i)^2}}(x_1^{(i)})\right\} \quad \text{in } \Omega(\varepsilon),$$

$$\frac{\partial T_0}{\partial n} = \varepsilon^{-1}\frac{\partial}{\partial n_{\mathbf{X}}}\mathcal{W}_0(\mathbf{X}) \quad \text{on } \partial\Omega_\mathcal{N}(\varepsilon), \quad (1.6.21)$$

$$T_0 = T_0(\mathbf{x},\varepsilon) \quad \text{on } \partial\Omega_D^{(i)}(\varepsilon). \quad (1.6.22)$$

Choosing $\mathcal{W}_0 = 0$ and

$$w_0^{(i)}(0) = 0 \quad (1.6.23)$$

we remove the leading order discrepancy (of order $O(\varepsilon^{-2})$) in (1.6.20) and obtain the homogeneous Neumann condition (1.6.21) for the function T_0.

The discrepancies of order $O(\varepsilon^{-1})$, $O(\varepsilon^0)$, etc., associated with the first two terms of the sum (1.6.20) and concentrated in the vicinity of the junction point will be compensated by the junction solutions in the next steps of the asymptotic algorithm. To compensate for the last term in (1.6.20) we require

$$\frac{d^2w_0^{(i)}}{dx_1^{(i)^2}}(x_1^{(i)}) = 0, \quad 0 < x_1^{(i)} < l_i, \quad i = 1,2,3,4. \quad (1.6.24)$$

With the above choice of $\mathcal{W}_0$ the condition (1.6.22) implies (1.6.3) provided

$$w_0^{(i)}(l_i) + C_0 = T_i, \quad i = 1,2,3,4. \quad (1.6.25)$$

It follows from (1.6.23)–(1.6.25) that $w_0^{(i)}$ is linear, and it is given by

$$w_0^{(i)}(x_1^{(i)}) = \frac{T_i - C_0}{l_i}x_1^{(i)}. \quad (1.6.26)$$

The constant C_0 will be specified in the next step of the asymptotic algorithm.

The relation (1.6.20) finally has the form

$$\Delta T_0 = \varepsilon^{-1}\sum_{i=1}^{4}\frac{dw_0^{(i)}}{dx_1^{(i)}}(0)(\chi''(X_1^{(i)})X_1^{(i)} + 2\chi'(X_1^{(i)})). \quad (1.6.27)$$

1.6.3 The next-order approximation

Now we shall approximate the solution by

$$\mathcal{T}_0 + \varepsilon \mathcal{T}_1, \qquad (1.6.28)$$

where

$$\mathcal{T}_1(\mathbf{x}, \varepsilon) = C_1 + \mathcal{W}_1(\mathbf{X}) + \sum_{i=1}^{4} w_1^{(i)}(x_1^{(i)}) \chi(x_1^{(i)}/\varepsilon). \qquad (1.6.29)$$

Substituting (1.6.28) in (1.6.1)–(1.6.3) we obtain

$$-\Delta(\mathcal{T}_0 + \varepsilon \mathcal{T}_1) = -\Delta \mathcal{T}_0 - \varepsilon^{-1} \Delta_{\mathbf{X}} \mathcal{W}_1(\mathbf{X})$$

$$- \sum_{i=1}^{4} \left\{ \varepsilon^{-1} \chi''(X_1^{(i)}) w_1^{(i)}(x_1^{(i)}) + 2\chi'(X_1^{(i)}) \frac{dw_1^{(i)}}{dx_1^{(i)}}(x_1^{(i)}) \right.$$

$$\left. + \varepsilon \chi(X_1^{(i)}) \frac{d^2 w_1^{(i)}}{dx_1^{(i)^2}}(x_1^{(i)}) \right\} \quad \text{in } \Omega(\varepsilon), \qquad (1.6.30)$$

$$\frac{\partial}{\partial n}(\mathcal{T}_0 + \varepsilon \mathcal{T}_1) = \frac{\partial}{\partial n_{\mathbf{X}}} \mathcal{W}_1(\mathbf{X}) \quad \text{on } \partial \Omega_{\mathcal{N}}(\varepsilon), \qquad (1.6.31)$$

$$\mathcal{T}_0 + \varepsilon \mathcal{T}_1 = T_i + \varepsilon \mathcal{T}_1 \quad \text{on } \partial \Omega_{\mathcal{D}}^{(i)}(\varepsilon). \qquad (1.6.32)$$

To compensate for the discrepancies of order $O(\varepsilon^{-1})$ in the right-hand side of (1.6.30) and keeping in mind formula (1.6.27) we set

$$-\Delta_{\mathbf{X}} \mathcal{W}_1(\mathbf{X}) = \sum_{i=1}^{4} \frac{dw_0^{(i)}}{dx_1^{(i)}}(0)(\chi''(X_1^{(i)}) X_1^{(i)} + 2\chi'(X_1^{(i)})) \quad (1.6.33)$$

$$+ \sum_{i=1}^{4} q_1^{(i)} \chi''(X_1^{(i)}), \quad \text{in } \Xi$$

where

$$w_1^{(i)}(0) = q_1^{(i)}, \qquad (1.6.34)$$

and $q_1^{(i)}$ are unknown constants. Comparing (1.6.31) with (1.6.2) we obtain the boundary condition for $\mathcal{W}_1$:

$$\frac{\partial \mathcal{W}_1}{\partial n_{\mathbf{X}}} = 0 \quad \text{on } \partial \Xi. \qquad (1.6.35)$$

By Theorem 1.6, problem (1.6.33), (1.6.35) has a unique solution $(\mathcal{W}_1, q_1^{(1)}, q_1^{(2)}, q_1^{(3)}, q_1^{(4)})$, where $\mathcal{W}_1$ decays exponentially at infinity, if and only if

$$\sum_{i=1}^{4} \frac{dw_0^{(i)}}{dx_1^{(i)}}(0) = 0. \qquad (1.6.36)$$

Here, we used

$$\int_\Xi (\chi''(X_1^{(i)})X_1^{(i)} + 2\chi'(X_1^{(i)}))d\mathbf{X} = \int_\Xi (\chi(X_1^{(i)})X_1^{(i)})'' d\mathbf{X} = h. \qquad (1.6.37)$$

Relation (1.6.34) gives the boundary condition for $w_1^{(i)}$ at $x_1^{(i)} = 0$. The discrepancies of order $O(1)$, $O(\varepsilon)$, etc., in the right-hand side of (1.6.30) concentrated in the vicinity of the junction point will be compensated by the junction solutions in the following steps of the asymptotic algorithm. To compensate for the last term in (1.6.30) we set

$$\frac{d^2 w_1^{(i)}}{dx_1^{(i)2}}(x_1^{(i)}) = 0, \quad 0 < x_1^{(i)} < l_i, \quad i = 1, 2, 3, 4. \qquad (1.6.38)$$

To obtain a good approximation for (1.6.3) we set

$$C_1 + w_1^{(i)}(l_i) = 0. \qquad (1.6.39)$$

Then, the right-hand sides in (1.6.32) and (1.6.3) coincide up to exponentially small terms (with respect to ε^{-1}).

It follows from (1.6.36) and (1.6.26) that

$$C_0 = \left(\sum_{i=1}^{4} l_i^{-1}\right)^{-1} \sum_{i=1}^{4} T_i l_i^{-1}. \qquad (1.6.40)$$

We note that this formula coincides with (1.6.17) (with C replaced by C_0).

It follows from (1.6.34), (1.6.38) and (1.6.39) that

$$w_1^{(i)}(x_1^{(i)}) = -\frac{C_1 + q_1^{(i)}}{l_i} x_1^{(i)} + q_1^{(i)}, \qquad (1.6.41)$$

where the constant C_1 will be specified in the next step of the asymptotic algorithm. With the above choice of $\mathcal{W}_1$ and $w_1^{(i)}$ equation (1.6.30) becomes

$$\Delta(\mathcal{T}_0 + \varepsilon\mathcal{T}_1) = \sum_{i=1}^{4} \frac{dw_1^{(i)}}{dx_1^{(i)}}(0)(\chi''(X_1^{(i)})X_1^{(i)} + 2\chi'(X_1^{(i)}))$$

(compare with (1.6.27).

1.6.4 The complete asymptotic expansion

We seek an asymptotic expansion of the solution in the form

$$u \sim \sum_{k=0}^{\infty} \varepsilon^k \mathcal{T}_k, \qquad (1.6.42)$$

where

$$\mathcal{T}_k = C_k + \mathcal{W}_k(\mathbf{X}) + \sum_{i=1}^{4} w_k^{(i)}(x_1^{(i)})\chi(X_1^{(i)}), \qquad (1.6.43)$$

and $\mathcal{W}_k$ are the junction layer terms.

Let

$$u^{(N)} = \sum_{k=0}^{N} \varepsilon^k \mathcal{T}_k. \qquad (1.6.44)$$

Assume that we have constructed all terms $\mathcal{T}_k$, $0 \le k \le N$, up to the constant C_N, with linear functions $w_0^{(i)}$ given by (1.6.26) and

$$w_k^{(i)} = -\frac{C_k + q_k^{(i)}}{l_i} x_1^{(i)} + q_k^{(i)}, \ 1 \le k \le N. \qquad (1.6.45)$$

Also assume that

$$\Delta u^{(N)} = \varepsilon^{N-1} \sum_{i=1}^{4} \frac{dw_N^{(i)}}{dx_1^{(i)}}(0)(\chi''(X_1^{(i)})X_1^{(i)} + 2\chi'(X_1^{(i)})) \quad \text{in } \Omega(\varepsilon),$$

(1.6.46)

$$\frac{\partial u^{(N)}}{\partial n} = 0 \quad \text{on } \partial\Omega_\mathcal{N}(\varepsilon), \qquad (1.6.47)$$

and

$$u^{(N)} = T_j + O(e^{-\beta/\varepsilon}) \quad \text{on } \partial\Omega_\mathcal{D}^{(j)}(\varepsilon),$$

where β is some positive number.

Let us make the next step where we specify the constant C_N and the next term $\mathcal{T}_{N+1}$ up to the constant C_{N+1}.

Substituting $u^{(N+1)}$ in (1.6.1)–(1.6.3) we find

$$-\Delta u^{(N+1)} = -\Delta u^{(N)} - \varepsilon^{N-1}\Delta_{\mathbf{X}}\mathcal{W}_{N+1}(\mathbf{X})$$
$$- \varepsilon^N \sum_{i=1}^{4}\left\{\varepsilon^{-1}\chi''(X_1^{(i)})w_{N+1}^{(i)}(x_1^{(i)}) + 2\chi'(X_1^{(i)})\frac{dw_{N+1}^{(i)}}{dx_1^{(i)}}(x_1^{(i)})\right.$$
$$\left. + \varepsilon\chi(X_1^{(i)})\frac{d^2 w_{N+1}^{(i)}}{dx_1^{(i)^2}}(x_1^{(i)})\right\} \text{ in } \Omega(\varepsilon), \qquad (1.6.48)$$

$$\frac{\partial}{\partial n}u^{(N+1)} = \varepsilon^N \frac{\partial}{\partial n_{\mathbf{X}}}\mathcal{W}_{N+1} \text{ on } \partial\Omega_{\mathcal{N}}(\varepsilon), \qquad (1.6.49)$$

$$u^{(N+1)} = T_i + \varepsilon^{N+1}T_{N+1} + O(e^{-\beta/\varepsilon}) \text{ on } \partial\Omega_{\mathcal{D}}^{(i)}. \qquad (1.6.50)$$

To compensate for the terms of order $O(\varepsilon^{N-1})$ in the right-hand side of (1.6.48) and keeping in mind (1.6.46) we set

$$-\Delta_{\mathbf{X}}\mathcal{W}_{N+1}(\mathbf{X}) = \sum_{i=1}^{4}\frac{dw_N^{(i)}}{dx_1^{(i)}}(0)(\chi''(X_1^{(i)})X_1^{(i)} + 2\chi'(X_1^{(i)}))$$
$$+ \sum_{i=1}^{4}q_{N+1}^{(i)}\chi''(X_1^{(i)}) \text{ on } \Xi \qquad (1.6.51)$$

and
$$w_{N+1}^{(i)}(0) = q_{N+1}^{(i)}, \qquad (1.6.52)$$

where $q_{N+1}^{(i)}$ are unknown constants. As before, the higher-order discrepancies in the vicinity of the junction point are compensated in the next step of the asymptotic algorithm. To compensate for the last term in (1.6.48) we set

$$\frac{d^2 w_{N+1}^{(i)}}{dx_1^{(i)^2}}(x_1^{(i)}) = 0, \quad 0 < x_1^{(i)} < l_i, \quad i = 1, 2, 3, 4. \qquad (1.6.53)$$

To obtain the homogeneous boundary condition (1.6.49) we need

$$\frac{\partial \mathcal{W}_{N+1}}{\partial n_{\mathbf{X}}} = 0 \text{ on } \partial\Xi. \qquad (1.6.54)$$

By Theorem 1.6, there exists a solution

PROBLEM OF JUNCTION BETWEEN THIN BODIES

$$(\mathcal{W}_{N+1}, q_{N+1}^{(1)}, q_{N+1}^{(2)}, q_{N+1}^{(3)}, q_{N+1}^{(4)}) \in E_0(\Xi) \times \mathbb{R}^4$$

of problem (1.6.51), (1.6.54) if and only if

$$\sum_{i=1}^{4} \frac{dw_N^{(i)}}{dx_1^{(i)}}(0) = 0. \tag{1.6.55}$$

The function $\mathcal{W}_{N+1}$ satisfies

$$\mathcal{W}_{N+1}(\mathbf{X}) = O(e^{-\pi h^{-1} X_1^{(i)}}), \quad \text{as } X_1^{(i)} \to +\infty.$$

It follows from (1.6.45) for $k = N$ and (1.6.55) that the constant C_N is specified by

$$C_N = -\left(\sum_{i=1}^{4} l_i^{-1}\right)^{-1} \sum_{i=1}^{4} q_N^{(i)} l_i^{-1}. \tag{1.6.56}$$

Assuming that

$$C_{N+1} + w_{N+1}^{(i)}(l_i) = 0, \quad i = 1, 2, 3, 4, \tag{1.6.57}$$

we obtain from (1.6.50)

$$S_{N+1} = T_i + O(e^{-\beta/\varepsilon}) \quad \text{on } \partial\Omega_{\mathcal{D}}^{(i)}(\varepsilon).$$

It follows from (1.6.52), (1.6.53), (1.6.57) that

$$w_{N+1}^{(i)}(x_1^{(i)}) = -\frac{C_{N+1} + q_{N+1}^{(i)}}{l_i} x_1^{(i)} + q_{N+1}^{(i)}.$$

Thus, we have evaluated the constant C_N and constructed all terms of $\mathcal{T}_{N+1}$ up to the constant C_{N+1}. Moreover, S_{N+1} satisfies (1.6.46), (1.6.47) with the index N being replaced by $N+1$. Thus, by induction we have constructed all the terms of the asymptotic expansion.

1.6.5 The remainder estimate

Let $u^{(N)}$ be the finite sum (1.6.44) and let

$$\tilde{u}^{(N)} = u - u^{(N)}.$$

By (1.6.51), the function $\tilde{u}^{(N)}$ satisfies the boundary value problem

$$-\Delta \tilde{u}^{(N)} = \mathcal{F}_N \quad \text{in } \Omega(\varepsilon) \tag{1.6.58}$$

$$\frac{\partial \tilde{u}^{(N)}}{\partial n} = 0 \quad \text{on } \partial\Omega_N(\varepsilon) \tag{1.6.59}$$

$$\tilde{u}^{(N)} = h_N \quad \text{on } \partial\Omega_D(\varepsilon) \tag{1.6.60}$$

where

$$\mathcal{F}_N = \varepsilon^{N-1} \sum_{i=1}^{4} \frac{dw_N^{(i)}}{dx_1^{(i)}}(0)\{2\chi'(X_1^{(i)}) + X_1^{(i)}\chi''(X_1^{(i)})\},$$

and

$$h_N = -\sum_{k=0}^{N} \varepsilon^k \mathcal{W}_k(\mathbf{X}).$$

Here and in what follows we shall use the standard notation $H^1(G)$ for the Sobolev space of functions which are defined on a domain G and are square summable together with all first-order derivatives. The norm of a function v in $H^1(G)$ is given by

$$\|v\|_{H^1(G)} = \left(\int_\Omega (v^2 + \|\nabla v\|^2) dx \right)^{1/2}.$$

THEOREM 1.7

The remainder term $\tilde{u}^{(N)}$ satisfies the estimate

$$\|\tilde{u}^{(N)}\|_{H^1(\Omega(\varepsilon))} \leq \text{Const } \varepsilon^N. \tag{1.6.61}$$

Proof

Let $\tilde{u}^{(N)}$ be written as

$$\tilde{u}^{(N)}(\mathbf{x}, \varepsilon) = R(\mathbf{x}, \varepsilon) - \eta(x_1)\eta(x_2) \sum_{k=0}^{N} \varepsilon^k \mathcal{W}_k(\mathbf{X});$$

$\eta(t)$ is a smooth function which is equal to zero for $|t| \leq \min_j l_j/4$, and to one for $|t| \geq \min_j l_j/2$. Since the functions $\mathcal{W}_k$ and their derivatives are $O(e^{-\beta/\varepsilon})$, $\beta = \text{Const} > 0$, outside any neighbourhood of $x = 0$ independent of ε, it is sufficient to prove (1.6.61) for R. This function

satisifies the homogeneous Dirichlet and Neumann conditions on $\partial\Omega_\mathcal{D}(\varepsilon)$ and $\partial\Omega_\mathcal{N}(\varepsilon)$. Moreover,
$$-\Delta R = \mathcal{F},$$
where
$$\mathcal{F} = \mathcal{F}_N + O(e^{-\beta_1/\varepsilon}) \quad \text{with} \quad \beta_1 > 0.$$

Since $R = 0$ on $\partial\Omega_\mathcal{D}(\varepsilon)$, a standard application of the Newton–Leibniz formula leads to the estimate
$$\int_{\Omega(\varepsilon)} R^2 d\mathbf{x} \leq \text{Const} \int_{\Omega(\varepsilon)} |\nabla R|^2 d\mathbf{x}. \tag{1.6.62}$$

Using
$$\int_{\Omega(\varepsilon)} |\nabla R|^2 d\mathbf{x} = \int_{\Omega(\varepsilon)} R\mathcal{F} d\mathbf{x} \leq \|R\|_{L_2(\Omega(\varepsilon))} \|\mathcal{F}\|_{L_2(\Omega(\varepsilon))},$$

together with (1.6.62), we obtain
$$\|R\|_{H^1(\Omega(\varepsilon))} \leq \text{Const} \, \varepsilon^N. \tag{1.6.63}$$

Here we used that the function $\mathcal{F}_N$ vanishes outside the ε-neighbourhood of the junction point. $\square$

2

A BOUNDARY VALUE PROBLEM FOR THE LAPLACIAN IN A MULTI-STRUCTURE

In the present chapter we consider a multi-structure Ω_ε defined as a union of a three-dimensional domain and a number of thin cylinders of normalised thickness ε parallel to each other (see Fig. 2.1). We deal with the asymptotic analysis of a solution u of a mixed boundary value problem for the Laplacian operator in Ω_ε. The Dirichlet data are prescribed on the bases of thin cylinders; on the remaining part of the surface $\partial \Omega_\varepsilon$ we have the Neumann boundary condition. We note that the solution of this problem describes the temperature field in the multi-structure for the case when the heat flux is specified on the boundary surface outside the base regions of the thin cylinder, where the values of

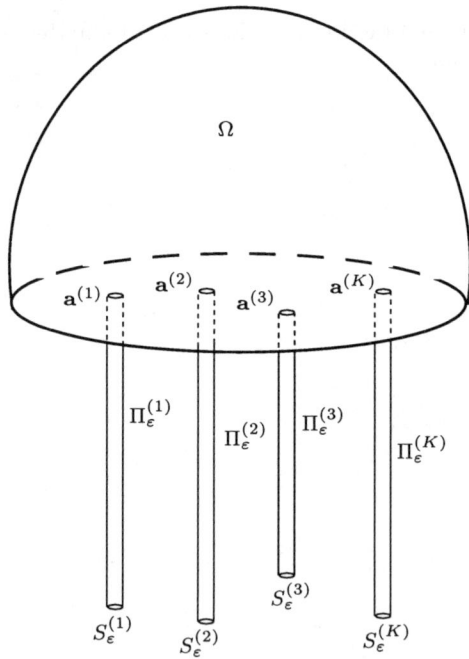

FIG. 2.1. A 3D–1D multi-structure Ω_ε.

the temperature are given.

Our objective is to obtain a complete asymptotic expansion for the solution:

$$u(x,\varepsilon) \sim \sum_{j=0}^{\infty} \varepsilon^j u_j(x,\varepsilon). \qquad (2.0.1)$$

We use the technique of compound asymptotic expansions outlined in Chapter 1.

We show that the functions u_j in (2.0.1) involve solutions to model problems which are formulated in original or scaled variables and posed in limit domains independent of ε: a domain Ω, a union of a half-space and a semi-infinite cylinder, a semi-infinite cylinder, the cross-section and the axis of a thin cylinder, the skeleton of the multi-structure.

The behaviour of the solution in Ω outside neighbourhoods of the junction points $\mathbf{a}^{(1)},\ldots,\mathbf{a}^{(K)}$ is described by the model Neumann problem in Ω with point sources applied at $\mathbf{a}^{(1)},\ldots,\mathbf{a}^{(K)}$. The asymptotics of fields near junctions are specified by means of solutions of the Neumann boundary value problem in the union of a half-space and a semi-infinite cylinder. These solutions vanish at infinity and hence play the role of junction layers. The terms of the asymptotic expansion, describing the solution in thin cylinders $\Pi_\varepsilon^{(1)},\ldots,\Pi_\varepsilon^{(K)}$, are obtained by means of the following limit problems: an ordinary differential equation on a segment and the Neumann boundary value problem on the scaled cross-section of $\Pi_\varepsilon^{(j)}$. Finally, the 'bottom layer' near the base $S_\varepsilon^{(j)}$ of a cylinder is defined as a solution of a mixed boundary value problem in a semi-infinite cylinder with the Dirichlet data on the cylinder base and the Neumann boundary condition on the lateral part of the surface.

In Section 2.1 we formulate the boundary value problem in Ω_ε. Section 2.2 deals with the analysis of model problems. An auxiliary asymptotic solution to a problem in a thin cylinder is constructed in Section 2.2.6.

We describe the right-hand sides of the boundary value problem in Ω_ε in Section 2.3. In Section 2.4 we construct the principal term u_0 of expansion (2.0.1). It includes the junction layer and the bottom layer terms and special fields in thin cylinders, as well as solutions of the model problem in Ω_0 with possible singularities at junction points. It also includes the solution of the algebraic system associated with the skeleton of the multi-structure.

In Section 2.5 we construct the higher-order terms of asymptotics (2.0.1). We solve the same sequence of model problems as for the leading approximation but with different right-hand sides. Section 2.6 is devoted to the justification of the whole asymptotic procedure.

As an example, in Section 2.7 we describe the principal term of the asymptotics of the solution of the mixed boundary value problem in Ω_ε for the Poisson equation with a constant right-hand side. Up to an additive constant, this leading term can be approximated in Ω, outside the vicinity of the junction points $\mathbf{a}^{(j)}$, by a sum

$$\sum_{j=1}^{K} T_j N(\mathbf{x}, \mathbf{a}^{(j)}),$$

where N is the Neumann function and T_j, $j = 1, \ldots, K$, are explicitly determined constants. In Chapter 6 we show that this leading order term approximates the first eigenfunction for the mixed boundary value problem for the Laplacian in the multi-structure.

In Section 2.8 we analyse some particular right-hand sides and obtain the asymptotics of the energy integral.

At the end of the chapter we briefly discuss the case of 1D–3D multi-structures of a general configuration as well as an example of 2D–3D multi-structure including a massive 3D body with an attached thin-walled tube.

2.1 Formulation of the problem

Let Ω be a bounded three-dimensional domain with Lipschitz boundary $\partial\Omega$, which means that $\partial\Omega$ can be locally represented in some Cartesian system by a graph of the Lipschitz function. We suppose that a part of $\partial\Omega$ is located in the plane $x_3 = 0$. The axis Ox_3 will be directed downwards. Let us consider the domain

$$\Omega_\varepsilon = \Omega \cup \Pi_\varepsilon^{(1)} \cup \cdots \cup \Pi_\varepsilon^{(K)},$$

where

$$\Pi_\varepsilon^{(j)} = \{x : 0 \leq x_3 < l^{(j)}, ((x_1 - a_1^{(j)})/\varepsilon, (x_2 - a_2^{(j)})/\varepsilon) \in g^{(j)} \subset \mathbb{R}^2\},$$

$$j = 1, \ldots, K,$$

are 'thin' cylinders compiled with the flat part of the surface $\partial\Omega_0$ at the points $(a_1^{(j)}, a_2^{(j)}, 0), j = 1, \ldots, K$. Here $g^{(j)}$ is a finite two-dimensional domain with Lipschitz boundary, and ε is a small parameter. Without loss of generality, we assume that the closure $\bar{g}_j$ lies in the unit ball centred at the origin. The base region of the thin cylinder $\Pi_\varepsilon^{(j)}$ will be denoted by $S_\varepsilon^{(j)}$, i.e.

MODEL PROBLEMS

$$S_\varepsilon^{(j)} = \{\mathbf{x} : x_3 = l^{(j)}, ((x_1 - a_1^{(j)})/\varepsilon, (x_2 - a_2^{(j)})/\varepsilon) \in g^{(j)}\}$$

(see Fig. 2.1). Also, we use the notation

$$S_\varepsilon = \cup_{j=1}^K S_\varepsilon^{(j)}.$$

We consider the following boundary value problem:

$$-\Delta_x u(\mathbf{x}, \varepsilon) = F(\mathbf{x}, \varepsilon), \quad \mathbf{x} \in \Omega_\varepsilon, \tag{2.1.1}$$

$$\frac{\partial u}{\partial n}(\mathbf{x}, \varepsilon) = P(\mathbf{x}, \varepsilon), \quad \mathbf{x} \in \partial\Omega_\varepsilon \setminus \bar{S}_\varepsilon, \tag{2.1.2}$$

$$u(\mathbf{x}, \varepsilon) = \Phi^{(j)}(\mathbf{x}, \varepsilon), \quad \mathbf{x} \in S_\varepsilon^{(j)}, \ j = 1, \ldots, K, \tag{2.1.3}$$

where $\partial/\partial n$ is the outward normal derivative on $\partial\Omega_\varepsilon$. We suppose that F, P and $\Phi^{(j)}$ are functions from $L_2(\Omega), L_2(\partial\Omega_\varepsilon \setminus \cup_j S_\varepsilon^{(j)})$ and $H^{1/2}(S_\varepsilon^{(j)})$, where we use the standard notation $H^{1/2}(S_\varepsilon^{(j)})$ for the space of traces (limit values) on $S_\varepsilon^{(j)}$ of functions in $H^1(\Pi_\varepsilon^{(j)})$. It is well known that the problem (2.1.1)–(2.1.3) is uniquely solvable in the class of functions $H^1(\Omega_\varepsilon)$ with the finite energy integral (see Section 9 of Chapter 2 in Lions and Magenes (1968) and also Theorem 2.1 in the text below). Our aim is to construct an asymptotic expansion, uniform in Ω_ε, for the solution.

2.2 Model problems

The asymptotic representation of u will contain six types of terms which can be found by solving the several model problems analysed in the present section.

2.2.1 Limit domains

We assign to the multi-structure Ω_ε the following collection of sets independent of ε:

(1) The *bounded domain* Ω with the Lipschitz boundary (see Fig. 2.2). We recall that $\partial\Omega$ includes a plane part containing the junction points $\mathbf{a}^{(j)} = (a_1^{(j)}, a_2^{(j)}, 0),\ j = 1, \ldots, K$.

(2) *The domain G* is the union of the half-space $\mathbb{R}^3_- = \{\mathbf{X} : X_3 < 0\}$ and a semi-infinite cylinder

$$\Pi_+ = \{\mathbf{X} : X_3 \geq 0, (X_1, X_2) \in g\},$$

where g is a two-dimensional bounded domain with Lipschitz boundary (see Fig. 2.3). This limit domain describes a scaled junction region of the multi-structure, where the junction layer problem will be posed.

(3) *The domain* Π_- is the semi-infinite cylinder:

$$\Pi_- = \{\mathbf{Y} : Y_3 < 0, (Y_1, Y_2) \in g \subset \mathbb{R}^2\},$$

which is shown in Fig. 2.4. The bottom layer will be characterised by a solution of a mixed boundary value problem with the Dirichlet data on the base region S of the boundary $\partial\Pi_-$.

(4) The Neumann problem on the *scaled cross-section* g (see Fig. 2.5(a)) occurs in the asymptotic analysis of the solution in the thin cylinders.

(5) The Dirichlet problem for an ordinary differential equation on the *axis of a thin cylinder* (see Fig. 2.5(b)) is also used to describe the asymptotics of the solution in the thin cylinders.

(6) Finally, we introduce the *skeleton* of the multi-structure, i.e. the union of the domain Ω and one-dimensional segments connected to Ω at the junction points (see Fig. 2.6).

2.2.2 Model problem in Ω

Formulation of the problem

It is natural to try to approximate the solution of (2.1.1)–(2.1.3) in the domain Ω by the function u_Ω such that

$$-\Delta_x u_\Omega(\mathbf{x}) = f(\mathbf{x}), \quad \mathbf{x} \in \Omega, \qquad (2.2.1)$$

$$\frac{\partial u_\Omega}{\partial n_x}(\mathbf{x}) = p(\mathbf{x}) + \sum_{j=1}^{K} T^{(j)} \delta(\mathbf{x} - \mathbf{a}^{(j)}), \quad \mathbf{x} \in \partial\Omega, \qquad (2.2.2)$$

where $T^{(j)}, j = 1, \ldots, K$, are constants, and f and p are functions from $L_2(\Omega)$ and $L_2(\partial\Omega)$ (compare with Section 1.4).

MODEL PROBLEMS

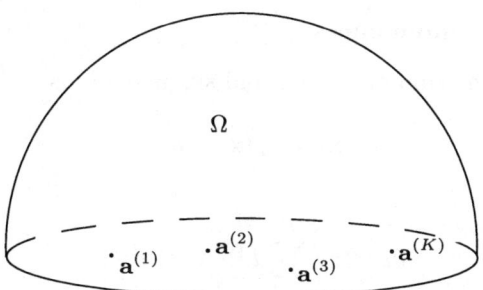

FIG. 2.2. Three-dimensional 'cap' of the multi-structure Ω_ε.

The Neumann function

The Neumann function N is defined as a solution of

$$-\Delta_x N(\mathbf{x}, \mathbf{y}) = \delta(\mathbf{x} - \mathbf{y}) - (\mathrm{mes}_3 \Omega)^{-1}, \quad \mathbf{x} \in \Omega, \qquad (2.2.3)$$

$$\frac{\partial N}{\partial n_x}(\mathbf{x}, \mathbf{y}) = 0, \quad \mathbf{x} \in \partial \Omega, \qquad (2.2.4)$$

with the orthogonality condition

$$\int_\Omega N(\mathbf{x}, \mathbf{y}) d\mathbf{x} = 0. \qquad (2.2.5)$$

The function N is symmetric, i.e. $N(\mathbf{x}, \mathbf{y}) = N(\mathbf{y}, \mathbf{x})$. Furthermore, $N(\mathbf{x}, \mathbf{a}^{(j)})$ satisfies the boundary value problem

$$\Delta_x N(\mathbf{x}, \mathbf{a}^{(j)}) = (\mathrm{mes}_3 \Omega)^{-1}, \quad \mathbf{x} \in \Omega, \qquad (2.2.6)$$

$$\frac{\partial N}{\partial n_x}(\mathbf{x}, \mathbf{a}^{(j)}) = \delta(\mathbf{x} - \mathbf{a}^{(i)}), \quad \mathbf{x} \in \partial \Omega. \qquad (2.2.7)$$

If Ω is a half-space, then the function

$$\mathbf{x} \to \frac{1}{2\pi \|\mathbf{x} - \mathbf{a}^{(j)}\|}$$

is harmonic and (2.2.7) holds. Hence, $N(\mathbf{x}, \mathbf{a}^{(j)}) - (2\pi \|\mathbf{x} - \mathbf{a}^{(j)}\|)^{-1}$ solves the Neumann boundary value problem in Ω with the right-hand sides smooth in a neighbourhood of $\mathbf{x} = \mathbf{a}^{(j)}$. Therefore,

$$N(\mathbf{x}, \mathbf{a}^{(j)}) = \frac{1}{2\pi \|\mathbf{x} - \mathbf{a}^{(j)}\|} + m_j(\mathbf{x}), \qquad (2.2.8)$$

where m_j is a smooth function.

Solvability and uniqueness

We seek u_Ω as the sum of regular and singular terms:

$$u_\Omega(\mathbf{x}) = u_r(\mathbf{x}) + u_s(\mathbf{x}), \qquad (2.2.9)$$

where

$$u_s(\mathbf{x}) = \sum_{j=1}^{K} T^{(j)} N(\mathbf{x}, \mathbf{a}^{(j)}).$$

Substituting (2.2.9) into (2.2.1), (2.2.2) we obtain the boundary value problem for u_r:

$$-\Delta_x u_r(\mathbf{x}) = f(\mathbf{x}) + \sum_{j=1}^{K} T^{(j)} (\text{mes}_3 \Omega)^{-1} \quad \text{in } \Omega, \qquad (2.2.10)$$

$$\frac{\partial u_r}{\partial n_x}(\mathbf{x}) = p(\mathbf{x}) \quad \text{on } \partial\Omega. \qquad (2.2.11)$$

It is known that this problem is solvable in the space $H^1(\Omega)$ (in a generalised sense) if and only if

$$\int_\Omega f(\mathbf{x})d\mathbf{x} + \int_{\partial\Omega} p(\mathbf{x})ds + \sum_{j=1}^{K} T^{(j)} = 0 \qquad (2.2.12)$$

(see Chapter 7 of Dautray and Lions (1988)). The solution is specified up to an arbitrary additive constant, and, to provide the uniqueness, we impose the orthogonality condition

$$\int_\Omega u_r(\mathbf{x})d\mathbf{x} = 0.$$

Using Green's formula

$$\int_\Omega (u\Delta v - v\Delta u)d\mathbf{x} = \int_{\partial\Omega} \left(u\frac{\partial v}{\partial n} - v\frac{\partial u}{\partial n} \right) ds \qquad (2.2.13)$$

one can verify that u_r admits the representation

$$u_r(\mathbf{x}) = \int_\Omega N(\mathbf{x},\mathbf{y})f(\mathbf{y})d\mathbf{y} + \int_{\partial\Omega} N(\mathbf{x},\mathbf{y})p(\mathbf{y})ds_y. \qquad (2.2.14)$$

2.2.3 Model problem for the junction region

Formulation of the problem

We consider the domain $G = \mathbb{R}^3_- \cup \Pi_+$, where $\mathbb{R}^3_-$ is a half-space $\{\mathbf{X} : X_3 < 0\}$ and Π_+ is a semi-infinite cylinder

$$\{\mathbf{X} : X_3 \geq 0, (X_1, X_2) \in g\}$$

(see Fig. 2.3); g is a bounded domain in $\mathbb{R}^2$ with Lipschitz boundary.

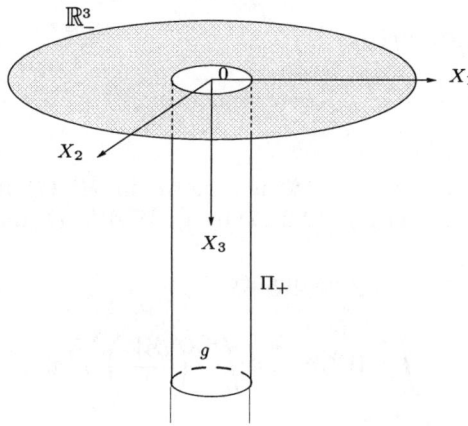

FIG. 2.3. The junction layer region.

Let W satisfy the Neumann problem

$$-\Delta_X W(\mathbf{X}) = F(\mathbf{X}), \quad \mathbf{X} \in G, \quad (2.2.15)$$

$$\frac{\partial W}{\partial n_X}(\mathbf{X}) = P(\mathbf{X}), \quad \mathbf{X} \in \partial G, \quad (2.2.16)$$

where F and P are functions from $L_2(G)$ and $L_2(\partial G)$ with bounded supports.

The space $\mathcal{H}^1(G)$

A function W is said to belong to the space $\mathcal{H}^1(G)$ if

$$\int_G \|\nabla W\|^2 d\mathbf{X} + \int_{\mathbb{R}^3_-} W^2 \frac{d\mathbf{X}}{\|\mathbf{X}\|^2} < \infty. \quad (2.2.17)$$

LEMMA 2.1

The functional
$$W \to \|\nabla W\|_{L_2(G)} \quad (2.2.18)$$
defines a norm in the space $\mathcal{H}^1(G)$. The estimate

$$\int_{\mathbb{R}^3_-} W^2 \frac{d\mathbf{X}}{\|\mathbf{X}\|^2} + \sup_{X_3>0} \frac{\|W(\cdot, X_3)\|^2_{L_2(g)}}{1 + X_3}$$

$$\leq \text{Const} \int_G \|\nabla W\|^2 d\mathbf{X} \quad (2.2.19)$$

is valid for all $W \in \mathcal{H}^1(G)$.

Proof

In order to show that (2.2.18) is a norm in $\mathcal{H}^1(G)$ it is sufficient to estimate the second term in (2.2.17) by $\int_G \|\nabla W\|^2 d\mathbf{X}$ multiplied by some positive constant.

Integrating the Hardy inequality

$$\int_0^\infty W^2 dr \leq 4 \int_0^\infty \left(\frac{\partial W}{\partial r}\right)^2 r^2 dr$$

over the lower unit half-sphere we obtain

$$\int_{\mathbb{R}^3_-} W^2 \frac{d\mathbf{X}}{\|\mathbf{X}\|^2} \leq 4 \int_G \|\nabla W\|^2 d\mathbf{X}. \quad (2.2.20)$$

Thus, we have proved that (2.2.18) gives the norm in $\mathcal{H}^1(G)$ and the first integral in (2.2.19) is estimated by the square of this norm.

Now, we estimate the second term in the left-hand side of (2.2.19). The application of the Newton–Leibniz formula and the triangle inequality yield

$$\|W(\cdot, X_3)\|_{L_2(g)} \leq \|W(\cdot, -t)\|_{L_2(g)} + \int_{-t}^{X_3} \left\|\frac{\partial}{\partial \tau} W(\cdot, \tau)\right\|_{L_2(g)} d\tau,$$

for any $t \in (0, 1)$ and $X_3 > 0$. It follows that

$$\|W(\cdot, X_3)\|_{L_2(g)} \leq \inf_{t \in (0,1)} \|W(\cdot, -t)\|_{L_2(g)}$$

$$+ \int_{-1}^{X_3} \left\| \frac{\partial}{\partial \tau} W(\cdot, \tau) \right\|_{L_2(g)} d\tau. \qquad (2.2.21)$$

We note that

$$\inf_{t \in (0,1)} \|W(\cdot, -t)\|_{L_2(g)} \leq \left(\int_{-1}^{0} \|W(\cdot, -t)\|_{L_2(g)}^2 dt \right)^{1/2}. \qquad (2.2.22)$$

Applying Cauchy's inequality to the second integral in the right-hand side of (2.2.21) we obtain

$$\int_{-1}^{X_3} \left\| \frac{\partial}{\partial \tau} W(\cdot, \tau) \right\|_{L_2(g)} d\tau$$

$$\leq \left(\int_{-1}^{X_3} \left\| \frac{\partial}{\partial \tau} W(\cdot, \tau) \right\|_{L_2(g)}^2 d\tau \right)^{1/2} (X_3 + 1)^{1/2}. \qquad (2.2.23)$$

Using estimates (2.2.22) and (2.2.23) together with (2.2.20) we deduce from (2.2.21) that

$$\|W(\cdot, X_3)\|_{L_2(g)} \leq \operatorname{Const} \left\{ \int_G \|\nabla W\|^2 d\mathbf{X} \right\}^{1/2} (1 + X_3)^{1/2}.$$

Thus, we have proved inequality (2.2.19). $\square$

By a solution of (2.2.15), (2.2.16) we mean the function $W \in \mathcal{H}^1(G)$ such that

$$\int_G \nabla W \cdot \nabla V d\mathbf{X} = \int_{\partial G} PV ds + \int_G FV d\mathbf{X}, \qquad (2.2.24)$$

for any $V \in \mathcal{H}^1(G)$.

Solvability in $\mathcal{H}^1(G)$ and asymptotics

Now, we turn to the solvability of problem (2.2.15), (2.2.16) in the space $\mathcal{H}^1(G)$.

LEMMA 2.2

Let F and P be functions in $L_2(G)$ and $L_2(\partial G)$ with bounded supports. Then

(i) problem (2.2.15), (2.2.16) has a unique solution in $\mathcal{H}^1(G)$;

(ii) this solution is represented as the convergent series

$$W(\mathbf{X}) = \sum_{k=0}^{\infty} \frac{p_k(\mathbf{X}/\|\mathbf{X}\|)}{\|\mathbf{X}\|^{k+1}}, \quad \|\mathbf{X}\| \to \infty, \ \mathbf{X} \in \mathbb{R}^3_-, \qquad (2.2.25)$$

where p_k are homogeneous harmonic polynomials of order k that include only even powers of X_3, and

$$W(\mathbf{X}) = \text{Const} + O(e^{-\alpha X_3}), \quad X_3 \to +\infty, \qquad (2.2.26)$$

where α is a positive constant.

Proof

(i) **Existence and uniqueness.** Consider the functional

$$V \to \int_{\partial G} PV\,ds + \int_G FV\,d\mathbf{X}.$$

From estimate (2.2.19) this functional is continuous on $\mathcal{H}^1(G)$. The Riesz theorem gives the existence of a unique function $W \in \mathcal{H}^1(G)$ satisfying (2.2.24).

(ii) **Asymptotics at infinity in $\mathbb{R}^3_-$.** We introduce Neumann's function $\mathcal{N}(\mathbf{X}, \mathbf{Y})$ for the half-space $\mathbb{R}^3_-$:

$$\Delta \mathcal{N}(\mathbf{X}, \mathbf{Y}) + \delta(\mathbf{X} - \mathbf{Y}) = 0, \quad \mathbf{X}, \mathbf{Y} \in \mathbb{R}^3_-,$$

$$\frac{\partial \mathcal{N}}{\partial X_3} = 0 \ \text{ for } \ X_3 = 0.$$

One can check that

$$\mathcal{N}(\mathbf{X}, \mathbf{Y}) = \frac{1}{4\pi}(\|\mathbf{X} - \mathbf{Y}\|^{-1} + \|\mathbf{X} - \mathbf{Y}^*\|^{-1}), \qquad (2.2.27)$$

where $\mathbf{Y}^* = (Y_1, Y_2, -Y_3)$. Let $\theta = \theta(\tau)$ denote a smooth cut-off function such that

MODEL PROBLEMS

$$\theta = \begin{cases} 0, & \tau < r, \\ 1, & \tau > r+1, \end{cases}$$

where r is a positive number chosen in such a way that the ball B_r contains the supports of F and P. Then, the function $\mathcal{V}(\mathbf{X}) = W(\mathbf{X})\theta(\|\mathbf{X}\|)$ satisfies

$$-\Delta\mathcal{V} = f := -W\Delta\theta - 2\nabla W \cdot \nabla\theta \quad \text{in } \mathbb{R}^3_-,$$

$$\left.\frac{\partial\mathcal{V}}{\partial X_3}\right|_{X_3=0} = 0.$$

Using Neumann's function (2.2.27) we obtain

$$\mathcal{V}(\mathbf{X}) = \int_{\mathbb{R}^3_-} f(\mathbf{Y})\mathcal{N}(\mathbf{X},\mathbf{Y})d\mathbf{Y}. \qquad (2.2.28)$$

For large values of $\|\mathbf{X}\|$ and for $\|\mathbf{Y}\| < r+1$ the function $\mathcal{N}(\mathbf{X},\mathbf{Y})$ can be expanded as

$$\mathcal{N}(\mathbf{X},\mathbf{Y}) = \sum_{k=0}^{\infty} \|\mathbf{X}\|^{-k-1} Q_k(\mathbf{Y}; \mathbf{X}\|\mathbf{X}\|^{-1}), \qquad (2.2.29)$$

where $Q_k(\mathbf{Y},\mathbf{X})$ are homogeneous harmonic polynomials (in $\mathbf{X}$) of order k that satisfy the homogeneous Neumann boundary condition when $X_3 = 0$ and whose coefficients depend polynomially on $\mathbf{Y}$. The asymptotic expansion (2.2.29) is uniform for all $\mathbf{Y} \in B_{r+1}$. Using the asymptotic formula (2.2.29) and the integral representation (2.2.28) we obtain (2.2.25).

Asymptotics at infinity, as $X_3 \to +\infty$. Let θ be the same cut-off function as the one given above. The function $\mathcal{V}(\mathbf{X}) := W(\mathbf{X})\theta(X_3)$ satisfies the boundary value problem

$$-\Delta\mathcal{V} = f := -W\theta'' - 2\frac{\partial W}{\partial X_3}\theta' \quad \text{in } \Pi_+, \qquad (2.2.30)$$

$$\frac{\partial\mathcal{V}}{\partial n} = 0 \quad \text{on } \partial g \text{ for } X_3 > 0, \qquad (2.2.31)$$

$$\mathcal{V} = 0 \quad \text{for } X_3 = 0. \qquad (2.2.32)$$

This problem can be solved by the Fourier method which implies the asymptotics

$$\mathcal{V} = AX_3 + B + O(e^{-\alpha X_3}) \quad \text{as } X_3 \to +\infty,$$

where A and B are constants and α^2 is the first positive eigenvalue of the operator $-\Delta$ on the cross-section g with homogeneous Neumann boundary data on ∂g. The condition $\mathcal{V} \in \mathcal{H}^1(G)$ gives $A = 0$. □

Solvability in the class of functions with linear growth at infinity

In the text below we shall use the following solvability theorem for the problem (2.2.15), (2.2.16) in the class of functions that are allowed to grow linearly in X_3 as $X_3 \to +\infty$.

LEMMA 2.3

The problem (2.2.15), (2.2.16) has a unique solution with finite energy integral over any bounded subdomain of G, which decays like $O(\|\mathbf{X}\|^{-2})$ in $\mathbb{R}^3_-$ as $\|\mathbf{X}\| \to +\infty$, and may grow at most linearly in X_3 as $X_3 \to +\infty$. This solution admits the series representation

$$W(\mathbf{X}) = \sum_{k=1}^{\infty} \frac{p_k(\mathbf{X}/\|\mathbf{X}\|)}{\|\mathbf{X}\|^{k+1}} \quad as \ \|\mathbf{X}\| \to +\infty, \ \mathbf{X} \in \mathbb{R}^3_-, \quad (2.2.33)$$

where $p_k(\mathbf{X})$ are harmonic polynomials, homogeneous of order k, and including only even powers of X_3. Furthermore,

$$W(\mathbf{X}) = DX_3 + q + O(e^{-\alpha X_3}), \quad X_3 \to +\infty, \quad (2.2.34)$$

where α is a positive constant,

$$D = -(\mathrm{mes}_2 g)^{-1} \int_G F(\mathbf{X}) d\mathbf{X} - \int_{\partial G} P(\mathbf{X}) ds, \quad (2.2.35)$$

and $q = \mathrm{Const}$.

Proof

Uniqueness. Let a function W satisfy the homogeneous problem (2.2.15), (2.2.16). Assume that θ is the same cut-off function as in the proof of Lemma 2.2. The function $\mathcal{V}(\mathbf{X}) = W(\mathbf{X})\theta(X_3)$ satisfies problem (2.2.30)–(2.2.32). Using the Fourier method and the constraint on the behaviour of the field W at infinity we derive the asymptotic formula (2.2.34).

Let $G_R = (\mathbb{R}^3_- \cap B_R) \cup \{\mathbf{X} : \mathbf{X} \in \Pi_+, X_3 < R\}$. We apply Green's formula to W and a constant field in G_R:

$$\int_{\{\|\mathbf{X}\|=R,\ X_3<0\}} \frac{\partial W}{\partial R} ds = \int_{\{X_3=R,\ \mathbf{X}\in G_R\}} \frac{\partial W}{\partial X_3} ds. \quad (2.2.36)$$

As $R \to +\infty$, the left-hand side of (2.2.36) vanishes, whereas the right-hand side tends to D. Therefore, $D = 0$. Hence, $W \in \mathcal{H}^1(G)$, and it follows from Lemma 2.2 that $W = 0$.

Existence and asymptotics. First, we shall construct a special solution Υ of the homogeneous problem (2.2.15), (2.2.16) such that

$$\Upsilon(\mathbf{X}) = \frac{X_3}{\mathrm{mes}_2 g} + \mathrm{Const} + O(e^{-\alpha X_3}) \quad \text{as } X_3 \to +\infty,\ \mathbf{X} \in \Pi_+,$$

$$\Upsilon(\mathbf{X}) = \frac{1}{2\pi\|\mathbf{X}\|} + O\left(\frac{1}{\|\mathbf{X}\|^2}\right) \quad \text{as } \|\mathbf{X}\| \to \infty,\ \mathbf{X} \in \mathbb{R}^3_-.$$

We seek the solution of (2.2.15), (2.2.16) in the form

$$\Upsilon = \frac{X_3}{\mathrm{mes}_2 g}\theta(X_3) + w(\mathbf{X}),$$

where $w \in \mathcal{H}^1(G)$ and satisfies the boundary value problem

$$-\Delta w(\mathbf{X}) = 2\frac{\theta'}{\mathrm{mes}_2 g} + \frac{\theta'' X_3}{\mathrm{mes}_2 g} \quad \text{in } G,$$

$$\left.\frac{\partial w}{\partial n}\right|_{\partial G} = 0.$$

Applying Lemma 2.2 we obtain the existence of w and the asymptotic representation (2.2.26), as $X_3 \to +\infty$, and

$$w = \frac{\mathrm{Const}}{\|\mathbf{X}\|} + O\left(\frac{1}{\|\mathbf{X}\|^2}\right) \quad \text{as } \|\mathbf{X}\| \to +\infty,\ \mathbf{X} \in \mathbb{R}^3_-. \quad (2.2.37)$$

Replacing W in (2.2.36) by Υ and taking the limit as $R \to +\infty$, we obtain that the constant coefficient in (2.2.37) is equal to $(2\pi)^{-1}$.

By Lemma 2.2 there exists a solution of (2.2.15), (2.2.16) from the class $\mathcal{H}^1(G)$, and the asymptotic formulae (2.2.25), (2.2.26) hold. Subtracting Υ multiplied by the appropriate constant we obtain that there exists a solution of (2.2.15), (2.2.16) with the asymptotic behaviour (2.2.33), (2.2.34) at infinity.

It remains to prove (2.2.35). Applying Green's formula to W and the constant field in G_R and taking the limit as $R \to \infty$, we obtain that the constant coefficient D from (2.2.34) is given by (2.2.35). □

REMARK 2.1

Let Υ be the function constructed in Lemma 2.3. Applying Green's formula to the functions W and Υ in $G \cap B_r(O)$ as $r \to \infty$, we obtain that the coefficient q in formula (2.2.34) is given by

$$q = \int_G \Upsilon(\mathbf{X})\mathcal{F}(\mathbf{X})d\mathbf{X} + \int_{\partial G} \Upsilon(\mathbf{X})\mathcal{P}(\mathbf{X})ds. \qquad (2.2.38)$$

2.2.4 Junction layer

In what follows an important place will be allocated for a solution of the Neumann problem in G decreasing exponentially as $X_3 \to +\infty$ and decreasing as $O(\|\mathbf{X}\|^{-2})$ when $\|\mathbf{X}\| \to +\infty$, $X_3 \leq 0$. It will be called the *junction layer* (compare with Section 1.6.1). We begin with problem (2.2.15), (2.2.16) and consider its solution W defined by Lemma 2.3. Let

$$\mathcal{W}(\mathbf{X}) = W(\mathbf{X}) - (DX_3 + q)\eta(X_3),$$

where η is a smooth function on $\mathbb{R}$ such that $\eta(t) = 0$ for $t < 1$ and $\eta(t) = 1$ for $t > 2$. By Lemma 2.3, the function $\mathcal{W}$ is the junction layer solution.

LEMMA 2.4

The triple $(\mathcal{W}, D, q)$ can be uniquely found by solving the Neumann problem

$$-\Delta_X \mathcal{W}(\mathbf{X}) = F(\mathbf{X}) + q\eta''(X_3)$$
$$+ D(X_3\eta''(X_3) + 2\eta'(X_3)) \quad \text{for } \mathbf{X} \in G, \qquad (2.2.39)$$
$$\frac{\partial \mathcal{W}}{\partial n_X}(\mathbf{X}) = P(\mathbf{X}), \quad \mathbf{X} \in \partial G, \qquad (2.2.40)$$

in the class of functions $\mathcal{W}$ subject to

$$\mathcal{W}(\mathbf{X}) = O(\|\mathbf{X}\|^{-2}) \quad \text{as } \|\mathbf{X}\| \to +\infty \text{ and } X_3 \leq 0, \qquad (2.2.41)$$

MODEL PROBLEMS

$$\mathcal{W}(\mathbf{X}) = O(e^{-\alpha X_3}) \quad \text{as } X_3 \to +\infty, \qquad (2.2.42)$$

where α is a positive number.

Proof

Since it is sufficient to check only the uniqueness of $(\mathcal{W}, D, q)$ we may suppose that $F = 0$ and $P = 0$. By Green's formula (2.2.13) applied to the functions $\mathcal{W}$ and 1,

$$0 = \int_G \{(DX_3 + q)\eta''(X_3) + 2D\eta'(X_3)\}d\mathbf{X}$$

$$= D \int_G \eta'(X_3)d\mathbf{X} = D \ \text{mes}_2 g,$$

which implies $D = 0$. Similarly,

$$0 = q \int_G \Upsilon(\mathbf{X})\eta''(X_3)d\mathbf{X}.$$

Consider the integral

$$I(N) = \int_G \Upsilon(\mathbf{X})\Delta_X \eta(X_3 - N)d\mathbf{X}.$$

Since Υ is harmonic and has zero normal derivative on ∂G, the integral $I(N)$ does not depend on N. Hence,

$$0 = q \int_G \left(\frac{X_3}{\text{mes}_2 g} + C\right)\eta''(X_3 - N)d\mathbf{X} + O(e^{-\alpha N})$$

with $\alpha > 0$ and $q = 0$. $\qquad \square$

2.2.5 Model problem for the bottom region

Formulation of the problem, space $E(\Pi_-)$

We introduce the local scaled variables (Y_1, Y_2, Y_3) in the vicinity of the bottom region (see Fig. 2.4), and let Π_- denote the semi-infinite cylinder

$$\Pi_- = \{\mathbf{Y} : Y_3 < 0, \ (Y_1, Y_2) \in g\},$$

and let
$$S = \{\mathbf{Y} : Y_3 = 0, (Y_1, Y_2) \in g \subset \mathbb{R}^2\}$$
denote the base of Π_-.

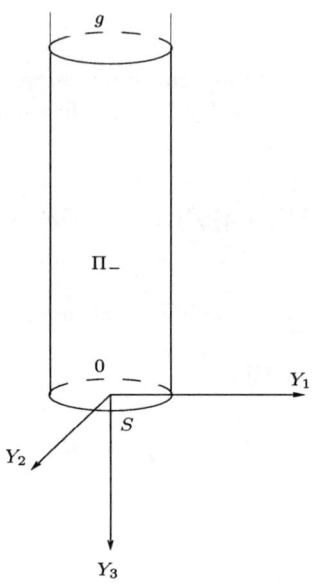

FIG. 2.4. Limit bottom region.

The asymptotic procedure will require the following model boundary value problem:

$$-\Delta_Y v(\mathbf{Y}) = \mathcal{G}(\mathbf{Y}), \quad \mathbf{Y} \in \Pi_-, \qquad (2.2.43)$$

$$\frac{\partial v}{\partial n_Y}(\mathbf{Y}) = H(\mathbf{Y}), \quad \mathbf{Y} \in \partial\Pi_- \setminus S, \qquad (2.2.44)$$

$$v(\mathbf{Y}) = \Phi(\mathbf{Y}), \quad \mathbf{Y} \in S. \qquad (2.2.45)$$

We seek a solution which belongs to the space $E(\Pi_-)$ of functions v with the finite norm

$$\|v\|_{E(\Pi_-)} = \left(\int_{\Pi_-} \|\nabla v\|^2 d\mathbf{Y} + \int_{\Pi_- \cap B_1} |v|^2 d\mathbf{Y}\right)^{1/2}. \qquad (2.2.46)$$

Using the arguments similar to those presented at the end of the proof of Lemma 2.1, we deduce

$$\sup_{Y_3<0} \frac{\|v(\cdot,Y_3)\|^2_{L_2(g)}}{1+|Y_3|} \leq \text{Const } \|v\|^2_{E(\Pi_-)}.$$

We assume that the function Φ in (2.2.45) belongs to the space $H^{1/2}(S)$ of traces on S of functions in $E(\Pi_-)$. The norm in $H^{1/2}(S)$ is defined as

$$\|\Phi\|_{H^{1/2}(S)} = \inf\,\{\|V\|_{E(\Pi_-)} : V \in E(\Pi_-), V = \Phi \text{ on } S\}.$$

It is well known that $H^{1/2}(S)$ can be supplied with the equivalent norm

$$\|\Phi\| = \left(\int_S\int_S \frac{(\Phi(\mathbf{X})-\Phi(\mathbf{Y}))^2}{|\mathbf{X}-\mathbf{Y}|^2} ds_{\mathbf{X}} ds_{\mathbf{Y}} + \int_S \Phi(\mathbf{X})^2 ds_{\mathbf{X}}\right)^{1/2}$$

(see, for example, Hörmander (1983), Section 7.9). Let G and H be functions with bounded supports defined on Π_- and $\partial\Pi_-\setminus S$ that belong to $L_2(\Pi_-)$ and $L_2(\partial\Pi_-\setminus S)$, respectively.

A solution of problem (2.2.43)–(2.2.45) is understood as a function $v \in E(\Pi_-)$ such that $v = \Phi$ on S, and

$$\int_{\Pi_-} \nabla v \cdot \nabla \mathcal{V} d\mathbf{Y} = \int_{\Pi_-} \mathcal{G}\mathcal{V} d\mathbf{Y} + \int_{\partial\Pi_-} H\mathcal{V} ds, \qquad (2.2.47)$$

for all $\mathcal{V} \in E(\Pi_-)$ subject to $\mathcal{V} = 0$ on S.

Solvability and asymptotics

LEMMA 2.5

There exists a unique solution of the problem (2.2.43)–(2.2.45). It satisfies

$$v(\mathbf{Y}) = C + O(e^{\alpha Y_3}) \quad \text{as } Y_3 \to -\infty, \qquad (2.2.48)$$

where

$$C = (\text{mes}_2 g)^{-1}\left(\int_S \Phi(\mathbf{Y}) ds - \int_{\Pi_-} Y_3 \mathcal{G}(Y) d\mathbf{Y}\right.$$

$$\left. - \int_{\partial\Pi_-\setminus S} Y_3 H(\mathbf{Y}) ds\right), \qquad (2.2.49)$$

and α is a positive constant.

Proof

Existence. Let $E^0(\Pi_-; S)$ be the subspace in $E(\Pi_-)$ of functions that vanish on S. Since, for a function $v \in E^0(\Pi_-; S)$, the second integral in (2.2.46) can be estimated by the first one, the functional

$$v \to \left(\int_{\Pi_-} \|\nabla v\|^2 d\mathbf{Y} \right)^{1/2}$$

can be regarded as a norm on the space $E^0(\Pi_-; S)$ which is equivalent to the norm (2.2.46).

We seek a solution of (2.2.43)–(2.2.45) in the form

$$v = v_0 + v_1, \tag{2.2.50}$$

where $v_1 \in E(\Pi_-)$, $v_1|_S = \Phi$, and $v_0 \in E^0(\Pi_-; S)$. Substituting (2.2.50) in (2.2.47) we obtain the equation for v_0:

$$\int_{\Pi_-} \nabla v_0 \cdot \nabla \mathcal{V} d\mathbf{X} = -\int_{\Pi_-} \nabla v_1 \cdot \nabla \mathcal{V} d\mathbf{X}$$

$$+ \int_{\Pi_-} \mathcal{G}\mathcal{V} d\mathbf{X} + \int_{\partial \Pi_-} H\mathcal{V} ds, \tag{2.2.51}$$

which should hold for any $\mathcal{V} \in E^0(\Pi_-; S)$. Since the right-hand side in (2.2.51) can be regarded as a linear continuous functional on $E^0(\Pi_-; S)$, we apply the Riesz representation theorem and obtain the existence of v_0 and, hence, the existence of the solution v.

Uniqueness. Let v be a solution of the homogeneous problem (2.2.43)–(2.2.45). Then, $v \in E^0(\Pi_-; S)$, and it follows from (2.2.47) with $\mathcal{V} = v$ that $\nabla v = 0$. Hence, $v = 0$.

The asymptotic representation (2.2.48) is obtained, for example, by the Fourier method.

Using Green's formula for the functions v and Y_3 in $\{\mathbf{Y} \in \Pi_- : |Y_3| < r\}$ we obtain (as $r \to +\infty$) the formula (2.2.49). □

REMARK 2.2

By (2.2.48), (2.2.49) we have that the solution v decays exponentially at infinity if and only if the right-hand sides of (2.2.43)–(2.2.45) satisfy the orthogonality condition

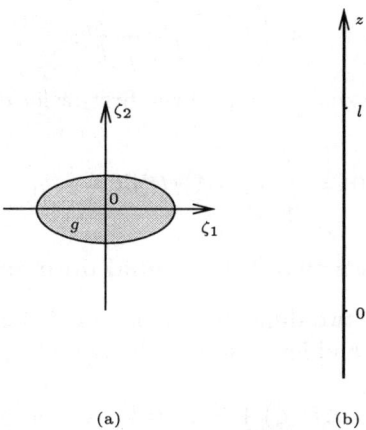

FIG. 2.5. Model regions for a thin cylinder.

$$\int_S \Phi(\mathbf{Y})ds - \int_{\Pi_-} Y_3 \mathcal{G}(\mathbf{Y})d\mathbf{Y} - \int_{\partial\Pi_-\setminus S} Y_3 H(\mathbf{Y})ds = 0. \qquad (2.2.52)$$

We shall call a solution v of problem (2.2.43)–(2.2.45) a *bottom layer* if it decays exponentially as $X_3 \to -\infty$. By Lemma 2.5, v is a bottom layer if and only if (2.2.52) holds.

2.2.6 Two model problems for a thin cylinder

The asymptotic procedure requires solutions of model problems on a one-dimensional segment (on the axis of the cylinder) and in a two-dimensional bounded domain (a scaled cross-section of the cylinder).

Model problem on a segment

We consider the Dirichlet problem on the segment $(0, l)$:

$$\frac{d^2 W}{dz^2}(z) = r(z), \quad z \in (0, l),$$

$$W(0) = b_0, \quad W(l) = b_1.$$

If $f \in L_1(0, l)$ then the solution is given by

$$W(z) = \left(\frac{z}{l} - 1\right) \int_0^z r(t) t\, dt + \frac{z}{l} \int_z^l (t - l) r(t)\, dt$$

$$+ \left(1 - \frac{z}{l}\right)b_0 + \frac{z}{l}b_1. \qquad (2.2.53)$$

In the text below we shall also need the first-order derivative of w at the origin:

$$W'(0) = \int_0^l (\frac{t}{l} - 1)r(t)dt + \frac{1}{l}(b_1 - b_0). \qquad (2.2.54)$$

Model problem in a two-dimensional domain

Let g be the same two-dimensional bounded domain with Lipschitz boundary as before, and let U satisfy the boundary value problem

$$\Delta U(\zeta) + \mathcal{F}(\zeta) = 0, \quad \zeta \in g, \qquad (2.2.55)$$

$$\frac{\partial U}{\partial n}(\zeta) = \mathcal{P}(\zeta), \quad \zeta \in \partial g, \qquad (2.2.56)$$

where $\mathcal{F}$ and $\mathcal{P}$ are functions, which belong to $L_2(g)$ and $L_2(\partial g)$ respectively. The problem (2.2.55), (2.2.56) is solvable in $H^1(g)$ provided

$$\int_g \mathcal{F}(\zeta)d\zeta + \int_{\partial g} \mathcal{P}(\zeta)ds = 0,$$

and the solution is defined up to an arbitrary additive constant. To specify a unique solution we require that

$$\int_g U(\zeta)d\zeta = 0.$$

2.2.7 Algebraic system for the skeleton

By the skeleton of the multi-structure we understand the union of Ω and one-dimensional segments r_j of length $l^{(j)}$ (the axes of thin cylinders $\Pi_\varepsilon^{(j)}$ attached to Ω at the junction points $\mathbf{a}^{(j)}$).

With the skeleton we associate a certain algebraic system which can be solved explicitly.

Suppose that the body Ω has an unknown constant temperature $C\varepsilon^{-2}$ and that the total heat flux created by internal and boundary sources is equal to I. To each r_j we assign the area $\varepsilon^2 \mathrm{mes}_2 g^{(j)}$ of the cross-section of $\Pi_\varepsilon^{(j)}$. Let us denote by $T^{(j)}$ the flux through the junction points $\mathbf{a}^{(j)}$, $j = 1, \ldots, K$, coming to the corresponding segment. We

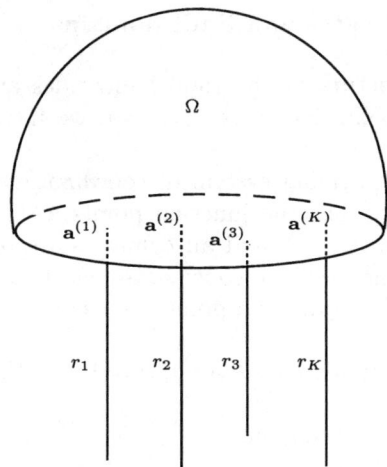

FIG. 2.6. Skeleton of the multi-structure Ω_ε.

assume that the distribution law for the temperature in the segments is linear and that the temperature at the j-th end point is equal to $J^{(j)}\varepsilon^{-2}$. The continuity of the temperature and the balance condition for the heat flux lead to the system for the unknowns $C, T^{(1)}, \ldots, T^{(K)}$:

$$T^{(j)} = (l^{(j)})^{-1}(J^{(j)} - C)\,\mathrm{mes}_2 g^{(j)}, \qquad (2.2.57)$$

$$\sum_{j=1}^{K} T^{(j)} + I = 0. \qquad (2.2.58)$$

Then

$$C = \left(\sum_{j=1}^{K} \frac{\mathrm{mes}_2 g^{(j)}}{l^{(j)}}\right)^{-1}\left(I + \sum_{j=1}^{K} \frac{\mathrm{mes}_2 g^{(j)}}{l^{(j)}} J^{(j)}\right).$$

The substitution of C into (2.2.57) gives an explicit representation for $T^{(j)}$.

We shall see that the system (2.2.57), (2.2.58) appears naturally in the construction of asymptotics of the solution $u(\mathbf{x}, \varepsilon)$.

2.3 Right-hand sides

Here the right-hand sides in (2.1.1)–(2.1.3) will be specified as multi-scaled asymptotic series. It is done in order to standardise a recurrent construction of terms in the asymptotic expansion of the solution.

2.3.1 Local coordinates and limit domains

To describe the structure of the right-hand sides and also the multi-scaled asymptotic expansion of the solution we introduce local scaled coordinates.

We recall that the global system of coordinates $\mathbf{x} = (x_1, x_2, x_3)$ is chosen in such a way that the junction points $\mathbf{a}^{(j)}$, $j = 1, \ldots, K$, are located on the plane $x_3 = 0$, and all cylinders are parallel to the Ox_3 axis directed downwards. These coordinates will be used in Ω.

In the vicinity of the junction points we shall use the coordinates

$$\mathbf{X}^{(j)} = \varepsilon^{-1}(\mathbf{x} - \mathbf{a}^{(j)}), \quad j = 1, \ldots, K, \qquad (2.3.1)$$

and introduce the model domains

$$G^{(j)} = \mathbb{R}^3_- \cup \Pi^{(j)}_+,$$

where

$$\mathbb{R}^3_- = \{\mathbf{X}^{(j)} : X^{(j)}_3 < 0\},$$

and

$$\Pi^{(j)}_+ = \{\mathbf{X}^{(j)} : X^{(j)}_3 \geq 0, (X^{(j)}_1, X^{(j)}_2) \in g^{(j)}\}.$$

The coordinate system

$$(\zeta^{(j)}_1, \zeta^{(j)}_2, z) = (X^{(j)}_1, X^{(j)}_2, x_3), \qquad (2.3.2)$$

is introduced in $\Pi^{(j)}_\varepsilon$.

Scaled variables

$$\mathbf{Y}^{(j)} = (X^{(j)}_1, X^{(j)}_2, \varepsilon^{-1}(x_3 - l^{(j)})) \qquad (2.3.3)$$

are used in the vicinity of the end of $\Pi^{(j)}_\varepsilon$. We shall also need the sets

$$\Pi^{(j)}_- = \{\mathbf{Y}^{(j)} : Y^{(j)}_3 < 0, (Y^{(j)}_1, Y^{(j)}_2) \in g^{(j)}\}$$

and

$$S^{(j)} = \{\mathbf{Y}^{(j)} : (Y^{(j)}_1, Y^{(j)}_2) \in g^{(j)}, Y^{(j)}_3 = 0\}.$$

2.3.2 Cut-off functions

We define the cut-off function for Ω as

$$\mathfrak{X}(\mathbf{x},\varepsilon) = \prod_{j=1}^{K}\chi(\mathbf{X}^{(j)}),\ \mathbf{x}\in\Omega, \qquad (2.3.4)$$

where χ is smooth, $\chi(\mathbf{X}) = 0$ for $\|\mathbf{X}\| < 1$ and $\chi(\mathbf{X}) = 1$ for $\|\mathbf{X}\| > 2$; $\mathfrak{X} = 0$ in the cylinders $\Pi_\varepsilon^{(j)}$.

Also, in thin cylinders we use the smooth cut-off function $\eta(Z)$ which equals 0 for $Z < 1$, and $\eta(Z) = 1$ for $Z > 2$.

The third smooth cut-off function Ξ for the junction layer is defined in $\mathbb{R}^2$ in such a way that $\Xi(\mathbf{x}) = 0$ for $(x_1^2+x_2^2)^{1/2} > d/2$, and $\Xi(\mathbf{x}) = 1$ as $(x_1^2 + x_2^2)^{1/2} < d/4$, where d is a sufficiently small constant independent of ε.

2.3.3 Asymptotic representation of the right-hand sides

Let us consider problem (2.1.1)–(2.1.3) with the right-hand sides defined by the multi-scaled asymptotic series. The right-hand side in (2.1.1) is given by

$$F(\mathbf{x},\varepsilon) \sim \sum_{k=0}^{\infty}\varepsilon^k\Big\{\mathfrak{X}(\mathbf{x},\varepsilon)f_k(\mathbf{x}) \qquad (2.3.5)$$
$$+\sum_{j=1}^{K}[\varepsilon^{-3}F_k^{(j)}(\mathbf{X}^{(j)}) + \varepsilon^{-2}\mathcal{F}_k^{(j)}(\xi^{(j)},z)\eta(\varepsilon^{-1}z) + \varepsilon^{-4}\mathcal{G}_k^{(j)}(\mathbf{Y}^{(j)})]\Big\}$$

where $f_k \in L_2(\Omega), F_k^{(j)} \in L_2(G^{(j)}),\ \mathcal{G}_k^{(j)} \in L_2(\Pi_-^{(j)})$ and f_k are smooth in fixed neighbourhoods of junction points; we also assume that $F_k^{(j)}, \mathcal{G}_k^{(j)}$ are supported by balls of fixed radius centred at the origins of the corresponding coordinate systems and that $\mathcal{F}_k^{(j)} \in C^\infty([0, l^{(j)}]; L_2(g^{(j)}))$.

Here we have used the space $C^\infty([0,l]; L_2(g))$ of smooth (with respect to z) functions taking values in $L_2(g)$. In other words, for these functions

$$\sup_{0<z<l}\left\|\frac{d^k}{dz^k}v(\cdot,z)\right\|_{L_2(g)} < \infty,\quad k = 0, 1, \ldots.$$

Replacing g by ∂g we introduce the space $C^\infty([0,l]; L_2(\partial g))$.

The asymptotic expansion (2.3.5) is understood in the following sense: the $L_2(\Omega_\varepsilon)$ norm of the difference between F and the partial sum of the series (2.3.5) extended over $k = 0,\ldots,N$ is estimated by Const $\varepsilon^{N-3/2}$. In fact, this is a majorant for the $L_2(\Omega_\varepsilon)$ norm of the next term ($k = N + 1$) in (2.3.5).

The right-hand side in the Neumann boundary condition (2.1.2) is defined as

$$P(\mathbf{x}, \varepsilon) \sim \sum_{k=0}^{\infty} \varepsilon^k \Big\{ \mathfrak{X}(\mathbf{x}, \varepsilon) p_k(\mathbf{x}) \qquad (2.3.6)$$

$$+ \sum_{j=1}^{K} [\varepsilon^{-2} P_k^{(j)}(\mathbf{X}^{(j)}) + \varepsilon^{-1} \mathcal{P}_k^{(j)}(\xi^{(j)}, z) \eta(\varepsilon^{-1} z) + \varepsilon^{-3} H_k^{(j)}(\mathbf{Y}^{(j)})] \Big\},$$

where $p_k \in L_2(\partial\Omega), P_k^{(j)} \in L_2(\partial G^{(j)}), H_k^{(j)} \in L_2(\partial\Pi_-^{(j)} \setminus S^{(j)})$, and p_k are smooth in fixed neighbourhoods of junction points; $P_k^{(j)}, H_k^{(j)}$ are supported by fixed balls centred at the origins of the corresponding coordinate systems, and $\mathcal{P}_k^{(j)} \in C^\infty([0, l^{(j)}]; L_2(\partial g^{(j)}))$.

The sign $\sim$ in (2.3.6) means that the $L_2(\partial\Omega_\varepsilon \setminus S_\varepsilon)$ norm of the difference between P and the partial sum of the series (2.3.6) extended over $k = 0, \ldots, N$, is estimated by Const ε^{N-1}. This is a majorant for the $L_2(\partial\Omega_\varepsilon \setminus S_\varepsilon)$ norm of the next term ($k = N + 1$) in (2.3.6).

The right-hand side of the Dirichlet condition (2.1.3) has the form

$$\Phi^{(j)}(x, \varepsilon) \sim \varepsilon^{-2} \sum_{k=0}^{\infty} \varepsilon^k \Phi_k^{(j)}(Y^{(j)}), \qquad (2.3.7)$$

with $\Phi_k^{(j)} \in H^{1/2}(S^{(j)})$. To explain the meaning of the asymptotic expansion (2.3.7), we need the space $H^{1/2}(S_\varepsilon^{(j)})$ of traces on $S_\varepsilon^{(j)}$ for functions in $H^1(\Omega_\varepsilon)$. The norm in $H^{1/2}(S_\varepsilon^{(j)})$ is defined by

$$\|\Phi\|_{H^{1/2}(S_\varepsilon^{(j)})} = \inf \{\|u\|_{H^1(\Omega_\varepsilon)} : u \in H^1(\Omega_\varepsilon), u = \Phi \text{ on } S_\varepsilon^{(j)}\}. \quad (2.3.8)$$

Relation (2.3.7) means that the $H^{1/2}(S_\varepsilon^{(j)})$ norm of the difference between $\Phi^{(j)}$ and the sum of the first $N + 1$ terms in the right-hand side of (2.3.7) is estimated by Const $\varepsilon^{N-1/2}$. This is a majorant for the next term ($k = N + 1$) in (2.3.7).

We see that the asymptotic series (2.3.5)–(2.3.7) contain terms corresponding to

(1) the three-dimensional set Ω;
(2) small neighbourhoods of the junction points;
(3) thin cylinders $\Pi_\varepsilon^{(j)}$, $j = 1, \ldots, K$;
(4) small neighbourhoods of $S_\varepsilon^{(j)}, j = 1, \ldots, K$.

Writing the right-hand sides F, P and $\Phi^{(j)}$ in the above form proves to be important for the construction of the complete asymptotic expansion of

THE LEADING TERM OF THE SOLUTION   81

the solution. In fact, the discrepancies in the equations and the boundary conditions generated by subsequent approximations of the solution have this general form. Hence all terms of this asymptotic series can be constructed uniformly.

Particular examples of the right-hand sides will be considered in Section 2.8.

2.4 The leading term of the asymptotic solution

The leading approximation u_0 of the field u will be sought in the form

$$u_0(\mathbf{x}, \varepsilon) = \varepsilon^{-2} C_0 + u_{0,\Omega}(\mathbf{x}) \mathfrak{X}(\mathbf{x}, \varepsilon) + \varepsilon^{-1} \sum_{j=1}^{K} \mathcal{W}_0^{(j)}(\mathbf{X}^{(j)}) \Xi(\mathbf{x} - \mathbf{a}^{(j)})$$

$$+ \sum_{j=1}^{K} (\varepsilon^{-2} W_0^{(j)}(z) + U_0^{(j)}(\zeta^{(j)}, z) + \varepsilon^{-2} v_0^{(j)}(\mathbf{Y}^{(j)})) \eta(\varepsilon^{-1} z), \quad (2.4.1)$$

where $u_{0,\Omega}$ is a solution of the boundary value problem (2.2.1), (2.2.2) in Ω with sources of unknown intensity at the junction points; $\mathcal{W}_0^{(j)}$ are junction layer terms; $W_0^{(j)}, U_0^{(j)}$ are terms allocated for the cylinder $\Pi_\varepsilon^{(j)}$; and $v_0^{(j)}$ is the bottom layer. These three types of terms have been defined in Sections 2.2.4–2.2.6.

To simplify formulae, we shall use the notations $\mathbf{X}, \mathbf{Y}$ and ζ without the superscript index j, and we hope that it will not cause any confusion.

2.4.1 Domain Ω

To find the function $u_{0,\Omega}$ one should take into account the leading terms f_0 and p_0 in (2.3.5), (2.3.6) which correspond to Ω. The boundary value problem is

$$-\Delta u_{0,\Omega} = f_0 \quad \text{in } \Omega, \qquad (2.4.2)$$

$$\frac{\partial u_{0,\Omega}}{\partial n} = p_0 + \sum_{j=1}^{K} T_0^{(j)} \delta(\mathbf{x} - \mathbf{a}^{(j)}) \quad \text{on } \partial\Omega. \qquad (2.4.3)$$

Here $T_0^{(j)}$ are constant coefficients which will be specified later. According to Section 2.2.2, problem (2.4.2), (2.4.3) has a solution of the form

$$u_{0,\Omega} = u_{0,r} + u_{0,s},$$

where $u_{0,r} \in H^1(\Omega)$ solves

$$-\Delta u_{0,\Omega} = f_0 + (\text{mes}_3\Omega)^{-1} \sum_{j=1}^{K} T_0^{(j)} \quad \text{in } \Omega, \qquad (2.4.4)$$

$$\frac{\partial u_{0,\Omega}}{\partial n} = p_0 \quad \text{on } \partial\Omega, \qquad (2.4.5)$$

and

$$u_{0,s} = \sum_{j=1}^{K} T_0^{(j)} N(\mathbf{x}, \mathbf{a}^{(j)}).$$

Problem (2.4.2), (2.4.3) is solvable if and only if

$$\int_\Omega f_0(\mathbf{x})d\mathbf{x} + \int_{\partial\Omega} p_0(\mathbf{x})ds + \sum_{j=1}^{K} T_0^{(j)} = 0. \qquad (2.4.6)$$

As above, to achieve uniqueness we require

$$\int_\Omega u_{0,r}(\mathbf{x})d\mathbf{x} = 0. \qquad (2.4.7)$$

Note that equation (2.4.6) is one of the relations required to specify $T_0^{(j)}$.

Owing to the smoothness of f_0 and p_0 in the vicinity of junction points, the field $u_{0,r}$ is also smooth in neighbourhoods of junction points $\mathbf{a}^{(j)}$, $j = 1, \ldots, K$.

By (2.2.8), the singular part $u_{0,s}$ admits the representation

$$u_{0,s}(\mathbf{x}) = \frac{T_0^{(j)}}{2\pi \|\mathbf{x} - \mathbf{a}^{(j)}\|} + \mu^{(j)}(\mathbf{x}) \qquad (2.4.8)$$

as $\mathbf{x} \to \mathbf{a}^{(j)}$; here $\mu^{(j)}$ is smooth in the vicinity of $\mathbf{a}^{(j)}$.

Multiplying $u_{0,\Omega}$ by the cut-off function $\mathcal{X}$ we obtain the field which produces a discrepancy in the right-hand sides (2.4.2), (2.4.3) in the vicinity of junction points:

$$-\Delta(\mathcal{X} u_{0,\Omega}) = \mathcal{X} f_0 - u_{0,\Omega}\Delta\mathcal{X} - 2\nabla\mathcal{X} \cdot \nabla u_{0,\Omega} \quad \text{in } \Omega, \qquad (2.4.9)$$

$$\frac{\partial(\mathcal{X} u_{0,\Omega})}{\partial n} = \mathcal{X} p_0 \quad \text{on } \partial\Omega. \qquad (2.4.10)$$

Here, we use the fact that $\partial\mathcal{X}/\partial n = 0$ on $\partial\Omega$ in the vicinity of junction points.

THE LEADING TERM OF THE SOLUTION

Owing to the smoothness of $u_{0,r}$ at the junction points and formula (2.4.8) for $u_{0,s}$, we can re-expand the discrepancy terms in (2.4.9), (2.4.10) in the scaled variables $\mathbf{X}$:

$$u_{0,\Omega}\Delta\mathfrak{X} + 2\nabla\mathfrak{X}\cdot\nabla u_{0,\Omega} \sim \frac{T_0^{(j)}}{2\pi}\varepsilon^{-3}\Big\{\|\mathbf{X}\|^{-1}\Delta_X\chi(\mathbf{X})$$

$$- 2\|\mathbf{X}\|^{-3}\mathbf{X}\cdot\nabla\chi(\mathbf{X})\Big\} + \sum_{k=0}^{\infty}\varepsilon^{k-2}F_{0,k}^{(j)}(\mathbf{X}), \qquad (2.4.11)$$

where $F_{0,k}^{(j)}$ are smooth functions with bounded support. We note that the last sum begins with the term of order $O(\varepsilon^{-2})$, whereas the corresponding series in expansion (2.3.5) begins with a term of order $O(\varepsilon^{-3})$.

2.4.2 Junction layer

In order to construct the junction layer $\mathcal{W}_0^{(j)}$ we should take into account the right-hand sides $F_0^{(j)}, P_0^{(j)}$ in (2.3.5), (2.3.6) and the leading discrepancy in (2.4.11).

The junction layer $\mathcal{W}_0^{(j)}$ together with the constants $D_0^{(j)}, q_0^{(j)}$ (see Section 2.2.4) satisfies the boundary value problem

$$-\Delta_X \mathcal{W}_0^{(j)}(\mathbf{X}) = F_0^{(j)} + \frac{T_0^{(j)}}{2\pi}\Big\{\|\mathbf{X}\|^{-1}\Delta_X\chi(\mathbf{X})$$

$$- 2\|\mathbf{X}\|^{-3}\mathbf{X}\cdot\nabla_X\chi(\mathbf{X})\Big\} + q_0^{(j)}\eta''(X_3)$$

$$+ D_0^{(j)}(X_3\eta''(X_3) + 2\eta'(X_3)), \quad \mathbf{X}\in G^{(j)}, \qquad (2.4.12)$$

$$\frac{\partial\mathcal{W}_0^{(j)}}{\partial n_X}(\mathbf{X}) = P_0^{(j)}(\mathbf{X}), \quad \mathbf{X}\in\partial G^{(j)}. \qquad (2.4.13)$$

According to Lemma 2.4, this problem has a unique solution $(\mathcal{W}_0^{(j)}, D_0^{(j)}, q_0^{(j)})$ such that

$$\mathcal{W}_0^{(j)}(\mathbf{X}) = O(\|\mathbf{X}\|^{-2}) \text{ as } \|\mathbf{X}\|\to+\infty, \ X_3<0,$$

$$\mathcal{W}_0^{(j)}(\mathbf{X}) = O(e^{-\alpha X_3}) \text{ as } X_3\to+\infty,$$

where α is a positive constant.

The coefficient $D_0^{(j)}$ can be evaluated by (2.2.35). To do this, we prove the following assertion:

LEMMA 2.6

The identity

$$\frac{1}{2\pi}\int_{G^{(j)}} \{\|\mathbf{X}\|^{-1}\Delta_X\chi(\mathbf{X}) - 2\|\mathbf{X}\|^{-3}\mathbf{X}\cdot\nabla_X\chi(\mathbf{X})\}d\mathbf{X} = -1 \quad (2.4.14)$$

is valid for the cut-off function χ introduced in Section 2.3.2.

Proof

We represent the integrand as

$$\Delta(\|\mathbf{X}\|^{-1}\chi(\mathbf{X})) - \chi(\mathbf{X})\Delta(\|\mathbf{X}\|^{-1}).$$

Since the cut-off function vanishes in the vicinity of the origin, the second term is equal to zero. Hence, the left-hand side in (2.4.14) is equal to

$$\frac{1}{2\pi}\int_{G^{(j)}} \Delta(\|\mathbf{X}\|^{-1}\chi(\mathbf{X}))d\mathbf{X}. \quad (2.4.15)$$

Since the integrand has a compact support, the quantity (2.4.15) coincides with

$$\frac{1}{2\pi}\int_{B_R^-} \Delta(\|\mathbf{X}\|^{-1}\chi(\mathbf{X}))d\mathbf{X},$$

where B_R^- is the lower semi-ball of sufficiently large radius R. Integrating by parts we obtain that the last integral is equal to

$$-\frac{1}{2\pi}\int_{S_R^-} \frac{ds}{\|\mathbf{X}\|^2} = -1,$$

where S_R^- is the lower semi-sphere of radius R. This completes the proof. $\square$

By (2.2.35) and (2.4.14) we can specify $D_0^{(j)}$:

$$D_0^{(j)} = \frac{-1}{\operatorname{mes}_2 g_j}\left(\int_{G^{(j)}} F_0^{(j)}(\mathbf{X})d\mathbf{X} + \int_{\partial G^{(j)}} P_0^{(j)}(\mathbf{X})ds - T_0^{(j)}\right). \quad (2.4.16)$$

The constant $q_0^{(j)}$ is calculated in accordance with (2.2.38).

THE LEADING TERM OF THE SOLUTION

In the lower half-space, as $\|\mathbf{X}\| \to \infty$, the function $\mathcal{W}_0^{(j)}$ can be re-expanded in inverse powers of $\|\mathbf{X}\|$ (see Theorem 2.3). Hence, multiplying $\varepsilon^{-1}\mathcal{W}_0^{(j)}$ by the cut-off function Ξ we verify that discrepancies in the equation and boundary condition are concentrated in Ω and $\partial\Omega$, respectively, and can be presented as the asymptotic series

$$\sum_{k=1}^{\infty} \varepsilon^k f_{0,k}(\mathbf{x}), \quad \sum_{k=1}^{\infty} \varepsilon^k p_{0,k}(\mathbf{x}), \qquad (2.4.17)$$

where $f_{0,k} \in L_2(\Omega)$, $p_{0,k} \in L_2(\partial\Omega)$, and these functions vanish in the vicinity of the junction points. Note that the series (2.4.17) are one order higher than the series

$$\sum_{k=0}^{\infty} \varepsilon^k \mathfrak{X} f_k \text{ and } \sum_{k=0}^{\infty} \varepsilon^k \mathfrak{X} p_k$$

in (2.3.5), (2.3.6) which we started with.

It follows from (2.4.9), (2.4.10), (2.4.11) and (2.3.7) that the discrepancy produced in the Poisson equation by

$$u_{0,\Omega}\mathfrak{X} + \varepsilon^{-1}\sum_{j=1}^{K} \mathcal{W}_0^{(j)}(\mathbf{X})\Xi(\mathbf{x} - \mathbf{a}^{(j)})$$

in the vicinity of the junction point $\mathbf{a}^{(j)}$ is given in the form

$$\varepsilon^{-3}\left\{q_0^{(j)}\eta''(X_3) + D_0^{(j)}(X_3\eta''(X_3) + 2\eta'(X_3))\right\} + O(\varepsilon^{-2}), \qquad (2.4.18)$$

where the term $O(\varepsilon^{-2})$ is similar to that in the right-hand side of (2.4.11).

2.4.3 Thin cylinder

To define the leading part

$$\varepsilon^{-2}W_0^{(j)}(z) + U_0^{(j)}(\zeta, z)$$

of the asymptotic expansion in the cylinder $\Pi_\varepsilon^{(j)}$, we take into account the terms $\mathcal{F}_0^{(j)}, \mathcal{P}_0^{(j)}$ in the right-hand sides of (2.3.5), (2.3.6).

We have

$$-\Delta_x(\varepsilon^{-2}W_0^{(j)} + U_0^{(j)}) = -\varepsilon^{-2}\left(\frac{d^2}{dz^2}W_0^{(j)} + \Delta_\zeta U_0^{(j)}\right) - \frac{\partial^2}{\partial z^2}U_0^{(j)}$$

and
$$\frac{\partial}{\partial n_x}(\varepsilon^{-2}W_0^{(j)} + U_0^{(j)}) = \varepsilon^{-1}\frac{\partial}{\partial n_\zeta}U_0^{(j)}.$$

The function $U_0^{(j)}$ is required to satisfy

$$-\Delta_\zeta U_0^{(j)} = \mathcal{F}_0^{(j)} + \frac{d^2}{dz^2}W_0^{(j)} \quad \text{in } g,$$

$$\frac{\partial}{\partial n_\zeta}U_0^{(j)} = \mathcal{P}_0^{(j)} \quad \text{on } \partial g.$$

The solvability condition for this problem provides the ordinary differential equation for $W_0^{(j)}$:

$$\frac{d^2}{dz^2}W_0^{(j)} = -(\text{mes}_2 g)^{-1}\left(\int_g \mathcal{F}_0^{(j)} d\zeta + \int_{\partial g} \mathcal{P}_0^{(j)} ds\right). \quad (2.4.19)$$

In order to determine $U_0^{(j)}$ uniquely we impose the orthogonality condition

$$\int_g U_0^{(j)}(\zeta, z)dz = 0 \quad \text{for } z \in (0, l^{(j)}).$$

Clearly, $W_0^{(j)} \in C^\infty([0, l^{(j)}])$ and $U_0^{(j)} \in C^\infty([0, l^{(j)}]; H^1(g))$. Now, the discrepancy

$$-\Delta_x(\varepsilon^{-2}W_0^{(j)} + U_0^{(j)}) - \varepsilon^{-2}\mathcal{F}_0^{(j)} = -\frac{\partial^2}{\partial z^2}U_0^{(j)}$$

is smaller than the leading term $\varepsilon^{-2}\mathcal{F}_0^{(j)}$ in (2.3.5) corresponding to $\Pi_\varepsilon^{(j)}$.

Next, we derive the boundary condition for equation (2.4.19). The product

$$\eta(\varepsilon^{-1}z)(\varepsilon^{-2}W_0^{(j)} + U_0^{(j)})$$

gives a discrepancy in the right-hand side of the equation in the vicinity of the junction point

$$-\Delta\{\eta(\varepsilon^{-1}z)(\varepsilon^{-2}W_0^{(j)} + U_0^{(j)})\} = \eta(\varepsilon^{-1}z)\left(\varepsilon^{-2}\mathcal{F}_0^{(j)} - \frac{\partial^2}{\partial z^2}U_0^{(j)}\right)$$

$$- \varepsilon^{-2}(\varepsilon^{-2}W_0^{(j)} + U_0^{(j)})\eta'' - 2\varepsilon^{-1}\eta'\frac{\partial}{\partial z}(\varepsilon^{-2}W_0^{(j)} + U_0^{(j)}).$$

In the scaled coordinates $\mathbf{X}$ the discrepancy terms can be written as

$$-\varepsilon^{-4}W_0^{(j)}(0)\eta''(X_3) - \varepsilon^{-3}(X_3\eta''(X_3) + 2\eta'(X_3))\frac{\partial W_0^{(j)}}{\partial z}(0)$$

$$+ \varepsilon^{-2} \sum_{k=0}^{\infty} \varepsilon^k \mathcal{F}_{0,k}^{(j)}(\mathbf{X}), \qquad (2.4.20)$$

where $\mathcal{F}_{0,k}^{(j)}$ are smooth in the vicinity of $\mathbf{a}^{(j)}$ and have bounded supports.

We remark that the multiplication by $\eta(z/\varepsilon)$ does not leave any discrepancies in the Neumann boundary condition on the lateral surface of Π_ε. Comparing (2.4.18) and (2.4.20) we deduce that

$$W_0^{(j)}(0) = 0 \qquad (2.4.21)$$

and

$$\frac{\partial W_0^{(j)}}{\partial z}(0) = D_0^{(j)}. \qquad (2.4.22)$$

We note that (2.4.18) also includes the term $\varepsilon^{-3} q_0^{(j)} \eta''(X_3)$ which will be taken into account in the next step of the asymptotic algorithm. Namely, $W_1^{(j)}(0) = q_0^{(j)}$, where $W_1^{(j)}(z)$ is the quantity similar to $W_0^{(j)}(z)$ in the next step of the asymptotic procedure.

2.4.4 Bottom layer

The boundary value problem for $v_0^{(j)}$ will include the quantities $\mathcal{G}_0^{(j)}$, $H_0^{(j)}$, $\Phi_0^{(j)}$ from (2.3.5)–(2.3.7) and the discrepancy terms produced by the fields $\varepsilon^{-2} W_0^{(j)} + U_0^{(j)}$ in the thin cylinders $\Pi_\varepsilon^{(j)}$ and by the constant term $\varepsilon^{-2} C_0$ from (2.4.1). We remark that the junction layer also produces the discrepancy, but it is exponentially small and can be neglected.

Thus, the bottom layer $v_0^{(j)}$ should satisfy the boundary value problem

$$-\Delta_Y v_0^{(j)}(\mathbf{Y}) = \mathcal{G}_0^{(j)}(\mathbf{Y}) \text{ in } \Pi_-^{(j)}, \qquad (2.4.23)$$

$$\frac{\partial v_0^{(j)}(\mathbf{Y})}{\partial n} = H_0^{(j)}(\mathbf{Y}) \text{ on } \partial \Pi_-^{(j)} \setminus S^{(j)}, \qquad (2.4.24)$$

and

$$v_0^{(j)}(\mathbf{Y}) = \Phi_0^{(j)}(\mathbf{Y}) - C_0 - W_0^{(j)}(l^{(j)}) \text{ on } S^{(j)}. \qquad (2.4.25)$$

From Lemma 2.5 and Remark 2.2, this problem has a solution which decays exponentially at infinity if and only if

$$-C_0 + (\text{mes}_2 g^{(j)})^{-1} \left(\int_{S^{(j)}} \Phi_0^{(j)}(\mathbf{Y}) ds - \int_{\Pi_-^{(j)}} Y_3 \mathcal{G}_0^{(j)}(\mathbf{Y}) d\mathbf{Y} \right.$$

$$-\int_{\partial\Pi_-^{(j)}\setminus S^{(j)}} Y_3 H_0^{(j)}(\mathbf{Y})ds\bigg) = W_0^{(j)}(l^{(j)}). \qquad (2.4.26)$$

This yields the boundary condition for $W_0^{(j)}$ at the point $z = l^{(j)}$.

2.4.5 Evaluation of $W_0^{(j)}$

The function $W_0^{(j)}$ can be found from (2.4.19) and the boundary conditions (2.4.21), (2.4.26). It is given by (2.2.53), with $b_0 = 0$, $l = l^{(j)}$ and b_1 being equal to the right-hand side of (2.4.26); the function r is equal to the right-hand side of (2.4.19).

2.4.6 Evaluation of the constants $T_0^{(j)}$ and C_0

It follows from (2.2.54) and (2.4.16) that the boundary condition (2.4.22) is satisfied if and only if

$$T_0^{(j)}(\mathrm{mes}_2 g^{(j)})^{-1} = (l^{(j)})^{-1}(J_0^{(j)} - C_0), \qquad (2.4.27)$$

where

$$J_0^{(j)} = \frac{l^{(j)}}{\mathrm{mes}_2 g^{(j)}} \left(\int_{G^{(j)}} F_0^{(j)}(\mathbf{X}) d\mathbf{X} + \int_{\partial G^{(j)}} P_0^{(j)}(\mathbf{X}) ds \right)$$

$$+ \frac{1}{\mathrm{mes}_2 g^{(j)}} \int_0^{l^{(j)}} (l^{(j)} - z) \left(\int_{g^{(j)}} \mathcal{F}_0^{(j)}(\zeta; z) d\zeta + \int_{\partial g^{(j)}} \mathcal{P}_0^{(j)}(\zeta; z) ds \right) dz$$

$$+ \frac{1}{\mathrm{mes}_2 g^{(j)}} \left(\int_{S^{(j)}} \phi_0^{(j)}(\mathbf{Y}) ds - \int_{\Pi_-^{(j)}} Y_3 \mathcal{G}_0^{(j)}(\mathbf{Y}) d\mathbf{Y} \right.$$

$$\left. - \int_{\partial\Pi_-^{(j)}\setminus S^{(j)}} Y_3 H_0^{(j)}(\mathbf{Y}) ds \right). \qquad (2.4.28)$$

Let

$$I_0 = \int_\Omega f_0(\mathbf{x}) d\mathbf{x} + \int_{\partial\Omega} p_0(\mathbf{x}) ds. \qquad (2.4.29)$$

Then (2.4.6) can be written as

$$I_0 + \sum_{j=1}^K T_0^{(j)} = 0. \qquad (2.4.30)$$

COMPLETE ASYMPTOTIC EXPANSION

Equations (2.4.27), (2.4.30) coincide with (2.2.57), (2.2.58), and hence

$$C_0 = \left(\sum_{j=1}^{K} \frac{\mathrm{mes}_2 g^{(j)}}{l^{(j)}}\right)^{-1} \left(I_0 + \sum_{j=1}^{K} \frac{\mathrm{mes}_2 g^{(j)}}{l^{(j)}} J_0^{(j)}\right). \qquad (2.4.31)$$

Now, the constants $T_0^{(j)}$ are found from (2.4.27).

2.4.7 Concluding remarks on formal algorithm

We have described the algorithm required to construct the leading term (2.4.1) of the asymptotic expansion.

From this algorithm it follows that the difference $u - u_0$ satisfies the boundary value problem (2.1.1)–(2.1.3) with the right-hand sides of the form (2.3.5)–(2.3.7), where the summation starts with $k = 1$, and with the additional term

$$-\varepsilon^{-3} \sum_{j=0}^{K} q_0^{(j)} \eta''(X_3) \qquad (2.4.32)$$

in (2.1.1) which arose from (2.4.12). This term will be compensated in the next step of the algorithm owing to the choice of the boundary condition

$$W_1^{(j)}(0) = q_0^{(j)}, \qquad (2.4.33)$$

where $W_1^{(j)}$ is analogous to $W_0^{(j)}$.

When solving model boundary value problems we have to deal with right-hand sides involving unknown constants $C_0, T_0^{(j)}, j = 1, \ldots, K$. We have chosen this procedure because it seems to us more natural mathematically. However, it is possible to change the algorithm and to start with evaluation of the constants $C_0, T_0^{(j)}$ from the physically obvious system (2.4.27), (2.4.31). Then, the right-hand sides in the model problems are known from the very beginning.

2.5 Complete asymptotic expansion

2.5.1 Structure of the asymptotic expansion

We seek a complete asymptotic expansion for the solution of the problem (2.1.1)–(2.1.3) in the form

$$u(\mathbf{x},\varepsilon) \sim \sum_{k=0}^{\infty} \varepsilon^k u_k(\mathbf{x},\varepsilon), \qquad (2.5.1)$$

where

$$u_k(\mathbf{x},\varepsilon) = \varepsilon^{-2} C_k + u_{k,\Omega}(\mathbf{x})\mathfrak{X}(\mathbf{x},\varepsilon) + \varepsilon^{-1} \sum_{j=1}^{K} \mathcal{W}_k^{(j)}(\mathbf{X})\Xi(\mathbf{x}-\mathbf{a}^{(j)})$$

$$+ \sum_{j=1}^{K} (\varepsilon^{-2} W_k^{(j)}(z) + U_k^{(j)}(\zeta,z) + \varepsilon^{-2} v_k^{(j)}(\mathbf{Y}))\eta(\varepsilon^{-1}z).$$

Here, $C_k =$ Const, and $u_{k,\Omega}$ is a solution of the model problem in Ω discussed in Section 2.2.2. It has the representation

$$u_{k,\Omega} = u_{k,r} + u_{k,s}$$

with the regular term $u_{k,r}$ being smooth in the vicinity of junction points and with the singular term

$$u_{k,s} = \sum_{j=1}^{K} T_k^{(j)} N(\mathbf{x},\mathbf{a}^{(j)}).$$

The constant coefficients $T_k^{(j)}$ are subject to the balance condition of the form (2.4.6). The function $\mathcal{W}_k^{(j)}$ is the junction layer; together with the constants $q_k^{(j)}, D_k^{(j)}$ it satisfies the model problem of the form (2.2.39), (2.2.40). Properties of $\mathcal{W}_k^{(j)}$ were studied in Section 2.2.4. Here we mention that $\mathcal{W}_k^{(j)}$ decays at infinity (see (2.2.41), (2.2.42)).

The combination

$$\varepsilon^{-2} W_k^{(j)}(z) + U_k^{(j)}(\zeta,z)$$

describes the field in the thin cylinders, and these terms are defined as solutions of the ordinary differential equation and the Neumann boundary value problem for the Laplacian in $g^{(j)}$ (see Section 2.2.6). At a junction point

$$W_k^{(j)}(0) = q_{k-1}^{(j)}, \qquad (2.5.2)$$

and

$$\frac{d}{dz} W_k^{(j)}(0) = D_k^{(j)}. \qquad (2.5.3)$$

The function $v_k^{(j)}$ is the bottom layer. It decays exponentially at infinity, and the balance relation of the form (2.2.57) yields the boundary condition for $W_k^{(j)}$ at $z = l^{(j)}$. We note that the extra boundary condition (2.5.3) gives a relation between the constants $T_k^{(j)}$ and C_k. The constant coefficients $C_k, T_k^{(j)}$, $j = 1, \ldots, K$, are defined from a system of linear algebraic equations of the form (2.2.57), (2.2.58).

2.5.2 The asymptotic algorithm

In accordance with Section 2.4.7 we can construct, literally in the same way as u_0, the leading order approximation for $\varepsilon^{-1}(u - u_0)$. The only difference is that the function $W_1^{(j)}$ will satisfy the inhomogeneous boundary condition (2.4.33) to compensate for the discrepancy term (2.4.32). Next, we see that the difference $u - u_0 - \varepsilon u_1$ is subject to the boundary value problem (2.1.1)–(2.1.3) with right-hand sides of the form (2.3.5)–(2.3.7), with the index of summation $k \geq 2$, and the additional term

$$-\varepsilon^{-2} \sum_{j=1}^{K} q_1^{(j)} \eta''(X_3)$$

in equation (2.1.1). The constants $q_1^{(j)}$ were found when solving the junction layer problem (see Section 2.4.2). Suppose the coefficients $u_0, u_1, \ldots, u_{N-1}$ in (2.5.1) have been constructed and the function

$$\varepsilon^{-N}\left(u - \sum_{k=0}^{N-1} \varepsilon^k u_k\right) \qquad (2.5.4)$$

satisfies problem (2.1.1)–(2.1.3) with

$$F = F_* - \varepsilon^{-4} \sum_{j=1}^{K} q_{N-1}^{(j)} \eta''(X_3), \qquad (2.5.5)$$

and $F_*, P, \Phi^{(j)}$ subject to (2.3.5)–(2.3.7). The constants $q_{N-1}^{(1)}, \ldots, q_{N-1}^{(K)}$ are assumed to be known. It remains to construct the leading term for (2.5.4) by the procedure described in Section 2.4. To compensate for the additional terms of order $O(\varepsilon^{-4})$ in (2.5.5) we prescribe the boundary condition

$$W_N^{(j)}(0) = q_{N-1}^{(j)}$$

instead of (2.4.21).

2.6 Justification of the asymptotic expansion

2.6.1 Auxiliary estimates for functions in $H^1(\Omega_\varepsilon)$

We begin with an estimate of the L_2 norm of a function $v \in H^1(\Omega_\varepsilon)$ by its Dirichlet integral and the L_2 norms corresponding to the bottom regions S_ε. The notation $\Gamma_\varepsilon^{(j)}$ will be used for the lateral surface of the cylinder $\Pi_\varepsilon^{(j)}$.

LEMMA 2.7

Let $V \in H^1(\Omega_\varepsilon)$. Then

$$\int_{\Omega_\varepsilon} |V|^2 d\mathbf{x} + \int_{\partial\Omega_\varepsilon} |V|^2 ds$$
$$\leq \text{Const } \varepsilon^{-2} \left\{ \int_{\Omega_\varepsilon} \|\nabla V\|^2 d\mathbf{x} + \int_{S_\varepsilon} |V|^2 ds \right\}. \qquad (2.6.1)$$

Proof

(i) We begin with the estimate for the first integral in the left-hand side. From the Newton–Leibniz formula

$$V(\mathbf{x}', x_3) = V(\mathbf{x}', l^{(j)}) - \int_{x_3}^{l^{(j)}} \frac{\partial V}{\partial x_3}(\mathbf{x}', \tau) d\tau, \qquad (2.6.2)$$

for all $(x_1 - a_1^{(j)}, x_2 - a_2^{(j)}) \in \varepsilon g^{(j)}$. Let $\Pi_{\varepsilon,\delta}^{(j)}$ denote the thin cylinder

$$\{\mathbf{x}: (x_1 - a_1^{(j)}, x_2 - a_2^{(j)}) \in \varepsilon g^{(j)}, \ -\delta < x_3 < l^{(j)}\},$$

where δ is a non-negative constant chosen in such a way that this cylinder is located in Ω_ε.

It follows from (2.6.2) that

$$\|V\|^2_{L_2(\Pi_{\varepsilon,\delta}^{(j)})} \leq \text{Const } \Big\{ \|V\|^2_{L_2(S_\varepsilon^{(j)})}$$
$$+ \int_{\Pi_{\varepsilon,\delta}^{(j)}} \left| \frac{\partial V}{\partial \tau}(\mathbf{x}', \tau) \right|^2 d\tau d\mathbf{x}' \Big\}$$
$$\leq \text{Const } \{\|V\|^2_{L_2(S_\varepsilon^{(j)})} + \|\nabla V\|^2_{L_2(\Pi_{\varepsilon,\delta}^{(j)})}\}. \qquad (2.6.3)$$

Now, let
$$H^1_\perp(\Omega) = \{w \in H^1(\Omega) : \int_\Omega w(\mathbf{x})d\mathbf{x} = 0\}.$$
Then, by the Poincaré inequality,
$$\|w\|_{L_2(\Omega)} \leq \text{Const } \|\nabla w\|_{L_2(\Omega)}, \qquad (2.6.4)$$
for $w \in H^1_\perp(\Omega)$. Assume that $v \in H^1(\Omega)$, and let v be represented in the form
$$v = w + \beta, \qquad (2.6.5)$$
where $w \in H^1_\perp(\Omega)$, and β is constant. We have
$$\begin{aligned}
\|v\|^2_{L_2(\Omega)} &= \int_\Omega (w^2 + \beta^2)d\mathbf{x} \\
&\leq \text{Const } \{\|\nabla v\|^2_{L_2(\Omega)} + \beta^2 \text{mes } \Omega\} \\
&\leq \text{Const } \{\|\nabla v\|^2_{L_2(\Omega)} + \beta^2 \varepsilon^{-2} \text{mes } (\Pi_{\varepsilon,\delta} \setminus \Pi_\varepsilon)\}.
\end{aligned}$$

Note that by (2.6.5) and (2.6.4)
$$\begin{aligned}
\beta^2(\text{mes }(\Pi^{(j)}_{\varepsilon,\delta} \setminus \Pi^{(j)}_\varepsilon)) &\leq 2 \int_{\Pi^{(j)}_{\varepsilon,\delta} \setminus \Pi^{(j)}_\varepsilon} (v^2 + w^2)d\mathbf{x} \\
&\leq \text{Const } \left\{ \int_{\Pi^{(j)}_{\varepsilon,\delta} \setminus \Pi^{(j)}_\varepsilon} v^2 d\mathbf{x} + \int_\Omega \|\nabla v\|^2 d\mathbf{x} \right\}.
\end{aligned}$$

Hence
$$\|v\|^2_{L_2(\Omega)} \leq \varepsilon^{-2} \text{ Const } \left\{ \|\nabla v\|^2_{L_2(\Omega)} + \int_{\Pi^{(j)}_{\varepsilon,\delta} \setminus \Pi^{(j)}_\varepsilon} v^2 d\mathbf{x} \right\}.$$

This inequality, with $v = V$, and (2.6.3) give the required estimate for the first integral in (2.6.1).

(ii) Here we obtain the estimate for the second integral in the left-hand side of (2.6.1).

It is well known (see, for example, Section 1.8 of Lions and Magenes (1968)) that
$$\int_{\partial\Omega} |V|^2 ds \leq \text{Const } \int_\Omega (|\nabla V|^2 + |V|^2)d\mathbf{x}, \qquad (2.6.6)$$
and

$$\int_{\partial g^{(j)}} |V|^2 ds \leq \text{Const} \int_{g^{(j)}} (|\nabla_{X'} V|^2 + |V|^2) d\mathbf{X}'. \qquad (2.6.7)$$

(In fact, the integrals in the right-hand side dominate even the $H^{1/2}$ norm of V on $\partial \Omega$ and $\partial g^{(j)}$, but this will not be used in the sequel.) Integrating (2.6.7) along the interval $(0, l^{(j)})$ and changing the variables

$$x_1 - a_1^{(j)} = \varepsilon X_1, \quad x_2 - a_2^{(j)} = \varepsilon X_2$$

we obtain

$$\int_{\Gamma_\varepsilon^{(j)}} |V|^2 ds \leq \text{Const} \int_{\Pi_\varepsilon^{(j)}} (\varepsilon |\nabla_{\mathbf{x}'} V|^2 + \varepsilon^{-1} |V|^2) d\mathbf{x}'. \qquad (2.6.8)$$

From (2.6.3) and (2.6.8) we deduce that

$$\int_{\Gamma_\varepsilon^{(j)}} |V|^2 ds \leq \text{Const} \, \varepsilon^{-1} \left\{ \|V\|^2_{L_2(S_\varepsilon^{(j)})} + \|\nabla V\|^2_{L_2(\Pi_\varepsilon^{(j)})} \right\}. \qquad (2.6.9)$$

Using (2.6.9) and (2.6.6) and the estimate for the first norm on the right obtained in (i) we arrive at the required estimate for the second term in the left-hand side of (2.6.1). □

COROLLARY 2.1

Let $V \in H^1(\Omega_\varepsilon)$. Then

$$\int_{\Pi_\varepsilon^{(j)}} |V|^2 d\mathbf{x} + \varepsilon \int_{\Gamma_\varepsilon^{(j)}} |V|^2 ds_x \leq \text{Const} \left\{ \|V\|^2_{L_2(S_\varepsilon^{(j)})} + \|\nabla V\|^2_{L_2(\Pi_\varepsilon^{(j)})} \right\}.$$

Proof

The result follows directly from (2.6.3) and (2.6.9). □

2.6.2 Estimate for solutions

By the Newton–Leibniz formula and the Cauchy inequality we have

$$\|\varphi\|_{L_2(S_\varepsilon)} \leq C \|u\|_{H^1(\Omega_\varepsilon)}$$

for any extension u of φ. Hence, by (2.3.8)

$$\|\varphi\|_{L_2(S_\varepsilon)} \leq C\|\varphi\|_{H^{1/2}(S_\varepsilon)} \text{ for all } \varphi \in H^{1/2}(S_\varepsilon), \quad (2.6.10)$$

where the constant C does not depend on ε.

THEOREM 2.1

Let $f \in L_2(\Omega_\varepsilon)$, $p \in L_2(\partial\Omega_\varepsilon \setminus S_\varepsilon)$ and $\varphi \in H^{1/2}(S_\varepsilon)$. Then the problem

$$-\Delta u = f \text{ in } \Omega_\varepsilon, \quad (2.6.11)$$

$$\frac{\partial u}{\partial n} = p \text{ on } \partial\Omega_\varepsilon \setminus S_\varepsilon, \quad (2.6.12)$$

$$u = \varphi \text{ on } S_\varepsilon, \quad (2.6.13)$$

has a unique solution $u \in H^1(\Omega_\varepsilon)$, and the estimate

$$\int_{\Omega_\varepsilon} \|\nabla u\|^2 d\mathbf{x} + \varepsilon^2 \int_{\Omega_\varepsilon} u^2 d\mathbf{x} \leq \text{Const}\Big\{\varepsilon^{-2}(\|f\|^2_{L_2(\Omega_\varepsilon)}$$

$$+ \|p\|^2_{L_2(\partial\Omega_\varepsilon \setminus S_\varepsilon)}) + \|\varphi\|^2_{H^{1/2}(S_\varepsilon)}\Big\} \quad (2.6.14)$$

is valid.

Proof

The unique solvability of (2.6.11)–(2.6.13) is a standard fact (see, for example, Lions and Magenes (1968), Chapter 2, Section 9). We concentrate on the proof of (2.6.14).

Let the solution u be represented as

$$u = u_1 + u_2,$$

where u_1 solves the problem (2.6.11)–(2.6.13) with $\varphi = 0$, and u_2 is a solution of (2.6.11)–(2.6.13) where $f = g = 0$.

Then, multiplying $-\Delta u_1 = f$ by u_1 and integrating by parts we obtain

$$\|\nabla u_1\|^2_{L_2(\Omega_\varepsilon)} = \int_{\Omega_\varepsilon} fu_1 d\mathbf{x} + \int_{\partial\Omega_\varepsilon} pu_1 ds.$$

Using Cauchy's inequality and estimate (2.6.1) we obtain

$$\|\nabla u_1\|_{L_2(\Omega_\varepsilon)} \leq \text{Const } \varepsilon^{-1}\{\|f\|_{L_2(\Omega_\varepsilon)} + \|p\|_{L_2(\partial\Omega_\varepsilon)}\}. \quad (2.6.15)$$

Next, we estimate u_2. There exists a function h from $H^1(\Omega_\varepsilon)$ with support in thin cylinders $\overline{\Pi}_\varepsilon^{(i)}$ such that

$$h|_{S_\varepsilon^{(i)}} = \varphi,$$

and the estimate

$$\|\nabla h\|_{L_2(\Omega_\varepsilon)} \leq \text{Const } \|\varphi\|_{H^{1/2}(S_\varepsilon)} \tag{2.6.16}$$

holds.

Now, we seek u_2 in the form

$$u_2 = h + W.$$

Since

$$\int_{\Omega_\varepsilon} \nabla u_2 \cdot \nabla W \, d\mathbf{X} = 0,$$

we have

$$\int_{\Omega_\varepsilon} |\nabla W|^2 d\mathbf{X} = -\int_{\Omega_\varepsilon} \nabla h \cdot \nabla W \, d\mathbf{X}. \tag{2.6.17}$$

It follows from (2.6.16), (2.6.17) and Cauchy's inequality that

$$\|\nabla W\|_{L_2(\Omega_\varepsilon)} \leq \text{Const } \|\varphi\|_{H^{1/2}(S_\varepsilon)}. \tag{2.6.18}$$

Estimates (2.6.18) and (2.6.16) imply

$$\|\nabla u_2\|_{L_2(\Omega_\varepsilon)} \leq \text{Const } \|\varphi\|_{H^{1/2}(S_\varepsilon)}. \tag{2.6.19}$$

Thus, using in addition (2.6.15) we obtain

$$\|\nabla u\|_{L_2(\Omega_\varepsilon)} \leq \text{Const } \{\varepsilon^{-1}(\|f\|_{L_2(\Omega_\varepsilon)} + \|p\|_{L_2(\partial\Omega_\varepsilon)}) + \|\varphi\|_{H^{1/2}(S_\varepsilon)}\}. \tag{2.6.20}$$

The estimate for $\|u\|_{L_2(\Omega_\varepsilon)}$ follows from Lemma 2.7 and (2.6.10). This completes the proof. $\square$

2.6.3 Estimate for the remainder term

THEOREM 2.2

The solution $u(x,\varepsilon)$ of problem (2.1.1)–(2.1.3) with the right-hand sides (2.3.5)–(2.3.7) has the asymptotic representation (2.5.1): a finite sum

$$s_N = \sum_{k=0}^{N} \varepsilon^k u_k(\mathbf{x}, \varepsilon) \qquad (2.6.21)$$

of the series (2.5.1) approximates u in the following sense:

$$\int_{\Omega_\varepsilon} |\nabla(u - s_N)|^2 d\mathbf{x} + \varepsilon \int_{\Omega_\varepsilon} |u - s_N|^2 d\mathbf{x} \leq \text{Const } \varepsilon^{2N-1} \qquad (2.6.22)$$

where the constant does not depend on ε.

Proof

Let M be an integer, $M \geq N + 2$. We have

$$\int_{\Omega_\varepsilon} |\nabla(u - s_N)|^2 d\mathbf{x} + \varepsilon \int_{\Omega_\varepsilon} |u - s_N|^2 d\mathbf{x}$$

$$\leq 2 \left(\int_{\Omega_\varepsilon} |\nabla(u - s_M)|^2 d\mathbf{x} + \varepsilon \int_{\Omega_\varepsilon} |u - s_M|^2 d\mathbf{x} \right.$$

$$\left. + \int_{\Omega_\varepsilon} |\nabla(s_M - s_N)|^2 d\mathbf{x} + \varepsilon \int_{\Omega_\varepsilon} |s_M - s_N|^2 d\mathbf{x} \right). \qquad (2.6.23)$$

We apply inequality (2.6.14) to estimate the first two terms in the right-hand side. Let us denote by F_M, P_M, $\phi_M^{(i)}$ the right-hand sides of the problem (2.1.1)–(2.1.3) for $u - s_M$. In accordance with Section 2.5.2, these right-hand sides are represented by asymptotic series (2.3.5)–(2.3.7), where $k \geq M + 1$, and, in addition, the right-hand side (2.1.1) includes the terms

$$-\varepsilon^{M-3} q_M^{(j)} \eta''.$$

From the definition of asymptotic expansions given in Section 2.3.3 we obtain the estimate

$$\varepsilon^{-1} \|F_M\|_{L_2(\Omega_\varepsilon)} + \varepsilon^{-1} \|P_M\|_{L_2(\partial \Omega_\varepsilon)}$$

$$+ \sum_{i=1}^{K} \|\Phi_M^{(i)}\|_{H^1(S_\varepsilon^{(i)})} \leq \text{Const } \varepsilon^{M-5/2}. \tag{2.6.24}$$

By virtue of (2.6.14) (see Theorem 2.1) we obtain

$$\int_{\Omega_\varepsilon} |\nabla(u - s_M)|^2 d\mathbf{x} + \varepsilon \int_{\Omega_\varepsilon} |u - s_M|^2 d\mathbf{x} \leq \text{Const } \varepsilon^{2M-5}, \tag{2.6.25}$$

where the constant does not depend on ε.

To estimate the last two terms in (2.6.23) we define the order (with respect to ε) of the $(N+1)$-th term of the series (2.5.1):

$$\int_{\Omega_\varepsilon} |\nabla(s_M - s_N)|^2 d\mathbf{x} + \varepsilon \int_{\Omega_\varepsilon} |s_M - s_N|^2 d\mathbf{x} \leq \text{Const } \varepsilon^{2N-1}. \tag{2.6.26}$$

Choosing $M \geq N+3$ and taking into account (2.6.23), (2.6.25) and (2.6.26), we arrive at (2.6.22). □

2.7 A constant right-hand side

We illustrate the general results obtained in this chapter by an example. Consider the boundary value problem

$$-\Delta_x A(\mathbf{x}, \varepsilon) = (\text{mes}_3 \Omega)^{-1}, \quad \mathbf{x} \in \Omega_\varepsilon, \tag{2.7.1}$$

$$\frac{\partial A}{\partial n_x}(\mathbf{x}, \varepsilon) = 0, \quad \mathbf{x} \in \partial \Omega_\varepsilon \setminus S_\varepsilon^{(i)}, \tag{2.7.2}$$

$$A(\mathbf{x}, \varepsilon) = 0, \quad \mathbf{x} \in S_\varepsilon^{(i)}. \tag{2.7.3}$$

It will be shown in Chapter 6 that the function A represents an asymptotic approximation of the eigenfunction which corresponds to the smallest positive eigenvalue of the multi-structure.

The right-hand side of (2.7.1) allows for the asymptotic expansion (2.3.5). Indeed,

$$\begin{aligned}
1 &= \mathfrak{X}(\mathbf{x}, \varepsilon) + 1 - \mathfrak{X}(\mathbf{x}, \varepsilon) \\
&= \mathfrak{X}(\mathbf{x}, \varepsilon) + \sum_{j=1}^{K}(1 - \xi_j(\mathbf{X})) \\
&= \mathfrak{X}(\mathbf{x}, \varepsilon) + \sum_{j=1}^{K}\left[(1 - \xi_j(\mathbf{X}))(1 - \eta(X_3)) + \eta(z/\varepsilon)\right].
\end{aligned}$$

Thus, in (2.3.5), for the right-hand side of (2.7.1)

A CONSTANT RIGHT-HAND SIDE

$$f_0 = (\mathrm{mes}_3 \Omega)^{-1}, \tag{2.7.4}$$

$$F_3^{(j)}(\mathbf{X}) = (\mathrm{mes}_3 \Omega)^{-1}(1 - \xi_j(\mathbf{X}))(1 - \eta(X_3))$$

and

$$\mathcal{F}_2^{(j)} = (\mathrm{mes}_3 \Omega)^{-1}.$$

All other terms in expansion (2.3.5) are equal to zero.

We shall restrict ourselves to the leading order approximation of the solution as given by (2.4.1). We start with evaluation of the constants $T_0^{(j)}$ and C_0 (compare with the end of Section 2.4.7). The balance relation (2.4.30) becomes

$$1 + \sum_{j=1}^{K} T_0^{(j)} = 0. \tag{2.7.5}$$

Since $J_0^{(j)}$ (see (2.4.27)) is equal to zero,

$$T_0^{(j)}(\mathrm{mes}_2 g^{(j)})^{-1} = -C_0(l^{(j)})^{-1}, \quad j = 1, \ldots, K. \tag{2.7.6}$$

The solution of system (2.7.5), (2.7.6) is given by

$$C_0 = \left(\sum_{i=1}^{K} (l^{(i)})^{-1} \mathrm{mes}_2 g^{(i)} \right)^{-1} \tag{2.7.7}$$

and

$$T_0^{(j)} = -(l^{(j)})^{-1} \mathrm{mes}_2 g^{(j)} \left(\sum_{i=1}^{K} (l^{(i)})^{-1} \mathrm{mes}_2 g^{(i)} \right)^{-1}, \quad j = 1, \ldots, K.$$

By (2.7.4) and (2.7.5), the right-hand sides in (2.4.4), (2.4.5) are equal to zero, and hence $u_{0,r} = 0$. Therefore,

$$u_{0,\Omega} = u_{0,s} = \sum_{j=1}^{K} T_0^{(j)} N(\mathbf{x}, \mathbf{a}^{(j)}).$$

Next, we pass to the junction zone. The junction layer $\mathcal{W}_0^{(j)}$ and the constants $q_0^{(j)}, D_0^{(j)}$ satisfy the boundary value problem

$$-\Delta_X \mathcal{W}_0^{(j)}(\mathbf{X}) = \frac{T_0^{(j)}}{2\pi} \{ \|\mathbf{X}\|^{-1} \Delta_X \xi(\mathbf{X})$$

$$- 2\|\mathbf{X}\|^{-3}\mathbf{X} \cdot \nabla_X \xi(\mathbf{X})\}$$
$$+ q_0^{(j)} \eta''(X_3) + D_0^{(j)}(X_3 \eta''(X_3) + 2\eta'(X_3)), \quad \mathbf{X} \in G^{(j)},$$
$$\frac{\partial W_0^{(j)}}{\partial n_x}(\mathbf{X}) = 0, \quad \mathbf{X} \in \partial G^{(j)}.$$

By (2.4.16),
$$D_0^{(j)} = \frac{T_0^{(j)}}{\mathrm{mes}_2 g^{(j)}}$$

and by (2.2.38) $q_0^{(j)}$ is proportional to $T_0^{(j)}$. Therefore $W_0^{(j)}$ is also proportional to $T_0^{(j)}$.

The function $U_0^{(j)}$ is equal to zero in the thin cylinders, and $W_0^{(j)}$ is linear and vanishes at $z = 0$. By (2.4.26), $W_0^{(j)}(l^{(j)}) = -C_0$. Hence,

$$W_0^{(j)} = -\frac{C_0}{l^{(j)}} z.$$

Finally, the functions $v_0^{(j)}$ are equal to zero in the bottom region we all the right-hand sides in (2.4.23)–(2.4.25) vanish.

Thus, the leading order approximation for A is given by

$$\mathcal{U}(\mathbf{x}, \varepsilon) = \varepsilon^{-2} C_0 + \sum_{j=1}^{K} T^{(j)} N(\mathbf{x}, \mathbf{a}^{(j)}) \mathfrak{X}(\mathbf{x}, \varepsilon)$$

$$+ \varepsilon^{-1} \sum_{j=1}^{K} W_0^{(j)}(\mathbf{X}) \Theta(\mathbf{x} - \mathbf{a}^{(j)}) - \varepsilon^{-2} C_0 z \sum_{j=1}^{K} (l^{(j)})^{-1} \eta\left(\frac{z}{\varepsilon}\right), \quad (2.7.8)$$

where the constants $C_0, T_0^{(j)}, j = 1, \ldots, K$, are defined in (2.7.6), (2.7.7).

By Theorem 2.2

$$\int_{\Omega_\varepsilon} |\nabla(A - \mathcal{U})|^2 dx + \varepsilon \int_{\Omega_\varepsilon} |A - \mathcal{U}|^2 dx \leq \mathrm{Const}\, \varepsilon^{-1}, \quad (2.7.9)$$

which shows that $\mathcal{U}$ is an asymptotic approximation of A.

2.8 Application to the asymptotics of the energy integral

Here, we consider some particular cases of the right-hand sides of (2.3.5)–(2.3.7) and derive asymptotic formulae for the energy integral.

2.8.1 The case of the right-hand sides concentrated in $\overline{\Omega}$

Let us consider the boundary value problem

$$-\Delta_x u(\mathbf{x}, \varepsilon) = f(x), \quad \mathbf{x} \in \Omega_\varepsilon, \qquad (2.8.1)$$

$$\frac{\partial u}{\partial n_x}(\mathbf{x}, \varepsilon) = p(\mathbf{x}), \quad \mathbf{x} \in \partial\Omega_\varepsilon \setminus \overline{S}_\varepsilon, \qquad (2.8.2)$$

$$u(\mathbf{x}, \varepsilon) = 0, \quad \mathbf{x} \in S_\varepsilon. \qquad (2.8.3)$$

Here f is a smooth function in $\overline{\Omega}$, and p is the trace of the normal derivative of a smooth function in $\overline{\Omega}$. We suppose that f and p vanish near the points $\mathbf{x} = \mathbf{a}^{(j)}$, $j = 1, \ldots, K$, and on the cylinders $\Pi_\varepsilon^{(1)}, \ldots, \Pi_\varepsilon^{(K)}$. It is obvious that f and p can be represented in the form (2.3.5), (2.3.6), where $f_0 = f$, $p_0 = p$, and the remaining coefficients are equal to zero.

Non-balanced right-hand sides

Let us suppose first that the right-hand side in (2.8.1), (2.8.2) is non-balanced, i.e.

$$\int_\Omega f(\mathbf{x})d\mathbf{x} + \int_{\partial\Omega} p(\mathbf{x})ds \neq 0. \qquad (2.8.4)$$

The leading order approximation of the solution is given by (2.4.1). According to Section 2.5.6, we have the equations for C_0 and $T_0^{(j)}$, $j = 1, \ldots, K$:

$$I_0 + \sum_{j=1}^K T_0^{(j)} = 0,$$

$$T_0^{(j)} = -(l^{(j)})^{-1} C_0 \operatorname{mes}_2 g^{(j)},$$

where

$$I_0 = \int_\Omega f(\mathbf{x})d\mathbf{x} + \int_{\partial\Omega} p(\mathbf{x})ds.$$

Therefore,

$$C_0 = \left[\sum_{i=1}^K (l^{(i)})^{-1} \operatorname{mes}_2 g^{(i)}\right]^{-1} I_0. \qquad (2.8.5)$$

Similar to the case of the constant right-hand side (see Section 2.7) one shows that $U_0^{(j)} = 0$, $v_0^{(j)} = 0$, and

$$W_0^{(j)}(z) = -\frac{z}{l^{(j)}} C_0.$$

Thus, the leading order approximation (2.4.1) takes the form

$$u_0(\mathbf{x},\varepsilon) = \varepsilon^{-2}C_0 + u_{0,\Omega}(\mathbf{x})\mathfrak{X}(\mathbf{x},\varepsilon)$$

$$+ \varepsilon^{-1}\sum_{j=1}^{K} \mathcal{W}_0^{(j)}(\mathbf{X})\Xi(\mathbf{x}-\mathbf{a}^{(j)}) - \varepsilon^{-2}zC_0\sum_{j=1}^{K}(l^{(j)})^{-1}\eta(\varepsilon^{-1}z),$$

where $u_{0,\Omega}$ is described in Section 2.4.1, and the junction layer $\mathcal{W}_0^{(j)}$ is found as in Section 2.4.2.

We use Green's formula

$$\int_{\Omega_\varepsilon} \|\nabla u\|^2 d\mathbf{x} = \int_{\Omega_\varepsilon} u(\mathbf{x},\varepsilon)f(\mathbf{x})d\mathbf{x} + \int_{\partial\Omega_\varepsilon} u(\mathbf{x},\varepsilon)p(\mathbf{x})ds. \qquad (2.8.6)$$

Inserting u_0 instead of u into the right-hand side we note that only the term $\varepsilon^{-2}C_0$ is relevant. Hence, by (2.8.5), the energy integral for the solution of the problem (2.8.1)–(2.8.4) satisfies the asymptotic equality

$$\|\nabla u\|^2_{L_2(\Omega_\varepsilon)} \sim \varepsilon^{-2}I_0^2\left[\sum_{i=1}^{K}(l^{(i)})^{-1}\mathrm{mes}_2 g^{(i)}\right]^{-1}. \qquad (2.8.7)$$

Self-balanced right-hand sides

Now, let us consider the case

$$\int_{\Omega} f(\mathbf{x})d\mathbf{x} + \int_{\partial\Omega} p(\mathbf{x})ds = 0. \qquad (2.8.8)$$

The leading order asymptotic approximation of the solution is given by (2.4.1). Following Section 2.4.6 we obtain that

$$C_0 = T_0^{(1)} = \cdots = T_0^{(K)} = 0.$$

Then, by Sections 2.4.1 and 2.4.2,

$$u_{0,s} = 0, \ \mathcal{W}_0^{(j)} = 0, \ j = 1,\ldots,K,$$

and the constants $q_0^{(j)}$, $D_0^{(j)}$ also vanish. Furthermore, $W_0^{(j)} = U_0^{(j)} = 0$ and $v_0^{(j)} = 0$ (see Sections 2.4.3 and 2.4.4). Thus,

$$u_0(\mathbf{x},\varepsilon) = u_{0,r}(\mathbf{x})\mathfrak{X}(\mathbf{x},\varepsilon),$$

where $u_{0,r}$ is the only solution of the problem

$$-\Delta u_{0,r} = f \text{ in } \Omega,$$

$$\frac{\partial u_{0,r}}{\partial n} = p \text{ on } \partial\Omega,$$

subject to the orthogonality condition (2.4.7). Using formula (2.8.6) we obtain

$$\|\nabla u\|^2_{L_2(\Omega_\varepsilon)} \sim \int_\Omega u_{0,r} f d\mathbf{x} + \int_{\partial\Omega} u_{0,r} p ds = \|\nabla u_{0,r}\|^2_{L_2(\Omega)}.$$

2.8.2 The case of the Dirichlet data at the bases of thin cylinders

Let us consider problem (2.1.1)–(2.1.3), where the Dirichlet data at the bases of thin cylinders are the only non-zero part in the right-hand sides:

$$\Delta_x u(\mathbf{x}, \varepsilon) = 0, \ \mathbf{x} \in \Omega_\varepsilon, \tag{2.8.9}$$

$$\frac{\partial u}{\partial n_x}(\mathbf{x}, \varepsilon) = 0, \ \mathbf{x} \in \partial\Omega_\varepsilon \setminus \overline{S}_\varepsilon, \tag{2.8.10}$$

$$u(\mathbf{x}, \varepsilon) = \varepsilon^{-2} \phi^{(j)}(\mathbf{Y}), \ \mathbf{x} \in S_\varepsilon^{(j)}, \ j = 1, \ldots, K. \tag{2.8.11}$$

To evaluate the leading order approximation of energy, we need only the bottom layer solution. The functions $v_0^{(j)}$ satisfy the boundary value problem

$$\Delta_Y v_0^{(j)}(\mathbf{Y}) = 0, \ \mathbf{Y} \in \Pi_-^{(j)}, \tag{2.8.12}$$

$$\frac{\partial v_0^{(j)}}{\partial n_Y}(\mathbf{Y}) = 0, \ \mathbf{Y} \in \partial\Pi_-^{(j)} \setminus \overline{S}^{(j)}, \tag{2.8.13}$$

$$v_0^{(j)}(\mathbf{Y}) = \phi^{(j)}(\mathbf{Y}) - (\text{mes}_2 g^{(j)})^{-1} \int_{S^{(j)}} \phi^{(j)}(\mathbf{Y}) ds, \tag{2.8.14}$$

where $\mathbf{Y} \in S^{(j)}$, $j = 1, \ldots, K$. By Green's formula

$$\int_{\Omega_\varepsilon} |\nabla u|^2 d\mathbf{x} = \varepsilon^{-2} \sum_{j=1}^K \int_{S_\varepsilon^{(j)}} \frac{\partial u}{\partial x_3} \cdot \phi^{(j)} ds.$$

The solution $u(\mathbf{x}, \varepsilon)$ has the asymptotic form

$$u(\mathbf{x}, \varepsilon) \sim \varepsilon^{-2} \sum_{j=1}^{K} \eta\left(\frac{x_3}{\varepsilon}\right) v_0^{(j)}(\mathbf{Y})$$

$$+ D\left(\varepsilon^{-2} C + \sum_{j=1}^{K} T^{(j)} \{\chi(\mathbf{X}) N(\mathbf{x}, \mathbf{a}^{(i)})\right.$$

$$\left. + \varepsilon^{-1} \kappa(\mathbf{x} - \mathbf{a}^{(i)}) W^{(i)}(\mathbf{X})\}\right)$$

$$+ \varepsilon^{-1} \sum_{i=1}^{K} \kappa(\mathbf{x} - \mathbf{a}^{(i)}) b^{(i)} \Xi^{(i)}(\mathbf{X}), \qquad (2.8.15)$$

where

$$b^{(i)} = (l^{(i)})^{-1} \int_{S_i} \phi^{(i)}(\mathbf{Y}) ds_Y, \quad D = \sum_{i=1}^{K} b^{(i)};$$

the values $C, T^{(i)}$ are defined by (2.8.5).

It follows that the bottom layer gives the main contribution to the energy integral, and hence

$$\|\nabla u\|_{L_2(\Omega_\varepsilon)}^2 \sim \varepsilon^{-1} \sum_{j=1}^{K} \int_{S^{(j)}} \frac{\partial v_0^{(j)}}{\partial Y_3}(\mathbf{Y}) \phi^{(j)}(\mathbf{Y}) ds.$$

Since

$$\int_{S^{(j)}} \frac{\partial v_0^{(j)}}{\partial Y_3} ds = \int_{\Pi_-^{(j)}} \Delta v_0^{(j)} d\mathbf{x} = 0,$$

we have, by Green's formula,

$$\int_{S^{(j)}} \frac{\partial v_0^{(j)}}{\partial Y_3} \phi^{(j)} ds = \int_{\Pi_-^{(j)}} |\nabla v_0^{(j)}|^2 d\mathbf{Y}.$$

Thus, the relation

$$\|\nabla u\|_{L_2(\Omega_\varepsilon)}^2 \sim \varepsilon^{-1} \sum_{j=1}^{K} \int_{\Pi_-^{(j)}} |\nabla v_0^{(j)}|^2 d\mathbf{Y}$$

gives the desired asymptotics for the energy.

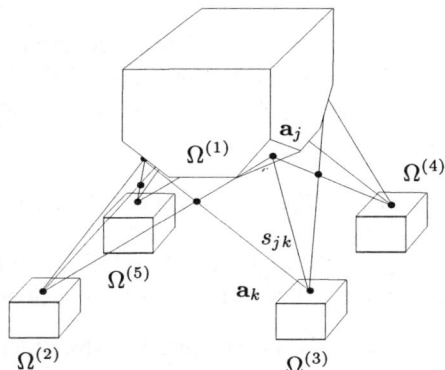

FIG. 2.7. Skeleton of a multi-structure, i.e. the union of three-dimensional bodies $\Omega^{(i)}$ and one-dimensional segments s_{jk}.

2.9 On a general 1D–3D multi-structure

Here we make some remarks on a possible extension of the previous asymptotic technique to 1D–3D multi-structures of a more general shape.

We assume that a multi-structure Ω_ε is a domain which consists of a finite number of three-dimensional bodies and several cylinders of a finite length and a small cross-section. These bodies and cylinders are connected by small junction elements. By 'small' we mean sets with a diameter of order ε (as above, ε is a positive non-dimensional small parameter).

Let three-dimensional parts $\Omega^{(i)}$ be domains with Lipschitz boundary and compact closure; $\overline{\Omega^{(i)}} \cap \overline{\Omega^{(j)}} = \emptyset$, when $i \neq j$. Consider a finite set $\{\mathbf{a}_k\}_{k=1}^M$ of points in $\mathbb{R}^3 \setminus \cup_i \Omega^{(i)}$ and connect some of them by segments, which do not have interior points in common. A segment with the end points $\mathbf{a}_k$, $\mathbf{a}_j$ will be denoted by s_{kj}. Skeleton of a multi-structure is shown in Fig. 2.7.

For each s_{kj} we introduce a local Cartesian system of coordinates $\mathbf{y} = (y_1, y_2, y_3)$ with the axis Oy_3 being directed along s_{kj} and construct a thin cylinder $\Pi_{kj}(\varepsilon) = s_{kj} \times g_{kj}(\varepsilon)$, where

$$g_{kj}(\varepsilon) = \{(y_1, y_2) : \varepsilon^{-1}(y_1, y_2) \in g_{kj}\},$$

with g_{kj} being two-dimensional Lipschitz bounded domains.

Assume that in the exterior of all balls $\mathcal{B}(\mathbf{a}_k, c\varepsilon) = \{\mathbf{x} \in \mathbb{R}^3 : |\mathbf{x} - \mathbf{a}_k| < c\varepsilon\}$, $c = \text{const}$, the sets

Ω_ε and $(\cup_j \Omega^{(j)}) \cup (\cup_{k,j} \Pi_{kj})(\varepsilon)$

are identical.

The parts of the domain Ω_ε which are located in the vicinity of $\mathbf{a}_k$ are described as follows. We place the origin of the Cartesian coordinate system at $\mathbf{a}_k$, and denote by δ a small positive quantity independent of ε. It is assumed that

$$\Omega_\varepsilon \cap \mathcal{B}(\mathbf{a}_k, \delta) = G_k(\varepsilon) \cap \mathcal{B}(\mathbf{a}_k, \delta),$$

where $G_k(\varepsilon) = \{\mathbf{y} : \varepsilon^{-1}\mathbf{y} \in G_k\}$ with G_k being a domain in $\mathbb{R}^3$, independent of ε, which belongs to one of the following three classes:

Domains of the first class correspond to the junctions between three-dimensional elements $\Omega^{(i)}$ and one or several thin rods. That is, G is called the domain of the first class if it is Lipschitz and if it coincides, in the exterior of a ball, with the union of the half-space $\mathbb{R}^3_-$ and semi-cylinders, located in $\mathbb{R}^3_+$, with axes transverse to $\partial \mathbb{R}^3_+$.

Domains of the second class correspond to junctions between rods. This class contains Lipschitz domains which coincide with the union of two or more semi-cylinders in the exterior of some ball.

By replacing a set of semi-cylinders in the last definition with a single cylindrical extension to infinity we introduce the *third class of domains*. This class corresponds to the end regions of thin rods which are not in a junction with other elements of the multi-structure.

As before, we consider a mixed Dirichlet–Neumann boundary value problem for the Laplacian. Similar to the previous sections of this chapter, the Neumann boundary conditions are prescribed at the surface of the bodies and on the lateral surface of thin rods. The Dirichlet data are prescribed at some ends of the rods, and in addition we introduce a finite number of rods subject to the Neumann boundary conditions at one of the ends.

Thus, the asymptotic analysis involves the following set of model problems which are ε-independent:

(1) The Neumann boundary value problems for the Laplacian in $\Omega^{(i)}$ whose solutions exhibit singular behaviour at junction points.

(2) The Neumann boundary value problems in the scaled region associated with the junction of a rod and one of $\Omega^{(i)}$ (the 'junction layer problem of the first kind').

(3) The model problems on a segment and the scaled cross-section of a thin cylinder.

(4) A mixed boundary value problem in a semi-infinite cylinder; the field describes the 'bottom layer' associated with the cylinder where the Dirichlet data are given at the end region.

(5) The Neumann boundary value problem in a semi-infinite cylinder; the corresponding field describes the 'bottom layer' associated with the cylinder where the Neumann boundary conditions are prescribed on the whole surface.

(6) The Neumann boundary value problem in a scaled junction region corresponding to the intersection of thin rods ('the junction layer problem of the second kind').

(7) A system of linear algebraic equations associated with the skeleton of the multi-structure.

The model problems (1)–(4) were discussed in detail in the earlier sections of the present chapter. The model problems (5) and (6) are new, but their analysis does not impose additional difficulties. Here, we consider the model problem (7) which generalises the one presented in Section 2.2.7.

We introduce the following physical model associated with the skeleton. We enumerate all three-dimensional bodies, the points of the junction between segments and the free end points of segments by the index m, and assign a constant temperature $C_m \varepsilon^{-2}$ to each element of this set Λ. The temperature is assumed to be known for the free ends of segments with Dirichlet boundary conditions; this quantity corresponds to $J\varepsilon^{-2}$ introduced in Section 2.2.7. The constants C_m are unknown for all other elements of this set. Also, for all elements, except for those where the Dirichlet data are given, the total heat flux $I^{(m)}$ produced by internal and boundary sources is specified.

Suppose that for all junction points on the surface of the same three-dimensional body the temperature is equal to the temperature of this body. The temperature value $C(\mathbf{a}^{(j)})\varepsilon^{-2}$ is allocated for each junction point $\mathbf{a}^{(j)}$, and it is equal to one of the temperatures $C_m \varepsilon^{-2}$. Assuming a linear distribution of temperature along the segments s_{ij} we define the heat flux at $\mathbf{a}^{(i)}$ as the quantity

$$T_{ij} = \frac{C(\mathbf{a}^{(j)}) - C(\mathbf{a}^{(i)})}{l_{ij}} \mathrm{mes}_2 g_{ij}. \tag{2.9.1}$$

For every element of the set Λ, except for those elements where the Dirichlet data are given, the following balance condition holds:

$$I^{(m)} + \sum_{j \in \Lambda} (C_j - C_m) \sum_{(k,q) \in \varkappa_{mj}} \frac{\mathrm{mes}_2 g_{kq}}{l_{kq}} = 0, \qquad (2.9.2)$$

where $\varkappa_{mj}$ denotes the set of pairs (k, q) of indices associated with the segments s_{kq} connecting the elements m and j of the set Λ. If there are no such segments then the corresponding terms in (2.9.2) are equal to zero.

Relations (2.9.2) form a system of linear algebraic equations with respect to C_m. The number of equations is equal to the number of unknowns. To prove the unique solvability we assume that $I^{(m)} = 0$ for all m and zero temperature is specified for those free ends where the Dirichlet data are given. Multiplying (2.9.2) by C_m and taking the sum over all m we obtain

$$\sum_{m,j \in \Lambda} (C_j - C_m)^2 \sum_{(k,q) \in \varkappa_{m,j}} \frac{\mathrm{mes}_2 g_{kq}}{l_{kq}} = 0.$$

Thus, the temperature at the ends of all the segments is equal to the same value; since the homogeneous Dirichlet data are prescribed at an end of at least one rod, this value is equal to zero.

When the constants C_m are known one can evaluate the heat flux by (2.9.1).

As in Section 2.5, the solution of the boundary value problem in the multi-structure can be represented by the asymptotic series in powers of ε. Compared with (2.5.1) it includes additional terms that correspond to model problems (5) and (6). However, the scheme of the asymptotic algorithm is the same.

2.10 A multi-structure with a thin-walled tube

A mixed boundary value problem for the Laplacian in a 2D–3D multi-structure Ω_ε which consists of a three-dimensional domain Ω and a thin-walled tube D_ε, has been considered in Åslund (1999). The thickness of the wall is equal to ε, where ε is a small parameter; see Fig. 2.8. The following boundary value problem is considered:

$$-\Delta_x u(\mathbf{x}, \varepsilon) = F(\mathbf{x}, \varepsilon), \quad \mathbf{x} \in \Omega_\varepsilon, \qquad (2.10.1)$$

$$\frac{\partial u}{\partial n}(\mathbf{x}, \varepsilon) = P(\mathbf{x}, \varepsilon), \quad \mathbf{x} \in \partial \Omega_\varepsilon \setminus \overline{S}_\varepsilon, \qquad (2.10.2)$$

$$u(\mathbf{x}, \varepsilon) = \Phi(\mathbf{x}, \varepsilon), \quad \mathbf{x} \in S_\varepsilon, \qquad (2.10.3)$$

where S_ε denotes the bottom surface. The skeleton of the multi-structure consists of the two-dimensional limit surface D_0 attached to the three-dimensional domain Ω along a curve denoted by Γ.

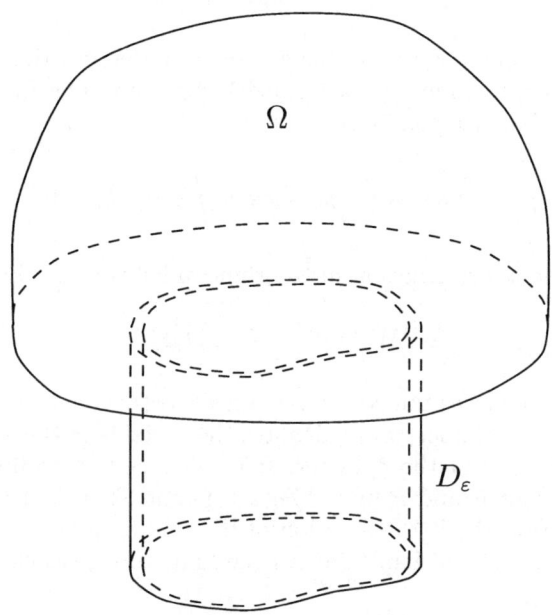

FIG. 2.8. The multi-structure.

A complete compound asymptotic expansion for the solution can be constructed using an approach similar to that in the previous sections. We restrict ourselves to the description of the principal term. Let f_0 and p_0 denote the leading terms in the expansions for the right-hand sides F and P, which correspond to the domain Ω. In this domain the principal term has the form

$$\varepsilon^{-1} C_0 + u_{0,\Omega}(\mathbf{x}), \qquad (2.10.4)$$

where C_0 is a constant, and $u_{0,\Omega}$ is the solution of the Neumann problem

$$-\Delta_x u_{0,\Omega}(\mathbf{x}) = f_0(\mathbf{x}), \quad \mathbf{x} \in \Omega,$$

$$\frac{\partial u_{0,\Omega}}{\partial n}(\mathbf{x}) = p_0(\mathbf{x}), \quad \mathbf{x} \in \partial \Omega \setminus \Gamma,$$

with the orthogonality condition

$$\int_\Omega u_{0,\Omega}(\mathbf{x})d\mathbf{x} = 0 \qquad (2.10.5)$$

and the prescribed singular behaviour

$$u_{0,\Omega} \sim T_0(t)\pi^{-1}\log(1/r), \quad r \to 0, \qquad (2.10.6)$$

along Γ. Here t denotes the arc length on Γ, r denotes the distance to Γ, and T_0 is a yet undetermined smooth function. The function T_0 is subject to the balance condition

$$\int_\Omega f_0(\mathbf{x})d\mathbf{x} + \int_{\partial\Omega} p_0(\mathbf{x})ds + \int_\Gamma T_0(t)dl_t = 0. \qquad (2.10.7)$$

The leading approximation in the thin-walled tube D_ε has the form

$$\varepsilon^{-1}W_0(z,t) + \varepsilon U_0(z,t,\xi), \qquad (2.10.8)$$

where $z \in [0,l]$ denotes the coordinate on D_0 orthogonal to Γ, l denotes the length of the cylinder, and ξ denotes the variable in the scaled cross-section $(-1/2, 1/2)$ of the cylinder. If we denote the leading terms in the expansions for F and P by $\varepsilon^{-1}\mathcal{F}_0(z,t,\xi)$ and $\mathcal{P}_0(z,t,\xi)$ respectively, then the function W_0 is the solution of the following Dirichlet problem for the two-dimensional Laplacian on the limit tube surface D_0:

$$-(\partial_z^2 + \partial_t^2)W_0(z,t) = \int_{-1/2}^{1/2} \mathcal{F}_0(z,t,\xi)d\xi + \sum_\pm \mathcal{P}_0(z,t,\pm 1/2) \quad \text{on } D_0,$$

$$W_0(0,t) = b_{0,0}(t),$$

$$W_0(l,t) = b_{0,1}(t).$$

The Dirichlet data $b_{0,0}$, $b_{0,1}$ will be specified when the bottom region and the junction region are analysed. The boundary value problem for the function U_0 is posed on the scaled cross-section of the tube, with the coordinates z and t as parameters:

$$-\partial_\xi^2 U_0(z,t,\xi) = \mathcal{F}_0(z,t,\xi)$$

$$-\int_{-1/2}^{1/2} \mathcal{F}_0(z,t,\xi)d\xi + \sum_\pm \mathcal{P}_0(z,t,\pm 1/2), \quad \xi \in (-1/2, 1/2),$$

$$\partial_\xi U_0(z,t,\pm 1/2) = \pm \mathcal{P}_0(z,t,\pm 1/2),$$

$$\int_{-1/2}^{1/2} U_0(z,t,\xi)d\xi = 0.$$

The functions in (2.10.4) and (2.10.8) are multiplied by cut-off functions, which give rise to discrepancies in the boundary value problem in an ε-neighbourhood of the junction region. The function $u_{0,\Omega}$ is multiplied by the cut-off function

$$\Psi(x,\varepsilon) = \chi(r/\varepsilon)$$

which vanishes near the junction region and equals one outside an ε-neighbourhood. The functions (2.10.8) are multiplied by the cut-off function η which is chosen so that $\eta(X_2) = 0$ for $X_2 < 1/2$, and $\eta(X_2) = 1$ for $X_2 > 1$, and the constant C_0 is multiplied by the function $(1 - \eta)$. One also has to take into account the discrepancies in the Dirichlet boundary conditions caused by (2.10.8). These discrepancies are compensated by a junction layer and a bottom layer.

First one considers the bottom region. The boundary value problem for the bottom layer $\varepsilon^{-1} v_0(\mathbf{Y},t)$ is posed on the scaled cross-section of the bottom region

$$\Pi^- = \{\mathbf{Y} : Y_2 < 0, Y_1 \in (-1/2, 1/2)\},$$

with the coordinate t as a parameter.

Let $\varepsilon^{-3} \mathcal{G}_0(\mathbf{Y},t)$, $\varepsilon^{-2} \mathcal{H}_0(\mathbf{Y},t)$ and $\varepsilon^{-1} \Phi_0(\mathbf{Y},t)$ denote the leading terms in the expansions for the right-hand sides in the boundary value problem (2.10.1)–(2.10.3). Then the function v_0 is the solution of the boundary value problem

$$-\Delta_Y v_0(\mathbf{Y},t) = \mathcal{G}_0(\mathbf{Y},t), \quad \mathbf{Y} \in \Pi^-,$$

$$\frac{\partial v_0}{\partial n}(\mathbf{Y},t) = \mathcal{H}_0(\mathbf{Y},t), \quad \mathbf{Y} \in \partial\Pi^- \setminus \overline{S},$$

$$v_0(\mathbf{Y},t) = \Phi_0(\mathbf{Y},t) - W_0(l,t), \quad \mathbf{Y} \in S,$$

where S denotes the base surface of the semi-strip. The solution is sought in a class of functions that vanish at infinity, and the condition

$$W_0(l,t) = \int_S \Phi_0(\mathbf{Y}) ds_Y - \int_{\Pi^-} Y_2 \mathcal{G}_0(\mathbf{Y}) d\mathbf{Y} - \int_{\partial\Pi^- \setminus \overline{S}} Y_2 \mathcal{H}_0(\mathbf{Y}) ds_Y$$

is necessary and sufficient for solvability. This relation gives the Dirichlet data $b_{0,1}$ in the boundary value problem for the function W_0.

Analogously, the junction layer $\mathcal{W}_0(\mathbf{X}, t)$ is defined as the solution of the Neumann problem posed in the scaled cross-section G of the junction region, which is the union of the half-plane $R_-^2 = \{\mathbf{X} : X_2 < 0\}$ and the semi-strip $\Pi^+ = \{\mathbf{X} : X_2 \geq 0, X_1 \in (-1/2, 1/2)\}$. In addition to the leading terms in the expansions for the functions F and P, which we denote by $\varepsilon^{-2} F_0(\mathbf{X})$ and $\varepsilon^{-1} P_0(\mathbf{X})$, one has to take into account the discrepancies mentioned above. In order to avoid the discrepancy of order $O(\varepsilon^{-3})$ in equation (2.10.1) we choose the Dirichlet data $b_{0,0}$ for the function W_0 as

$$b_{0,0}(t) = C_0.$$

This gives the representation

$$W_0(z,t) = C_0(1 - l^{-1} z) + R_0(z,t), \qquad (2.10.9)$$

where R_0 is the solution of the Dirichlet problem on the limit surface D_0 with the Dirichlet data $b_{0,0}$ replaced by zero. It remains to compensate for the discrepancy of order $O(\varepsilon^{-2})$, which has the form

$$\varepsilon^{-2}[\Delta_x, \Psi]\{T_0(t)\pi^{-1}(\log(1/\varrho) + \log(1/\varepsilon)) + d_0(z)\}$$

$$+ \varepsilon^{-2}[\partial_{X_2}^2, \eta]\{X_2 \partial_z W_0(0,t)\}.$$

Here $\varrho = r/\varepsilon$, and d_0 is the second term in the asymptotic expansion for $u_{0,\Omega}$ as $r \to 0$. The junction layer is multiplied by a cut-off function $\Xi(\mathbf{x}, \varepsilon)$ that equals one in a neighbourhood of the junction region. In order to find a solution that vanishes at infinity, the functions in the definitions of the right-hand sides have to satisfy the relation

$$T_0(t) = -C_0 l^{-1} + \partial_z R_0(0, z) + J_0(z), \qquad (2.10.10)$$

where

$$J_0(z) = \int_G F_0(\mathbf{X}) d\mathbf{X} + \int_{\partial G} P_0(\mathbf{X}) ds_X.$$

Here we used the representation (2.10.9). Moreover, one has to introduce an additional discrepancy term

$$\varepsilon^{-2} q_0(z) \eta''(X_2)$$

in equation (2.10.1), in the same way as in the previous sections. Using condition (2.10.7) and relation (2.10.10) one can evaluate the constant

$$C_0 = |\Gamma|^{-1} l \left\{ \int_\Omega f_0(\mathbf{x}) d\mathbf{x} + \int_{\partial \Omega} p_0(\mathbf{x}) ds \right.$$

$$+ \int_\Gamma J_0(t)dl_t + \partial_z\Big(\int_\Gamma R_0(z,t)dl_t\Big)\Big|_{z=0}\Big\}, \qquad (2.10.11)$$

and the function T_0 is given by (2.10.10).

The function
$$w_0(z) = \int_\Gamma R_0(z,t)dl_t$$
satisfies the Dirichlet problem
$$-\partial_z^2 w_0(z) = \int_\Gamma \overline{\mathcal{F}}_0(z,t)dl_t, \quad z \in (0,l),$$
$$w_0(0) = 0,$$
$$w_0(l) = \int_\Gamma b_{0,1}(t)dl_t,$$
where
$$\overline{\mathcal{F}}_0 = \int_{-1/2}^{1/2} \mathcal{F}_0(z,t,\xi)d\xi + \sum_\pm \mathcal{P}_0(z,t,\pm 1/2),$$
and one obtains the following representation for the last integral in (2.10.11):
$$\partial_z\Big(\int_\Gamma R_0(z,t)dl_t\Big)\Big|_{z=0} = \int_0^l (1-l^{-1}z) \int_\Gamma \overline{\mathcal{F}}_0(z,t)dl_t dz + l^{-1} \int_\Gamma b_{0,1}(t)dl_t.$$

Finally we define the leading approximation u_0 of the solution u by
$$u_0(\mathbf{x},\varepsilon) = \varepsilon^{-1}(1-\eta(\varepsilon^{-1}z))C_0$$
$$+ u_{0,\Omega}(\mathbf{x})\Psi(\mathbf{x},\varepsilon) + \mathcal{W}_0(\mathbf{X},t)\Xi(\mathbf{x}) + (\varepsilon^{-1}W_0(z,t)$$
$$+ \varepsilon U_0(z,t,\xi) + \varepsilon^{-1}v_0(\mathbf{Y},t))\eta(\varepsilon^{-1}z). \qquad (2.10.12)$$

The subsequent terms in the complete asymptotic expansion are constructed in a similar way. The main difference is that they, as well as the right-hand sides, depend polynomially on $\log\varepsilon$.

By using the principal term (2.10.12) in the asymptotics of u one can find, for example, an explicit asymptotic representation for the Dirichlet integral of u. Let us consider, in particular, homogeneous Dirichlet boundary conditions, and replace $F(\mathbf{x},\varepsilon)$ and $P(\mathbf{x},\varepsilon)$ by the smooth functions $f(\mathbf{x})$ and $p(\mathbf{x})$ that vanish on the tube D_ε and near the junction region. Then the asymptotic formula

$$\|\nabla u\|^2_{L_2(\Omega_\varepsilon)} \sim \varepsilon^{-1}|\Gamma|^{-1}l\left(\int_\Omega f(\mathbf{x})d\mathbf{x} + \int_{\partial\Omega} p(\mathbf{x})ds\right)^2, \qquad (2.10.13)$$

analogous to (2.8.7), is valid. This follows directly from Green's formula

$$\|\nabla u\|^2_{L_2(\Omega_\varepsilon)} = \int_\Omega f(\mathbf{x})u(\mathbf{x},\varepsilon)d\mathbf{x} + \int_{\partial\Omega} p(\mathbf{x})u(\mathbf{x},\varepsilon)ds$$

and from (2.10.12). The remainder term in (2.10.13) is $O(|\log\varepsilon|)$.

3
AUXILIARY FACTS FROM MATHEMATICAL ELASTICITY

3.1 Basic formulae of linear elasticity

In this section we present definitions and notations from linear elasticity. For more details we refer to the books by Westergaard (1952), Lurie (1970), Atkin and Fox (1980) and Gurtin (1981).

3.1.1 Stress and strain

Within an elastic continuum which occupies a domain $D \in \mathbb{R}^3$, we introduce the displacement field

$$\mathbf{u} = \begin{pmatrix} u_1 \\ u_2 \\ u_3 \end{pmatrix} \qquad (3.1.1)$$

and the strain tensor with components

$$\varepsilon_{ij} = \frac{1}{2}\left(\frac{\partial u_i}{\partial x_j} + \frac{\partial u_j}{\partial x_i}\right). \qquad (3.1.2)$$

These are called Cauchy's relations. The equalities $\varepsilon_{ij} = 0$, where $i, j = 1, 2, 3$, in a domain mean that

$$\mathbf{u}(\mathbf{x}) = \mathbf{a} + \mathbf{b} \times \mathbf{x}$$

with constant vectors $\mathbf{a}$ and $\mathbf{b}$; in other words, $\mathbf{u}$ is a rigid-body displacement.

Another characteristic of elastic media is the stress tensor (σ_{ij}), $i, j = 1, 2, 3$. For an isotropic material the stress and the strain components are related by Hooke's law:

$$\sigma_{ij} = 2\mu\varepsilon_{ij} + \lambda\delta_{ij}\sum_{k=1}^{3}\varepsilon_{kk}, \quad i,j = 1,2,3, \qquad (3.1.3)$$

where λ and μ are called the Lamé elastic moduli, and δ_{ik} is the Kronecker delta. In particular, the quantity μ is called the shear modulus. It is assumed in elasticity theory that $\mu > 0$ and $2\mu + 3\lambda > 0$. These inequalities guarantee

$$\frac{1}{2}\sum_{i,j=1}^{3}\sigma_{ij}\varepsilon_{ij} = \mu\sum_{i,j=1}^{3}\varepsilon_{ij}^{2} + \frac{1}{2}\lambda\left(\sum_{k=1}^{3}\varepsilon_{kk}\right)^{2} \geq c_0\sum_{i,j=1}^{3}\varepsilon_{ij}^{2}, \quad c_0 > 0.$$

The left-hand side is called the elastic energy density.

3.1.2 Equations of equilibrium and boundary conditions

The equations of equilibrium are written in the form

$$\sum_{j=1}^{3}\frac{\partial\sigma_{ij}}{\partial x_j} + F_i = 0, \quad i = 1,2,3, \qquad (3.1.4)$$

where F_i are components of the body-force density $\mathbf{F}$.

Substituting (3.1.3) in (3.1.4) we obtain the *Lamé system*

$$\mathbf{L}\left(\frac{\partial}{\partial \mathbf{x}}\right)\mathbf{u} := \mu\Delta\mathbf{u} + (\lambda+\mu)\nabla(\nabla\cdot\mathbf{u}) = -\mathbf{F}. \qquad (3.1.5)$$

The Lamé system with respect to components of displacement should be supplemented by certain boundary conditions. The main types of boundary conditions used in the formulations of problems of linear elasticity are:

(a) *displacement boundary conditions:*

$$\mathbf{u}(\mathbf{x}) = \varphi(\mathbf{x}), \quad \mathbf{x} \in \partial D, \qquad (3.1.6)$$

where the vector function φ is given;

(b) *traction boundary conditions:*

$$\sigma_i^{(n)} := \sum_{j=1}^{3}\sigma_{ij}n_j$$

$$= \sum_{j=1}^{3} \left\{ \lambda \delta_{ij} \sum_{k=1}^{3} \frac{\partial u_k}{\partial x_k} + \mu \left(\frac{\partial u_i}{\partial x_j} + \frac{\partial u_j}{\partial x_i} \right) \right\} n_j = p_i, \qquad (3.1.7)$$

where $i = 1, 2, 3$, $\mathbf{x} \in \partial D$, and n_j are components of the unit outward normal, and the corresponding vector form is

$$\boldsymbol{\sigma}^{(n)}(\mathbf{u}; \mathbf{x}) = \mathbf{p}(\mathbf{x}), \quad \mathbf{x} \in \partial D; \qquad (3.1.8)$$

here the vector function $\mathbf{p}(\mathbf{x}) = (p_1, p_2, p_3)$ is given, and $\boldsymbol{\sigma}^{(n)}$ is the vector with components $\sigma_i^{(n)}$, $i = 1, 2, 3$;

(c) *mixed boundary conditions* which involve displacements prescribed on a part of ∂D and tractions given on the remaining part.

It is well known (see, for example, Fichera (1984) and Oleinik, Shamaev and Yosifian (1992)) that the Dirichlet and mixed boundary value problems for the Lamé system are uniquely solvable in the class of vector functions $\mathbf{u}$ with finite elastic energy, provided D is a bounded domain with Lipschitz boundary. A solution of the problem with traction boundary conditions is specified up to an arbitrary rigid-body displacement. The solvability conditions for the static problem with prescribed tractions posed in a bounded region are expressed as the balance relations for the principal force and moment vectors of external load

$$\int_D \mathbf{F}(\mathbf{x}) d\mathbf{x} + \int_{\partial D} \mathbf{p}(\mathbf{x}) ds = 0,$$

and

$$\int_D \mathbf{x} \times \mathbf{F}(\mathbf{x}) d\mathbf{x} + \int_{\partial D} \mathbf{x} \times \mathbf{p}(\mathbf{x}) ds = 0.$$

The solution is uniquely determined up to a rigid-body displacement. For any elastic displacement fields $\mathbf{u}$ and $\mathbf{v}$ in a bounded domain D the Betti formula

$$\sum_{i,j=1}^{3} \int_D \sigma_{ij}(\mathbf{u}) \varepsilon_{ij}(\mathbf{v}) d\mathbf{x} = - \int_D \mathbf{L}\mathbf{u} \cdot \mathbf{v} d\mathbf{x} + \int_{\partial D} \boldsymbol{\sigma}^{(n)}(\mathbf{u}) \cdot \mathbf{v} ds \qquad (3.1.9)$$

is valid.

3.2 Two-dimensional problems of linear elasticity

3.2.1 Plane strain

Let the displacement vector **u** be parallel to the plane Ox_1x_2, and let the components u_i, $i = 1, 2, 3$, be independent of x_3:

$$u_i = u_i(\mathbf{x}'), \; i = 1, 2; \; u_3 = 0, \qquad (3.2.1)$$

where $\mathbf{x}' = (x_1, x_2)$. This assumption can be used in order to describe the deformation within an infinite cylindrical bar with the axis Ox_3 and with a cross-section $g \subset \mathbb{R}^2$, when tractions are applied on the lateral surface in such a way that $\sigma_1^{(n)}$ and $\sigma_2^{(n)}$ do not depend on x_3 and $\sigma_3^{(n)} = 0$, and when the body-force vector does not depend on x_3 and has the form

$$\mathbf{F} = (F_1, F_2, 0)^T. \qquad (3.2.2)$$

The basic relations of plane linear elasticity can be written as follows:

(a) *Cauchy's relations:*

$$\varepsilon_{11} = \frac{\partial u_1}{\partial x_1}, \; \varepsilon_{22} = \frac{\partial u_2}{\partial x_2}, \; \varepsilon_{12} = \frac{1}{2}\left(\frac{\partial u_1}{\partial x_2} + \frac{\partial u_2}{\partial x_1}\right), \qquad (3.2.3)$$

$$\varepsilon_{13} = \varepsilon_{23} = \varepsilon_{33} = 0.$$

(b) *Hooke's law:*

$$\sigma_{11} = (2\mu + \lambda)\varepsilon_{11} + \lambda\varepsilon_{22}, \; \sigma_{12} = 2\mu\varepsilon_{12},$$

$$\sigma_{22} = \lambda\varepsilon_{11} + (2\mu + \lambda)\varepsilon_{22}, \qquad (3.2.4)$$

$$\sigma_{33} = \nu(\sigma_{11} + \sigma_{22}), \; \sigma_{23} = \sigma_{13} = 0,$$

where $\nu = \lambda/2(\lambda + \mu)$ is the Poisson ratio which represents a measure of a lateral contraction compared with an axial extension for elastic solids.

(c) *Equations of equilibrium:*

$$\frac{\partial \sigma_{i1}}{\partial x_1} + \frac{\partial \sigma_{i2}}{\partial x_2} + F_i = 0, \; i = 1, 2. \qquad (3.2.5)$$

The components of the traction vector are given by

TWO-DIMENSIONAL PROBLEMS OF LINEAR ELASTICITY

$$\sigma_i^{(n)} = \sigma_{i1}n_1 + \sigma_{i2}n_2, \quad i = 1, 2, \tag{3.2.6}$$

where n_1, n_2 denote the components of the outward unit normal on the boundary of the elastic region.

For the case when the traction components, say p_i, $i = 1, 2$, are specified on the domain boundary ∂g, and when the components F_i, $i = 1, 2$, of the body force are given in the interior points of g, the displacement field (u_1, u_2) satisfies the system

$$\mu \Delta u_i + (\lambda + \mu) \frac{\partial}{\partial x_i} \nabla \cdot \mathbf{u} + F_i = 0, \quad i = 1, 2, \tag{3.2.7}$$

in g, and the boundary conditions

$$\left((2\mu + \lambda)\frac{\partial u_1}{\partial x_1} + \lambda \frac{\partial u_2}{\partial x_2}\right)n_1 + \mu\left(\frac{\partial u_1}{\partial x_2} + \frac{\partial u_2}{\partial x_1}\right)n_2 = p_1, \tag{3.2.8}$$

$$\mu\left(\frac{\partial u_1}{\partial x_2} + \frac{\partial u_2}{\partial x_1}\right)n_1 + \left((2\mu + \lambda)\frac{\partial u_2}{\partial x_2} + \lambda\frac{\partial u_1}{\partial x_1}\right)n_2 = p_2, \tag{3.2.9}$$

on ∂g.

The solvability conditions for the system (3.2.7)–(3.2.9) are

$$\int_g F_i d\mathbf{x}' + \int_{\partial g} p_i ds = 0, \quad i = 1, 2, \tag{3.2.10}$$

and

$$\int_g (x_2 F_1 - x_1 F_2) d\mathbf{x}' + \int_{\partial g} (x_2 p_1 - x_1 p_2) ds = 0. \tag{3.2.11}$$

A solution of (3.2.7)–(3.2.9) is defined up to an arbitrary rigid-body displacement, i.e. the vector field which admits the representation $(a_1, a_2)^T + b(-x_2, x_1)^T$, where a_1, a_2 and b are constant coefficients.

3.2.2 Anti-plane shear

Again, we consider an infinite cylinder with the axis Ox_3 and with the bounded cross-section $g \subset \mathbb{R}^2$. Assume that the external load is applied in such a way that the displacement field has the form

$$\mathbf{u} = (0, 0, u_3(\mathbf{x}'))^T. \tag{3.2.12}$$

The only non-zero components of stress are

$$\sigma_{i3} = \mu \frac{\partial u_3}{\partial x_i}, \quad i = 1, 2, \tag{3.2.13}$$

where μ is the shear modulus.

The Lamé system is reduced to the scalar Poisson equation

$$\mu \Delta u_3 + F_3 = 0 \quad \text{in } g, \tag{3.2.14}$$

where F_3 is the only non-zero component of the body-force density depending on $\mathbf{x}'$. The traction boundary conditions are represented by

$$\sigma_{13} n_1 + \sigma_{23} n_2 = p \quad \text{on } \partial g, \tag{3.2.15}$$

or, equivalently,

$$\mu \frac{\partial u_3}{\partial n} = p \quad \text{on } \partial g. \tag{3.2.16}$$

Here n_i, $i = 1, 2$, are components of the unit outward normal on the boundary ∂g, and p is a given traction.

Thus, we have reduced the formulation to the classical Neumann boundary value problem (3.2.14), (3.2.16) for the Poisson equation. In solid mechanics it is called the problem of *anti-plane shear*.

The solvability condition for (3.2.14), (3.2.16) is given by

$$\int_g F_3 d\mathbf{x}' + \int_{\partial g} p \, ds = 0, \tag{3.2.17}$$

which represents the balance relation for the external load.

3.3 Differential equations for engineering models of elastic rods

Here, we outline the differential equations used in the engineering theories of thin elastic rods subject to small deformations (see, for example, Westergaard (1952) and Gurtin (1981)). We distinguish between four modes:

- two transverse flexural modes;
- extension;
- torsion.

Consider a thin cylindrical elastic rod, and let z denote the longitudinal variable along its axis. Assume that, owing to the presence of applied load, the rod may bend, change its length and twist. To leading order, the displacement components and the angular rotation are

SPECIAL SOLUTIONS   121

approximated by functions which depend on the longitudinal variable z only. Let v_1 and v_2 denote the transverse components of displacement, v_3 be the longitudinal component along the rod axis, and the notation v_4 stand for the angular rotation of the cross-section.

For the *bending modes* of applied load, the tranverse displacement components satisfy the fourth-order differential equations

$$C_j \frac{d^4 v_j}{dz^4}(z) = p_j(z), \quad j = 1, 2,$$

where the quantities p_j represent the transverse components of the external load, and C_1 and C_2 are the flexural rigidity coefficients.

For the *extension mode*, the longitudinal displacement v_3 solves the second-order differential equation

$$C_3 \frac{d^2 v_3}{dz^2}(z) = p_3(z),$$

where p_3 is the external longitudinal load along the rod axis, and C_3 is the extensional rigidity.

Finally, for the *torsion mode*, the function v_4 satisfies the second-order differential equation

$$C_4 \frac{d^2 v_4}{dz^2}(z) = m(z),$$

where m is the external load consisting of a distribution of couples having their axes parallel to the axis of the rod, and C_4 is the torsional rigidity of the rod.

In practical applications, the above equations are supplied with the boundary conditions. In particular, when the rod is clamped at both ends, the functions v_j, $j = 1, 2, 3, 4$, and the derivatives v'_j, $j = 1, 2$, vanish at the ends of the rod.

3.4 Classical solutions of linear elasticity for a half-space

Here we refer to classical results (see, for example, Westergaard (1952)) for the linear elasticity problems in a half-space subject to a concentrated force applied on the boundary or at an internal point.

3.4.1 Boussinesq–Cerruti's solution

Let the displacement vector **u** satisfy the boundary value problem

$$\mathbf{L}(\partial_{\mathbf{x}})\mathbf{u}(\mathbf{x}) = \mathbf{0}, \quad \mathbf{x} \in \mathbb{R}^3_- = \{(x_1, x_2, x_3) : x_3 < 0\}, \qquad (3.4.1)$$

$$\boldsymbol{\sigma}^{(n)}(\mathbf{u};\mathbf{x}) = \mathbf{P}\delta(\mathbf{x}'), \quad x_3 = 0, \quad \mathbf{x}' = (x_1, x_2), \qquad (3.4.2)$$

where $\mathbf{P} = (P_1, P_2, P_3)^T$ is a constant vector, $\boldsymbol{\sigma}^{(3)} = (\sigma_{13}, \sigma_{23}, \sigma_{33})^T$. We seek a solution which decays at infinity. This field can be represented as a linear combination

$$\mathbf{u} = \sum_{j=1}^{3} P_j \boldsymbol{\mathcal{B}}^{(j)},$$

of the Boussinesq solution $\boldsymbol{\mathcal{B}}^{(3)}$, corresponding to a normal force, and of the Cerruti solutions $\boldsymbol{\mathcal{B}}^{(1)}$, $\boldsymbol{\mathcal{B}}^{(2)}$ for a tangential force ($i \neq 3$). We have no intention of performing the derivation of these solutions. We just present the result with reference to the book by Westergaard (1952).

The column $\boldsymbol{\mathcal{B}}^{(j)}$, $j = 1, 2, 3$, has the components

$$\mathcal{B}_i^{(j)}(\mathbf{x}) = \frac{1}{4\pi\mu R}\left[\frac{x_i x_j}{R^2} + 2(1-\nu)\delta_{ij} - (1-2\nu)p_i^{(j)}(\mathbf{x})\right],$$

where ν is the Poisson ratio, $R^2 = x_1^2 + x_2^2 + x_3^2$, and

$$p_3^{(3)} = 0,$$
$$p_j^{(3)}(\mathbf{x}) = -p_3^{(j)}(\mathbf{x}) = -\frac{x_j}{R - x_3}, \quad j = 1, 2,$$
$$p_2^{(1)}(\mathbf{x}) = p_1^{(2)}(\mathbf{x}) = \frac{x_1 x_2}{(R - x_3)^2},$$
$$p_j^{(j)}(\mathbf{x}) = -\frac{x_3}{R - x_3} + \frac{x_j^2}{(R - x_3)^2}, \quad j = 1, 2.$$

The matrix $\boldsymbol{\mathcal{B}}$ with columns $\boldsymbol{\mathcal{B}}^{(i)}$, $i = 1, 2, 3$, represents the well-known Boussinesq–Cerruti solution.

It should be mentioned that the fields

$$\partial_{x_1}\boldsymbol{\mathcal{B}}^{(3)}, \quad \partial_{x_2}\boldsymbol{\mathcal{B}}^{(3)}, \quad \frac{1}{2}(\partial_{x_1}\boldsymbol{\mathcal{B}}^{(2)} - \partial_{x_2}\boldsymbol{\mathcal{B}}^{(1)}),$$

correspond to concentrated moments applied at the boundary of the half-space, and the fields

$$\partial_{x_1}\boldsymbol{\mathcal{B}}^{(1)}, \quad \partial_{x_2}\boldsymbol{\mathcal{B}}^{(2)}, \quad \frac{1}{2}(\partial_{x_1}\boldsymbol{\mathcal{B}}^{(2)} + \partial_{x_2}\boldsymbol{\mathcal{B}}^{(1)}),$$

correspond to the concentrated expansion load and the shear load, respectively.

3.4.2 Mindlin's solution

Let a concentrated force $\mathbf{P} = (P_1, P_2, P_3)^T$ act at an internal point $\mathbf{x}_0 = (0, 0, y_3)$, $y_3 < 0$, of the half-space. We are looking for a solution of the boundary value problem

$$\mathbf{L}(\partial_\mathbf{x})\mathbf{u}(\mathbf{x}) + \mathbf{P}\delta(\mathbf{x} - \mathbf{x}_0) = 0, \quad \mathbf{x} \in \mathbb{R}^3_-, \quad (3.4.3)$$

$$\sigma^{(3)}(\mathbf{u}; \mathbf{x}) = 0, \quad x_3 = 0, \quad (3.4.4)$$

which vanishes at infinity. This problem was solved by Mindlin (1936). The displacement vector $\mathbf{u}$ admits the representation

$$\mathbf{u} = \sum_{j=1}^{3} P_j \mathcal{M}^{(j)},$$

where $\mathcal{M}^{(j)}$, $j = 1, 2$, correspond to a horizontal force applied at $\mathbf{x}_0$, and $\mathcal{M}^{(3)}$ gives the displacement field for the case of a vertical force. The columns $\mathcal{M}^{(j)}$, $j = 1, 2, 3$, are given by

$$\mathcal{M}_i^{(j)}(\mathbf{x}', x_3, y_3) = \frac{1}{16\pi\mu(1-\nu)} \left[\frac{\hat{x}_i \hat{x}_j}{R_1^3} + (3 - 4\nu)\frac{\hat{x}_i \hat{x}_j}{R_2^3} \right.$$
$$\left. + \left(\frac{1}{R_2} + \frac{3 - 4\nu}{R_1}\right)\delta_{ij} + t_i^{(j)}(\mathbf{x}', x_3, y_3) \right],$$

where $\mathbf{x}' = (x_1, x_2)$,

$$R_1 = [x_1^2 + x_2^2 + (x_3 - y_3)^2]^{1/2}, \quad R_2 = [x_1^2 + x_2^2 + (x_3 + y_3)^2]^{1/2},$$

$\hat{x}_i = x_i$, $i = 1, 2$, $\hat{x}_3 = x_3 - y_3$, and the terms $t_i^{(j)}$ have the form

$$t_j^{(3)} = -t_3^{(j)} = \frac{6y_3 x_3 (x_3 + y_3) x_j}{R_2^5} + \frac{4(1-\nu)(1-2\nu) x_j}{R_2(R_2 - x_3 - y_3)}, \quad j = 1, 2,$$

$$t_3^{(3)} = \frac{4(1-\nu)(1-2\nu)}{R_2} - \frac{2y_3 x_3}{R_2^3}\left(1 - \frac{3(x_3 + y_3)^2}{R_2^2}\right)$$

$$t_j^{(j)} = \frac{4(1-\nu)(1-2\nu)}{R_2 - x_3 - y_3}\left(1 - \frac{x_j^2}{R_2(R_2 - x_3 - y_3)}\right)$$
$$+ \frac{2y_3 x_3}{R_2^3}\left(1 - \frac{3x_j^2}{R_2^2}\right), \quad j = 1, 2,$$

$$t_1^{(2)} = t_2^{(1)} = -\frac{4(1-\nu)(1-2\nu) x_1 x_2}{R_2(R_2 - x_3 - y_3)^3} - \frac{6y_3 x_1 x_2 x_3}{R_2^5}.$$

The matrix function $\mathcal{M} = \mathcal{M}(\mathbf{x}', x_3, y_3)$ with the columns $\mathcal{M}^{(i)}$, $i = 1, 2, 3$, represents the Mindlin solution.

One can see that

$$\mathcal{M}^T(\mathbf{x}' - \mathbf{y}', x_3, y_3) = \mathcal{M}(\mathbf{y}' - \mathbf{x}', y_3, x_3). \tag{3.4.5}$$

3.4.3 Connection between the Boussinesq–Cerruti and Mindlin solutions,

The following statement gives a relation between the Boussinesq-Cerruti solution, the Mindlin solution and their derivatives, for $y_3 = 0$.

LEMMA 3.1

For any multi-index $\alpha = (\alpha_1, \alpha_2)$ and a non-negative integer k the formula

$$\partial_{\mathbf{y}'}^\alpha \partial_{y_3}^k \mathcal{M}(\mathbf{x}' - \mathbf{y}', x_3, y_3)|_{\mathbf{y}'=0,\ y_3=0}$$
$$= \sum_{|\beta|=k} (-\partial_{\mathbf{x}'})^\alpha \partial_{\mathbf{x}'}^\beta \mathcal{B}(\mathbf{x}', x_3) \mathbf{c}_\beta \tag{3.4.6}$$

is valid, where $\mathbf{c}_\beta$ are constant matrices. In particular, $\mathbf{c}_{(0,0)}$ is the identity matrix, and

$$\mathbf{c}_{(1,0)} = \begin{pmatrix} 0 & 0 & \lambda(2\mu+\lambda)^{-1} \\ 0 & 0 & 0 \\ 1 & 0 & 0 \end{pmatrix}, \quad \mathbf{c}_{(0,1)} = \begin{pmatrix} 0 & 0 & 0 \\ 0 & 0 & \lambda(2\mu+\lambda)^{-1} \\ 0 & 1 & 0 \end{pmatrix}.$$

Proof

(i) $k = 0$. Since

$$\mathcal{M}(\mathbf{x}', x_3, 0) = \mathcal{B}(\mathbf{x}', x_3), \tag{3.4.7}$$

we obtain (3.4.6) for $k = 0$.

(ii) $k = 1$. We represent the boundary operator of tractions $\sigma^{(3)}$ in the form

$$\sigma^{(3)}(\mathbf{u};\mathbf{x}) = \begin{pmatrix} \mu & 0 & 0 \\ 0 & \mu & 0 \\ 0 & 0 & \lambda+2\mu \end{pmatrix} (\partial_{x_3} + \mathbb{A}_1 \partial_{x_1} + \mathbb{A}_2 \partial_{x_2}) \mathbf{u}(\mathbf{x}),$$

where

$$\mathbb{A}_1 = \begin{pmatrix} 0 & 0 & 1 \\ 0 & 0 & 0 \\ \lambda(2\mu+\lambda)^{-1} & 0 & 0 \end{pmatrix}, \quad \mathbb{A}_2 = \begin{pmatrix} 0 & 0 & 0 \\ 0 & 0 & 1 \\ 0 & \lambda(2\mu+\lambda)^{-1} & 0 \end{pmatrix}.$$

Relation (3.4.4) yields

$$(\partial_{y_3} + \mathbb{A}_1 \partial_{y_1} + \mathbb{A}_2 \partial_{y_2})\mathcal{M}(\mathbf{y}' - \mathbf{x}', y_3, x_3)|_{y_3=0} = 0.$$

Using (3.4.5) we obtain

$$\partial^\alpha_{\mathbf{y}'} \partial_{y_3}\mathcal{M}(\mathbf{x}' - \mathbf{y}', x_3, y_3)|_{y_3=0} = (-\partial_{\mathbf{x}'})^\alpha \Big(\partial_{x_1}\mathcal{B}(\mathbf{x}' - \mathbf{y}', x_3)\mathbb{A}_1^T$$

$$+ \partial_{x_2}\mathcal{B}(\mathbf{x}' - \mathbf{y}', x_3)\mathbb{A}_2^T \Big), \qquad (3.4.8)$$

which gives (3.4.6) for $k=1$.

(iii) $k > 1$. The differential operator $\mathbf{L}$ can be represented in the form

$$\mathbf{L}(\partial_{\mathbf{x}}) = \begin{pmatrix} \mu & 0 & 0 \\ 0 & \mu & 0 \\ 0 & 0 & \lambda+2\mu \end{pmatrix} \Big(\partial^2_{x_3} + \sum_{j=1}^{2} \mathbb{A}_{j3}\partial_{x_j}\partial_{x_3} + \sum_{j=1}^{2}\sum_{k=1}^{2} \mathbb{A}_{jk}\partial_{x_j}\partial_{x_k} \Big),$$

where $\mathbb{A}_{j3}$, $\mathbb{A}_{jk}$ are constant matrices. By (3.4.5) and (3.4.3),

$$\partial^2_{y_3}\mathcal{M}(\mathbf{x}' - \mathbf{y}', x_3, y_3) \qquad (3.4.9)$$

$$= \sum_{j=1}^{2} \partial_{x_j}\partial_{y_3}\mathcal{M}(\mathbf{x}'-\mathbf{y}', x_3, y_3)\mathbb{A}_{j3}^T - \sum_{j=1}^{2}\sum_{k=1}^{2} \partial_{x_j}\partial_{x_k}\mathcal{M}(\mathbf{x}'-\mathbf{y}', x_3, y_3)\mathbb{A}_{jk}^T.$$

Using (3.4.9) m times we deduce that

$$\partial^{2m}_{y_3}\mathcal{M}(\mathbf{x}' - \mathbf{y}', x_3, y_3) \qquad (3.4.10)$$

$$= \sum_{|\beta|=2m-1} \partial^\beta_{\mathbf{x}'}\partial_{y_3}\mathcal{M}(\mathbf{x}'-\mathbf{y}', x_3, y_3)\mathbb{A}_{\beta 3}^T - \sum_{|\gamma|=2m} \partial^\gamma_{\mathbf{x}'}\mathcal{M}(\mathbf{x}'-\mathbf{y}', x_3, y_3)\mathbb{A}_\gamma^T,$$

where $\mathbb{A}_{\beta 3}$, $\mathbb{A}_\gamma$ are constant matrices. Then relations (3.4.10), (3.4.8) and (3.4.7) yield (3.4.6) for even k.

For odd values of k one can differentiate (3.4.10) with respect to y_3 and then use (3.4.9) to expand the term with $\partial^2_{y_3}\mathcal{M}$. As a result we obtain (3.4.6). □

The formula (3.4.6) implies the following relation between Mindlin's and Boussinesq's solutions:

$$\frac{1}{2}(\nabla_{\mathbf{y}} \times \mathcal{M}^T(\mathbf{x}' - \mathbf{y}', x_3, y_3)|_{\mathbf{y}=0})^T$$

$$= \left[-\partial_{x_2}\mathcal{B}^{(3)}(\mathbf{x}), \partial_{x_1}\mathcal{B}^{(3)}(\mathbf{x}), \frac{1}{2}\left(\partial_{x_2}\mathcal{B}^{(1)}(\mathbf{x}) - \partial_{x_1}\mathcal{B}^{(2)}(\mathbf{x})\right)\right]. \quad (3.4.11)$$

We recall that the right-hand side represents the displacement field corresponding to the concentrated moment applied at the origin. To verify (3.4.11), we use the equation

$$(\nabla_{\mathbf{y}} \times \mathcal{M}^T(\mathbf{x}' - \mathbf{y}', x_3, y_3))^T$$

$$= \left[\begin{pmatrix} 0 & -\partial_{y_3} & \partial_{y_2} \\ \partial_{y_3} & 0 & -\partial_{y_1} \\ -\partial_{y_2} & \partial_{y_1} & 0 \end{pmatrix} \mathcal{M}^T(\mathbf{x}' - \mathbf{y}', x_3, y_3) \right]^T,$$

together with the following two equations, which are consequences of (3.4.6):

$$\partial_{y_j}\mathcal{M}_{kl}(\mathbf{x}' - \mathbf{y}', x_3, y_3)\big|_{\mathbf{y}=0} = -\partial_{x_j}\mathcal{B}_{kl}(\mathbf{x}), \quad j = 1, 2,$$

and

$$\partial_{y_3}\mathcal{M}(\mathbf{x}' - \mathbf{y}', x_3, y_3)\big|_{\mathbf{y}=0} = \partial_{x_1}\begin{pmatrix} \mathcal{B}_{13}(\mathbf{x}) & 0 & (2\mu+\lambda)^{-1}\lambda\mathcal{B}_{11}(\mathbf{x}) \\ \mathcal{B}_{23}(\mathbf{x}) & 0 & (2\mu+\lambda)^{-1}\lambda\mathcal{B}_{21}(\mathbf{x}) \\ \mathcal{B}_{33}(\mathbf{x}) & 0 & (2\mu+\lambda)^{-1}\lambda\mathcal{B}_{31}(\mathbf{x}) \end{pmatrix}$$

$$+ \partial_{x_2}\begin{pmatrix} 0 & \mathcal{B}_{13}(\mathbf{x}) & (2\mu+\lambda)^{-1}\lambda\mathcal{B}_{12}(\mathbf{x}) \\ 0 & \mathcal{B}_{23}(\mathbf{x}) & (2\mu+\lambda)^{-1}\lambda\mathcal{B}_{22}(\mathbf{x}) \\ 0 & \mathcal{B}_{33}(\mathbf{x}) & (2\mu+\lambda)^{-1}\lambda\mathcal{B}_{32}(\mathbf{x}) \end{pmatrix}.$$

3.5 Special solutions for a bounded two-dimensional domain

Here we introduce solutions of some boundary value problems for the Laplace operator on the two-dimensional cross-section g of a cylinder which prove to be useful for the description of elastic fields within the cylinder. These solutions play the role of potential functions for the shear stress components, and for a detailed description we refer to the book by Lurie (1970), Chapter 6.

It is assumed that the origin coincides with the centre of inertia of g, and the coordinate axes OX_1, OX_2 are parallel to the principal inertia axes, i.e.

$$\int_g X_j d\mathbf{X}' = 0, \ j = 1, 2; \quad \int_g X_1 X_2 d\mathbf{X}' = 0. \tag{3.5.1}$$

We suppose also that the boundary of g is Lipschitz. Here, for simplicity of presentation we do not specify classes of solutions which should be understood in a generalised sense (see Chapter 1 of Oleinik, Shamaev and Yosifian (1992)).

3.5.1 The torsion potential

An important place in the further analysis will be taken by the function φ which solves the Neumann boundary value problem

$$\Delta\varphi(\mathbf{X}') = 0, \ \mathbf{X}' \in g, \tag{3.5.2}$$
$$\partial_n \varphi(\mathbf{X}') = X_2 n_1 - X_1 n_2, \ \mathbf{X}' \in \partial g. \tag{3.5.3}$$

Obviously,

$$\int_{\partial g} (X_2 n_1 - X_1 n_2) ds = 0,$$

and, therefore, the problem (3.5.2), (3.5.3) is solvable.

To provide the uniqueness, we assume that

$$\int_g \varphi(\mathbf{X}') d\mathbf{X}' = 0.$$

The function φ is used to represent the displacement vector within a cylindrical bar under the torsion load:

$$\mathbf{u}(\mathbf{X}) = \beta(-X_2 X_3, X_1 X_3, \varphi(X_1, X_2)),$$

where β is a constant characterising the twisting of the bar per unit length.

It should be mentioned that φ is closely related to the Prandtl potential function Φ for the shear stress components within an elastic rod subject to a torsion load. That is,

$$\Phi(\mathbf{X}') = \psi(\mathbf{X}') - \frac{1}{2}(X_1^2 + X_2^2),$$

where ψ is the conjugate function to φ with respect to the Cauchy–Riemann equations.

3.5.2 The bending potentials

Also, in Section 3.6 we shall use the functions χ_1, χ_2 which satisfy the following Neumann boundary value problems:

$$\Delta\chi_1(\mathbf{X}') = 0, \quad \mathbf{X}' \in g, \tag{3.5.4}$$

$$\partial_n\chi_1(\mathbf{X}') = X_1^2 n_1 + \frac{\lambda}{\lambda+\mu} X_1 X_2 n_2, \quad \mathbf{X}' \in \partial g, \tag{3.5.5}$$

and

$$\Delta\chi_2(\mathbf{X}') = 0, \quad \mathbf{X}' \in g, \tag{3.5.6}$$

$$\partial_n\chi_2(\mathbf{X}') = X_2^2 n_2 + \frac{\lambda}{\lambda+\mu} X_1 X_2 n_1, \quad \mathbf{X}' \in \partial g. \tag{3.5.7}$$

One can verify directly that

$$\int_{\partial g} \left(X_j^2 n_j + \frac{\lambda}{\lambda+\mu} X_1 X_2 n_{3-j} \right) ds = 0, \quad j = 1, 2,$$

which guarantees the solvability of the above-formulated Neumann boundary value problems.

To provide the uniqueness we assume that

$$\int_g \chi_j(\mathbf{X}')d\mathbf{X}' = 0, \quad j = 1, 2.$$

In the engineering literature χ_1, χ_2 are used as potential functions for the shear stress components within an elastic rod subject to a bending load (see Lurie (1970), Chapter 6).

3.5.3 Example

Consider the elliptical domain g bounded by

$$\frac{X_1^2}{a^2} + \frac{X_2^2}{b^2} = 1,$$

where a and b are non-zero constants. In this case the functions φ and χ_i, $i = 1, 2$, admit the representations

$$\varphi = \frac{b^2 - a^2}{a^2 + b^2} X_1 X_2,$$

$$\chi_1 = \frac{(1-2\nu)a^2}{3(3a^2+b^2)}(X_1^3 - 3X_1X_2^2) + \frac{b^2 + 2a^2(1+\nu)}{3a^2+b^2}a^2 X_1,$$

$$\chi_2 = \frac{(1-2\nu)b^2}{3(3b^2+a^2)}(X_2^3 - 3X_2X_1^2) + \frac{a^2 + 2b^2(1+\nu)}{3b^2+a^2}b^2 X_2.$$

3.6 Special solutions of linear elasticity for an infinite cylinder

Here we describe polynomial (in the axial direction) solutions of the homogeneous elasticity equations in an infinite cylinder with homogeneous traction boundary conditions.

Let $\Pi = \{\mathbf{X} \in \mathbb{R}^3 : X_3 \in \mathbb{R}, (X_1, X_2) \in g \subset \mathbb{R}^2\}$, where g is a domain with Lipschitz boundary. Also, we shall use the notation $\mathbf{X}' = (X_1, X_2)$ and denote by $\mathbf{e}^{(k)}$ the unit vector directed along the OX_k axis.

As in the previous section, it is assumed that the coordinate axes OX_1, OX_2 coincide with the principal inertia axes of g.

We seek vector functions, polynomial in X_3, which satisfy the homogeneous boundary value problem:

$$\mathbf{L}(\partial_X)\psi(\mathbf{X}) = 0, \ \mathbf{X} \in \Pi, \qquad (3.6.1)$$
$$\sigma^{(n)}(\psi; \mathbf{X}) = 0, \ \mathbf{X} \in \partial\Pi, \qquad (3.6.2)$$

where $\mathbf{n} = (n_1(\mathbf{X}'), n_2(\mathbf{X}'), 0)$ is the outward unit normal on $\partial\Pi$.

3.6.1 Representation of differential operators

We use the following representation for the operators of the problem:

$$\mathbf{L}(\partial_X)\psi = \mathbf{L}_0(\partial_{X'})\psi + \mathbf{L}_1(\partial_{X'})\partial_{X_3}\psi + \mathbf{L}_2 \partial_{X_3}^2 \psi, \qquad (3.6.3)$$

$$\sigma^{(n)}(\psi; \mathbf{X}) = \mathbf{T}_0(\mathbf{n}, \partial_{X'})\psi + \mathbf{T}_1(\mathbf{n})\partial_{X_3}\psi. \qquad (3.6.4)$$

Here

$$\mathbf{L}_0(\partial_{X'}) = \begin{pmatrix} (2\mu+\lambda)\partial_{X_1}^2 + \mu\partial_{X_2}^2 & (\lambda+\mu)\partial_{X_1}\partial_{X_2} & 0 \\ (\lambda+\mu)\partial_{X_1}\partial_{X_2} & (2\mu+\lambda)\partial_{X_2}^2 + \mu\partial_{X_1}^2 & 0 \\ 0 & 0 & \mu(\partial_{X_1}^2 + \partial_{X_2}^2) \end{pmatrix}; \qquad (3.6.5)$$

$$\mathbf{L}_1(\partial_{X'}) = (\lambda + \mu) \begin{pmatrix} 0 & 0 & \partial_{X_1} \\ 0 & 0 & \partial_{X_2} \\ \partial_{X_1} & \partial_{X_2} & 0 \end{pmatrix}; \qquad (3.6.6)$$

$$\mathbf{L}_2 = \begin{pmatrix} \mu & 0 & 0 \\ 0 & \mu & 0 \\ 0 & 0 & (2\mu + \lambda) \end{pmatrix}; \qquad (3.6.7)$$

$$\mathbf{T}_0(\mathbf{n}, \partial_{X'}) = n_1 \begin{pmatrix} (2\mu+\lambda)\partial_{X_1} & \lambda\partial_{X_2} & 0 \\ \mu\partial_{X_2} & \mu\partial_{X_1} & 0 \\ 0 & 0 & \mu\partial_{X_1} \end{pmatrix}$$

$$+ n_2 \begin{pmatrix} \mu\partial_{X_2} & \mu\partial_{X_1} & 0 \\ \lambda\partial_{X_1} & (2\mu+\lambda)\partial_{X_2} & 0 \\ 0 & 0 & \mu\partial_{X_2} \end{pmatrix}; \qquad (3.6.8)$$

$$\mathbf{T}_1(\mathbf{n}) = \begin{pmatrix} 0 & 0 & \lambda n_1 \\ 0 & 0 & \lambda n_2 \\ \mu n_1 & \mu n_2 & 0 \end{pmatrix}. \qquad (3.6.9)$$

3.6.2 The spectral problem

Construction of the exponential polynomial solutions of (3.6.1), (3.6.2) of the form

$$\psi(\mathbf{X}) = e^{\Lambda X_3} \sum_{j=0}^{N} \frac{X_3^j}{j!} \mathbf{\Phi}^{(N-j)}(\mathbf{X'}) \qquad (3.6.10)$$

leads to the eigenvalue problem

$$\mathbf{L}(\partial_{X'}, \Lambda)\mathbf{\Phi}^{(0)}(\mathbf{X'}) = 0, \ \mathbf{X'} \in g, \qquad (3.6.11)$$
$$\big(\mathbf{T}_0(\mathbf{n}, \partial_{X'}) + \Lambda \mathbf{T}_1(\mathbf{n})\big)\mathbf{\Phi}^{(0)}(\mathbf{X'}) = 0, \ \mathbf{X'} \in \partial g. \qquad (3.6.12)$$

It suffices for our purposes to describe only the solutions (3.6.10) which grow, as $X_3 \to \pm\infty$, slower than some polynomial. In the following theorem we show that there are no such solutions with non-zero pure imaginary Λ.

THEOREM 3.1

Let $\Lambda = i\tau$, where τ is a non-zero real number. Then the problem (3.6.11), (3.6.12) has only the trivial solution in the class $(H^1(g))^3$.

Proof

First, we note that the absence of non-trivial solutions of (3.6.11), (3.6.12) implies the absence of non-trivial solutions of the form (3.6.10).

Let a vector-valued function $\mathbf{u} \in (H^1(g))^3$ solve the problem (3.6.11), (3.6.12) with $\Lambda = i\tau$. Then $e^{i\tau X_3}\mathbf{u}$ is a solution of (3.6.1), (3.6.2). We shall use the notation

$$\Pi_1 = \{\mathbf{X} : (X_1, X_2) \in g, \ 0 < X_3 < 1\}.$$

Direct calculations show that

$$0 = -\int_g (\mathbf{L}(\partial_{\mathbf{X}'}, i\tau)\mathbf{u}) \cdot \overline{\mathbf{u}} d\mathbf{X}' = -\int_{\Pi_1} (\mathbf{L}(\partial_{\mathbf{X}})(e^{i\tau X_3}\mathbf{u})) \cdot \overline{(e^{i\tau X_3}\mathbf{u})} d\mathbf{X}$$

$$= \sum_{k,j=1}^{3} \int_{\Pi_1} \sigma_{kj}(e^{i\tau X_3}\mathbf{u}) \overline{\varepsilon_{kj}(e^{i\tau X_3}\mathbf{u})} d\mathbf{X}$$

$$= \int_{\Pi_1} \left\{ 2\mu \sum_{k,j=1}^{3} |\varepsilon_{kj}(e^{i\tau X_3}\mathbf{u})|^2 + \lambda \left|\sum_{k=1}^{3} \varepsilon_{kk}(e^{i\tau X_3}\mathbf{u})\right|^2 \right\} d\mathbf{X}.$$

Therefore,

$$\varepsilon_{kj}(e^{i\tau X_3}\mathbf{u}) = 0 \ \text{ for } \ k, j = 1, 2, 3.$$

This implies that all components of $\mathbf{u}$ are equal to zero. □

This result shows that we need to consider the eigenvalue $\Lambda = 0$. In this case one can check that the vector functions

$$\phi^{(k,0)} = \mathbf{e}^{(k)}, \ k = 1, 2, 3, \tag{3.6.13}$$

$$\phi^{(4,0)} = -X_2 \mathbf{e}^{(1)} + X_1 \mathbf{e}^{(2)} \tag{3.6.14}$$

are eigenfunctions.

The vector coefficients $\mathbf{\Phi}^{(1)}, \ldots, \mathbf{\Phi}^{(N)}$ are generalised eigenvectors of the same spectral problem corresponding to $\mathbf{\Phi}^{(0)}$, i.e. they satisfy

$$\mathbf{L}_0(\partial_{\mathbf{X}'})\mathbf{\Phi}^{(j)} = -(\mathbf{L}_1(\partial_{\mathbf{X}'})\mathbf{\Phi}^{(j-1)} + \mathbf{L}_2\mathbf{\Phi}^{(j-2)}), \tag{3.6.15}$$

$$\mathbf{T}_0(\mathbf{n}, \partial_{\mathbf{X}'})\mathbf{\Phi}^{(j)} = -\mathbf{T}_1(\mathbf{n})\mathbf{\Phi}^{(j-1)}, \tag{3.6.16}$$

where $j = 1, 2, \ldots, N$, and, by definition, $\mathbf{\Phi}^{(-1)} = 0$. In particular, for the case when $\Lambda = 0$ the eigenvectors $\phi^{(k,0)}$, $k = 3, 4$, generate the generalised eigenvectors

$$\phi^{(3,1)} = -\frac{\lambda}{2(\mu + \lambda)}(X_1 \mathbf{e}^{(1)} + X_2 \mathbf{e}^{(2)}), \tag{3.6.17}$$

$$\phi^{(4,1)} = \varphi(\mathbf{X}')\mathbf{e}^{(3)}, \tag{3.6.18}$$

where φ, introduced in Section 3.5.1, satisfies the Neumann boundary value problem (3.5.2), (3.5.3).

The eigenvectors $\phi^{(j,0)}$, $j = 1, 2$, give rise to generalised eigenvectors

$$\phi^{(j,1)} = -X_j \mathbf{e}^{(3)}, \tag{3.6.19}$$

$$\phi^{(1,2)} = \frac{\lambda}{4(\lambda + \mu)}[(2X_2 X_1 + k_1 X_1)\mathbf{e}^{(2)}$$
$$+ (X_1^2 - X_2^2 - k_1 X_2)\mathbf{e}^{(1)}], \tag{3.6.20}$$

$$\phi^{(2,2)} = \frac{\lambda}{4(\lambda + \mu)}[(2X_2 X_1 - k_2 X_2)\mathbf{e}^{(1)}$$
$$+ (X_2^2 - X_1^2 + k_2 X_1)\mathbf{e}^{(2)}], \tag{3.6.21}$$

$$\phi^{(j,3)} = \left(\frac{4\mu + 3\lambda}{12(\lambda + \mu)} X_j^3 + \frac{\lambda}{4(\lambda + \mu)} \frac{X_1^2 X_2^2}{X_j} - \chi_j(\mathbf{X}') \right.$$
$$\left. + k_j \frac{\lambda}{4(\lambda + \mu)} \varphi(\mathbf{X}') \right) \mathbf{e}^{(3)}, \tag{3.6.22}$$

where k_1, k_2 are constants which will be chosen later, and the functions χ_j, $j = 1, 2$, are solutions of (3.5.4)–(3.5.7).

3.6.3 X_3-polynomial solutions

According to the previous section, the vector functions

$$\sum_{j=0}^{l} \frac{X_3^j}{j!} \phi^{(k,l-j)}(\mathbf{X}'), \quad k = 1, 2, 3, 4, \; l = 0, \ldots, m_k - 1, \tag{3.6.23}$$

satisfy the boundary value problem (3.6.1), (3.6.2). Here

$$m_1 = m_2 = 4 \quad \text{and} \quad m_3 = m_4 = 2. \tag{3.6.24}$$

Expression (3.6.23) describes 12 linearly independent vector-valued functions

$$\boldsymbol{\psi}^{(i)} = \mathbf{e}^{(i)}, \; i = 1, 2, 3; \; \boldsymbol{\psi}^{(4)}(\mathbf{X}) = (-X_2, X_1, 0)^T,$$

$$\boldsymbol{\psi}^{(5)}(\mathbf{X}) = (X_3, 0, -X_1)^T, \; \boldsymbol{\psi}^{(6)}(\mathbf{X}) = (0, X_3, -X_2)^T,$$

$$\boldsymbol{\xi}^{(i)}(\mathbf{X}) = \frac{1}{6} X_3^3 \mathbf{e}^{(i)} + \left(-\frac{1}{2} X_3^2 X_i + \frac{4\mu + 3\lambda}{12(\lambda + \mu)} X_i^3 \right.$$

$$+\frac{\lambda}{4(\lambda+\mu)}\frac{X_1^2 X_2^2}{X_i}-\chi_i(\mathbf{X'})+k_i\frac{\lambda}{4(\lambda+\mu)}\varphi(\mathbf{X'})\Big)\mathbf{e}^{(3)}+X_3\phi^{(i,2)}(\mathbf{X'}), \ i=1,2,$$

$$\boldsymbol{\xi}^{(3)}(\mathbf{X}) = -\lambda[2(\lambda+\mu)]^{-1}(X_1\mathbf{e}^{(1)}+X_2\mathbf{e}^{(2)})+X_3\mathbf{e}^{(3)},$$

$$\boldsymbol{\xi}^{(4)}(\mathbf{X}) = (-X_2 X_3, X_1 X_3, \varphi(\mathbf{X'}))^T,$$

$$\boldsymbol{\xi}^{(5)}(\mathbf{X}) = (\tfrac{1}{2}X_3^2, 0, -X_3 X_1)^T + \boldsymbol{\phi}^{(1,2)}(\mathbf{X'}),$$

$$\boldsymbol{\xi}^{(6)}(\mathbf{X}) = (0, \tfrac{1}{2}X_3^2, -X_3 X_2)^T + \boldsymbol{\phi}^{(2,2)}(\mathbf{X'}). \qquad (3.6.25)$$

Note that $\boldsymbol{\psi}^{(i)}$, $i=1,2,\ldots,6$, are rigid-body displacements, whereas the vector polynomials $\boldsymbol{\xi}^{(i)}$, $i=1,2,\ldots,6$, have infinite elastic energy.

3.6.4 Biorthogonality conditions

Here and in what follows we choose k_1 and k_2 in (3.6.20) and (3.6.21) to satisfy

$$\frac{\lambda}{4(\lambda+\mu)}k_j\int_g(\|\mathbf{X'}\|^2-\|\nabla\varphi\|^2)d\mathbf{X'} = \int_g\Big(X_1\partial_{X_2}\chi_j - X_2\partial_{X_1}\chi_j$$

$$\pm\frac{\mu}{\lambda+\mu}X_j X_1 X_2\Big)d\mathbf{X'}, \ j=1,2, \qquad (3.6.26)$$

where the '+' sign corresponds to $j=1$, and '−' is related to the case $j=2$.

We note that the coefficient near k_j in (3.6.26) is positive. In fact,

$$\int_g \nabla\varphi\cdot(X_2\mathbf{e}^{(1)}-X_1\mathbf{e}^{(2)})d\mathbf{X'} = \int_g \|\nabla\varphi\|^2 d\mathbf{X'},$$

and therefore,

$$\int_g(\|\mathbf{X'}\|^2-\|\nabla\varphi\|^2)d\mathbf{X'} = \int_g\|\nabla\varphi - X_2\mathbf{e}^{(1)}+X_1\mathbf{e}^{(2)}\|^2 d\mathbf{X'} > 0.$$

Hence, the quantities k_1 and k_2 are uniquely defined by (3.6.26).

LEMMA 3.2

If k_1, k_2 satisfy (3.6.26) then the biorthogonality conditions

$$\left(\int_g \sigma^{(3)}(\boldsymbol{\xi}^{(k)}; \mathbf{X}) \cdot \boldsymbol{\psi}^{(i)}(\mathbf{X}) d\mathbf{X}'\right)_{k,i=1}^{6}$$

$$= \text{diag}\{-D_1, -D_2, D_3, D_4, D_1, D_2\} \qquad (3.6.27)$$

hold, where $\boldsymbol{\sigma}^{(3)} = (\sigma_{13}, \sigma_{23}, \sigma_{33})^T$, the integrals in (3.6.27) do not depend on X_3, and the elements of the diagonal matrix are

$$D_k = \mu \frac{2\mu + 3\lambda}{\lambda + \mu} \int_g X_k^2 d\mathbf{X}', \quad k = 1, 2,$$

$$D_3 = \mu \frac{2\mu + 3\lambda}{\lambda + \mu} \text{mes}_2 g,$$

$$D_4 = \mu \int_g \|\nabla \varphi - X_2 \mathbf{e}^{(1)} + X_1 \mathbf{e}^{(2)}\|^2 d\mathbf{X}'.$$

Proof

First, applying the Betti formula to the vector functions $\boldsymbol{\xi}^{(k)}$, $\boldsymbol{\psi}^{(i)}$, $i, k = 1, \ldots, 6$, in the region $\{\mathbf{x} \in \Pi, a < X_3 < b\}$ for some finite a, b, we obtain that the integral in the left-hand side of (3.6.27) does not depend on X_3.

Since

$$\partial_{X_3} \boldsymbol{\xi}^{(1)} = \boldsymbol{\xi}^{(5)} \quad \text{and} \quad \partial_{X_3} \boldsymbol{\xi}^{(2)} = \boldsymbol{\xi}^{(6)},$$

the relations (3.6.27) for $k = 5, 6$ follow from those obtained for $k = 1, 2$, by differentiation of (3.6.27) with respect to X_3.

Now, we prove (3.6.27) for $k = 1$. We have

$$\sigma_{13}(\boldsymbol{\xi}^{(1)}; \mathbf{X}) = \mu \left\{ X_1^2 - \partial_{X_1} \chi_1 + \frac{\lambda k_1}{4(\lambda + \mu)} (\partial_{X_1} \varphi - X_2) \right\},$$

$$\sigma_{23}(\boldsymbol{\xi}^{(1)}; \mathbf{X}) = \mu \left\{ \frac{\lambda}{\lambda + \mu} X_1 X_2 - \partial_{X_2} \chi_1 + \frac{\lambda k_1}{4(\lambda + \mu)} (\partial_{X_2} \varphi + X_1) \right\},$$

$$\sigma_{33}(\boldsymbol{\xi}^{(1)}; \mathbf{X}) = -\mu \frac{2\mu + 3\lambda}{\lambda + \mu} X_1 X_3.$$

By Green's formula,

$$\int_g \partial_{X_1} \chi_1 d\mathbf{X}' = \int_g \nabla \chi_1 \cdot \nabla X_1 d\mathbf{X}' = \int_{\partial g} X_1 (\partial_n \chi_1) ds,$$

and from the boundary condition (3.5.5),

$$\int_g \partial_{X_1}\chi_1 d\mathbf{X}' = \int_{\partial g} \left(X_1^3 n_1 + \frac{\lambda}{\lambda+\mu} X_1^2 X_2 n_2\right) ds$$

$$= \frac{4\lambda+3\mu}{\lambda+\mu} \int_{\partial g} X_1^2 d\mathbf{X}'.$$

Similarly,

$$\int_g \partial_{X_2}\chi_1 d\mathbf{X}' = \int_g \nabla\chi_1 \cdot \nabla X_2 ds = \int_{\partial g} X_2 \partial_n \chi_1 ds$$

$$= \int_{\partial g} \left(X_1^2 X_2 n_1 + \frac{\lambda}{\lambda+\mu} X_1 X_2^2 n_2\right) ds = 2\frac{2\lambda+\mu}{\lambda+\mu} \int_g X_1 X_2 d\mathbf{X}'.$$

Hence, using (3.5.1) we obtain

$$\int_g \partial_{X_2}\chi_1 d\mathbf{X}' = 0.$$

In the same way, it follows from (3.5.2), (3.5.3) that

$$\int_g \partial_{X_i}\varphi d\mathbf{X}' = 0, \ i = 1, 2.$$

Then, direct calculations give (3.6.27) for $k = 1$, $i = 1, 2, 3, 5, 6$. For $i = 4$ the relation (3.6.27) is valid owing to the choice of the coefficient k_1 (see (3.6.26)). The proof of (3.6.27) for $k = 2$ follows the same pattern. The verification of (3.6.27) for $k = 3, 4$ is trivial. □

3.6.5 The normalised stiffness coefficients

The quantities D_j, $1 \leq j \leq 4$, occur in the theory of elastic rods and can be interpreted as normalised stiffness coefficients, where D_1, D_2 are related to transverse displacements in the directions of the principal inertia axes of the cross-section, and D_3, D_4 correspond to the longitudinal displacement and axial rotation, respectively.

For example, for an elastic cylinder with a circular cross-section of radius a we have

$$D_1 = D_2 = \frac{\mu(2\mu+3\lambda)}{\lambda+\mu} \frac{\pi a^4}{4}, \quad D_3 = \frac{\mu(2\mu+3\lambda)}{\lambda+\mu} \pi a^2, \quad D_4 = \frac{\mu\pi a^4}{2}.$$

3.6.6 Biorthogonality relations for eigenvectors and generalised eigenvectors

Further, we shall need the following corollary of Lemma 3.2.

LEMMA 3.3

The biorthogonality relations hold:

$$\int_g (\mathbf{L}_1 \phi^{(k,m_k-1)} + \mathbf{L}_2 \phi^{(k,m_k-2)}) \cdot \phi^{(j,0)} d\mathbf{X}'$$
$$- \int_{\partial g} \mathbf{T}_1 \phi^{(k,m_k-1)} \cdot \phi^{(j,0)} dl = \mp \delta_{kj} \mathcal{D}_k, \; j = 1, 2, 3, 4, \quad (3.6.28)$$

where '$-$' corresponds to $k = 1, 2$, and '$+$' to $k = 3, 4$. Here, δ_{kj} denotes the Kronecker delta.

Proof

Let $\eta = \eta(X_3)$ be a smooth function which is equal to 1 for large X_3 and 0 for $X_3 < 0$. First, we show that

$$\left\{ \int_\Pi \mathbf{L}(\partial_\mathbf{x})(\eta \boldsymbol{\xi}^{(j)}) \cdot \boldsymbol{\psi}^{(k)} d\mathbf{X} - \int_{\partial \Pi} \sigma^{(n)}(\eta \boldsymbol{\xi}^{(j)}; \mathbf{X}) \cdot \boldsymbol{\psi}^{(k)} ds \right\}_{j,k=1}^{6}$$
$$= \text{diag}\{-D_1, -D_2, D_3, D_4, D_1, D_2\}. \quad (3.6.29)$$

Indeed, applying the Betti formula to $\eta \boldsymbol{\xi}^{(j)}$ and $\boldsymbol{\psi}^{(k)}$ in the domain $\{\mathbf{X} \in \Pi : 0 < X_3 < c\}$, where c is a large positive number, we obtain that the left-hand side (3.6.29) is equal to

$$\left\{ \int_g \sigma^{(3)}(\boldsymbol{\xi}^{(j)}; \mathbf{X}) \cdot \boldsymbol{\psi}^{(k)} d\mathbf{X}'|_{X_3=c} \right\}_{j,k=1}^{6}.$$

Hence, (3.6.29) follows from (3.6.27).

Using the representations (3.6.3) and (3.6.4) one has

$$\int_\Pi \mathbf{L}(\partial_\mathbf{x})(\eta \boldsymbol{\xi}^{(j)}) \cdot \boldsymbol{\psi}^{(k)} d\mathbf{X} - \int_{\partial \Pi} \sigma^{(n)}(\eta \boldsymbol{\xi}^{(j)}; \mathbf{X}) \cdot \boldsymbol{\psi}^{(k)} ds$$
$$= \int_\Pi \eta'(\mathbf{L}_1 \boldsymbol{\xi}^{(j)} + \mathbf{L}_2 \partial_{X_3} \boldsymbol{\xi}^{(j)}) \cdot \boldsymbol{\psi}^{(k)} d\mathbf{X} - \int_{\partial \Pi} \eta'(\mathbf{T}_1 \boldsymbol{\xi}^{(j)}) \cdot \boldsymbol{\psi}^{(k)} ds, \quad (3.6.30)$$

for $k = 1, 2, 3, 4$. The left-hand side of (3.6.30) does not depend on η. Hence, we can replace η by η_s given by $\eta_s(X_3) = s\eta(sX_3)$. Taking the limit $s \to \infty$, we arrive at

$$\int_g (\mathbf{L}_1 \boldsymbol{\xi}^{(j)} + \mathbf{L}_2 \partial_{X_3} \boldsymbol{\xi}^{(j)}) \cdot \boldsymbol{\psi}^{(k)} d\mathbf{X}'|_{X_3=0}$$

$$- \int_{\partial g} (\mathbf{T}_1 \boldsymbol{\xi}^{(j)}) \cdot \boldsymbol{\psi}^{(k)} dl|_{X_3=0} = \mp \delta_{kj} D_k, \qquad (3.6.31)$$

where '$-$' corresponds to $k = 1, 2$, and '$+$' to $k = 3, 4$. The left-hand sides (3.6.31) and (3.6.28) coincide. This completes the proof. $\square$

REMARK 3.1

(i) Similar to (3.6.29), one can show that

$$\int_\Pi \mathbf{L}(\partial_\mathbf{X})(\eta \boldsymbol{\psi}^{(j)}) \cdot \boldsymbol{\psi}^{(k)} d\mathbf{X} - \int_{\partial \Pi} \sigma^{(n)}(\eta \boldsymbol{\psi}^{(j)}; \mathbf{X}) \cdot \boldsymbol{\psi}^{(k)} ds = 0, \quad (3.6.32)$$

for $j, k = 1, \ldots, 6$.

(ii) We recall that according to (3.6.15) and (3.6.16) the generalised eigenvectors

$$\boldsymbol{\phi}^{(k,j)}, \ k = 1, 2, 3, 4; j = 1, \ldots, m_k - 1,$$

are defined as solutions of the boundary value problem

$$\mathbf{L}_0(\partial_{X'})\boldsymbol{\phi}^{(k,j)} = -(\mathbf{L}_1(\partial_{X'})\boldsymbol{\phi}^{(k,j-1)} + \mathbf{L}_2 \boldsymbol{\phi}^{(k,j-2)}), \quad (3.6.33)$$
$$\mathbf{T}_0(\mathbf{n}, \partial_{X'})\boldsymbol{\phi}^{(k,j)} = -\mathbf{T}_1(\mathbf{n})\boldsymbol{\phi}^{(k,j-1)}. \qquad (3.6.34)$$

Functions with negative indices are assumed to be zero. The solvability conditions (3.2.10), (3.2.11) and (3.2.17) for (3.6.33), (3.6.34) imply the orthogonality relation

$$\int_g (\mathbf{L}_1 \boldsymbol{\phi}^{(k,s)} + \mathbf{L}_2 \boldsymbol{\phi}^{(k,s-1)}) \cdot \boldsymbol{\phi}^{(j,0)} d\mathbf{X}'$$

$$- \int_{\partial g} (\mathbf{T}_1 \boldsymbol{\phi}^{(k,s)}) \cdot \boldsymbol{\phi}^{(j,0)} ds = 0, \qquad (3.6.35)$$

for $s < m_k - 1$, $j = 1, 2, 3, 4$, where $m_1, \ldots, m_4$ are given by (3.6.24).

3.6.7 There are no other polynomials

THEOREM 3.2

Any X_3-polynomial solution of (3.6.1), (3.6.2) with coefficients from $(\mathbf{H}^1(g))^3$ can be represented as a linear combination of the vector polynomials (3.6.23).

Proof

(a) First, consider a polynomial $\mathbf{p}^{(0)}(\mathbf{X}')$ of degree 0 (with respect to X_3) and assume that it satisfies (3.6.1), (3.6.2).

It follows from (3.6.3), (3.6.4) that

$$\mathbf{L}_0(\partial_{\mathbf{X}'})\mathbf{p}^{(0)}(\mathbf{X}') = 0, \quad \mathbf{X}' \in g, \tag{3.6.36}$$
$$\mathbf{T}_0(\mathbf{n}, \partial_{\mathbf{X}'})\mathbf{p}^{(0)}(\mathbf{X}') = 0, \quad \mathbf{X}' \in \partial g. \tag{3.6.37}$$

Thus, the vector $(p_1^{(0)}, p_2^{(0)})$ satisfies the homogeneous plane strain problem on g with traction boundary conditions, and $p_3^{(0)}$ solves the homogeneous Neumann boundary value problem for the Laplace operator on g. Hence, the vector function $\mathbf{p}^{(0)}$ can be represented as a linear combination of $\varphi^{(k,0)}$, $k = 1, 2, 3, 4$, given by (3.6.13), (3.6.14) (see Section 3.2).

(b) Let

$$\mathbf{p}(\mathbf{X}) = X_3 \mathbf{p}^{(0)}(\mathbf{X}') + \mathbf{p}^{(1)}(\mathbf{X}') \tag{3.6.38}$$

solve (3.6.1), (3.6.2). Then $\mathbf{p}^{(0)}$ satisfies the boundary value problem (3.6.36), (3.6.37), and, hence,

$$\mathbf{p}^{(0)} = \sum_{k=1}^{4} c_k^{(0)} \phi^{(k,0)} \tag{3.6.39}$$

with constant coefficients $c_k^{(0)}$.

The function $\mathbf{p}^{(1)}$ from (3.6.38) solves the boundary value problem

$$-\mathbf{L}_0(\partial_{\mathbf{X}})\mathbf{p}^{(1)}(\mathbf{X}') = \mathbf{L}_1(\partial'_{\mathbf{X}})\mathbf{p}^{(0)}(\mathbf{X}') \quad \text{in } g, \tag{3.6.40}$$
$$\mathbf{T}_0(\partial_{\mathbf{X}})\mathbf{p}^{(1)}(\mathbf{X}') = -\mathbf{T}_1(\partial'_{\mathbf{X}})\mathbf{p}^{(0)}(\mathbf{X}') \quad \text{on } \partial g. \tag{3.6.41}$$

In accordance with (3.6.15), (3.6.16) the function $\sum_{k=1}^{4} c_k^{(0)} \phi^{(k,1)}$ is a solution of (3.6.40), (3.6.41), and therefore

$$\mathbf{p}^{(1)} = \sum_{k=1}^{4} \{c_k^{(0)} \phi^{(k,1)} + c_k^{(1)} \phi^{(k,0)}\}. \qquad (3.6.42)$$

Hence, the vector function (3.6.38) can be represented as a linear combination of the polynomials (3.6.23) of degrees 0 and 1.

(c) Consider the vector function

$$\mathbf{p}(\mathbf{X}) = \frac{X_3^2}{2} \mathbf{p}^{(0)}(\mathbf{X}') + X_3 \mathbf{p}^{(1)}(\mathbf{X}') + \mathbf{p}^{(2)}(\mathbf{X}'), \qquad (3.6.43)$$

satisfying (3.6.3), (3.6.4).

The vector coefficient $\mathbf{p}^{(0)}$ solves (3.6.36), (3.6.37), and it can be written in the form (3.6.39).

For the vector function $\mathbf{p}^{(1)}$ equations (3.6.40), (3.6.41) hold, and, therefore, it can be represented by (3.6.42).

The vector $\mathbf{p}^{(2)}$ is a solution of the boundary value problem

$$-\mathbf{L}_0(\partial_\mathbf{X}')\mathbf{p}^{(2)}(\mathbf{X}') = \mathbf{L}_1(\partial_\mathbf{X}')\mathbf{p}^{(1)}(\mathbf{X}') + \mathbf{L}_2 \mathbf{p}^{(0)}(\mathbf{X}') \text{ in } g, \qquad (3.6.44)$$

$$\mathbf{T}_0(\partial_\mathbf{X}', \mathbf{n}) \mathbf{p}^{(2)}(\mathbf{X}') = -\mathbf{T}_1(\mathbf{n}) \mathbf{p}^{(1)}(\mathbf{X}') \text{ on } \partial g. \qquad (3.6.45)$$

According to Section 3.2, the problem (3.6.44), (3.6.45) is solvable if and only if the right-hand sides of (3.6.44), (3.6.45) are orthogonal to $\phi^{(j,0)}, j = 1, 2, 3, 4$, i.e.

$$\int_g \left(\mathbf{L}_1(\partial_\mathbf{X}')\mathbf{p}^{(1)}(\mathbf{X}') + \mathbf{L}_2 \mathbf{p}^{(0)}(\mathbf{X}') \right) \cdot \phi^{(j,0)}(\mathbf{X}') d\mathbf{X}'$$

$$- \int_{\partial g} \left(\mathbf{T}_1(\mathbf{n}) \mathbf{p}^{(1)}(\mathbf{X}') \right) \cdot \phi^{(j,0)} ds = 0. \qquad (3.6.46)$$

Using formulae (3.6.28) and (3.6.43) for $j = 3, 4$, we obtain that the left-hand side of (3.6.46) is equal to $c_j^{(0)} D_j$, and hence

$$c_3^{(0)} = c_4^{(0)} = 0. \qquad (3.6.47)$$

By (3.6.15), (3.6.16)

$$\mathbf{p}^{(2)}(\mathbf{X}') = \sum_{k=1}^{2} c_k^{(0)} \phi^{(k,2)} + \sum_{k=1}^{4} (c_k^{(1)} \phi^{(k,1)} + c_k^{(2)} \phi^{(k,0)}), \qquad (3.6.48)$$

and then (3.6.43) can be represented as a linear combination of polynomials (3.6.23) of degree $q \leq 2$.

(d) Now, consider a cubic polynomial in X_3

$$\mathbf{p}(\mathbf{X}) = \frac{X_3^3}{3!}\mathbf{p}^{(0)}(\mathbf{X}') + \frac{X_3^2}{2}\mathbf{p}^{(1)}(\mathbf{X}') + X_3\mathbf{p}^{(2)}(\mathbf{X}') + \mathbf{p}^{(3)}(\mathbf{X}), \quad (3.6.49)$$

which solves (3.6.1), (3.6.2).

The vector coefficients $\mathbf{p}^{(0)}, \mathbf{p}^{(1)}$ and $\mathbf{p}^{(2)}$ satisfy (3.6.36), (3.6.37), (3.6.40), (3.6.41) and (3.6.44), (3.6.45), respectively, and the relation (3.6.47) holds. The polynomials $\mathbf{p}^{(0)}, \mathbf{p}^{(1)}, \mathbf{p}^{(2)}$ are given by (3.6.39), (3.6.42), (3.6.48). For the term $\mathbf{p}^{(3)}(\mathbf{X}')$ we have

$$-\mathbf{L}_0(\partial'_\mathbf{X})\mathbf{p}^{(3)}(\mathbf{X}') = \mathbf{L}_1(\partial'_\mathbf{X})\mathbf{p}^{(2)}(\mathbf{X}') + \mathbf{L}_2\mathbf{p}^{(1)}(\mathbf{X}') \quad \text{in } g, \quad (3.6.50)$$

$$\mathbf{T}_0(\partial'_\mathbf{X}, \mathbf{n})\mathbf{p}^{(3)}(\mathbf{X}') = -\mathbf{T}_1(\mathbf{n})\mathbf{p}^{(2)}(\mathbf{X}') \quad \text{on } \partial g. \quad (3.6.51)$$

Similar to part (c) we derive

$$c_3^{(1)} = c_4^{(1)} = 0, \quad (3.6.52)$$

and

$$\mathbf{p}^{(3)}(\mathbf{X}') = \sum_{k=1}^{2}(c_k^{(0)}\phi^{(k,3)} + c_k^{(1)}\phi^{(k,2)})$$

$$+ \sum_{k=1}^{4}(c_k^{(2)}\phi^{(k,1)} + c_k^{(3)}\phi^{(k,0)}). \quad (3.6.53)$$

Hence, the polynomial $\mathbf{p}$ is represented as a linear combination of the vector polynomials (3.6.23).

(e) For the fourth-order polynomial

$$\mathbf{p}(\mathbf{X}) = \sum_{j=0}^{4} \frac{X_3^j}{j!}\mathbf{p}^{(4-j)}(\mathbf{X}'),$$

we have that the coefficients $\mathbf{p}^{(0)}$, $\mathbf{p}^{(1)}$, $\mathbf{p}^{(2)}$ and $\mathbf{p}^{(3)}$ are given by (3.6.39), (3.6.42), (3.6.48) and (3.6.53), where the constant coefficients

$$c_3^{(j)}, c_4^{(j)}, j = 0, 1,$$

vanish, and the term $\mathbf{p}^{(4)}$ satisfies the boundary value problem

$$-\mathbf{L}_0(\partial'_\mathbf{X})\mathbf{p}^{(4)}(\mathbf{X}') = \mathbf{L}_1(\partial'_\mathbf{X})\mathbf{p}^{(3)}(\mathbf{X}') + \mathbf{L}_2\mathbf{p}^{(2)}(\mathbf{X}') \quad \text{in } g, \quad (3.6.54)$$

$$\mathbf{T}_0(\partial'_\mathbf{X}, \mathbf{n})\mathbf{p}^{(4)}(\mathbf{X}') = -\mathbf{T}_1(\mathbf{n})\mathbf{p}^{(3)}(\mathbf{X}') \quad \text{on } \partial g. \quad (3.6.55)$$

The problem (3.6.54), (3.6.55) is solvable, provided the right-hand sides are orthogonal to $\phi^{(0,j)}$, $j = 1, 2, 3, 4$.

From (3.6.28) and (3.6.43) we deduce that

$$c_1^{(0)} = c_2^{(0)} = c_3^{(2)} = c_4^{(2)} = 0,$$

and hence $\mathbf{p}^{(0)} = 0$. Therefore, $\mathbf{p}$ is a cubic polynomial. □

3.7 Green's matrix in Ω

3.7.1 Definition

Let us define Green's matrix $\mathcal{N}(\mathbf{x}, \mathbf{y})$ of the traction problem for the Lamé system in Ω. We shall use the Cartesian system with coordinate axes which coincide with the principal inertia axes of the body Ω. The unit basis vectors are denoted by $\mathbf{e}^{(j)}$, $j = 1, 2, 3$. Then the columns $\mathcal{N}^{(j)}$ of $\mathcal{N}$ satisfy

$$-\mathbf{L}(\partial_\mathbf{x})\mathcal{N}^{(j)}(\mathbf{x},\mathbf{y}) = \left(\delta(\mathbf{x}-\mathbf{y}) - \frac{1}{\text{mes}_3\Omega}\right)\mathbf{e}^{(j)}$$
$$- J_j^{-1}\mathbf{x} \times (\mathbf{e}^{(j)} \times \mathbf{y}), \quad \mathbf{x} \in \Omega, \quad (3.7.1)$$
$$\sigma^{(n)}(\mathcal{N}^{(j)}(\cdot,\mathbf{y});\mathbf{x}) = 0, \quad \mathbf{x} \in \partial\Omega, \quad \mathbf{y} \in \Omega, \quad (3.7.2)$$

where J_j, $j = 1, 2, 3$, are the principal inertia moments

$$J_j = \int_\Omega \left(\|\mathbf{x}\|^2 - x_j^2\right) d\mathbf{x}. \qquad (3.7.3)$$

We note that the principal force and moment vectors, evaluated for the right-hand side in (3.7.1), are equal to zero; this implies the solvability of (3.7.1), (3.7.2). To provide the uniqueness of $\mathcal{N}^{(j)}$, we set the orthogonality conditions

$$\int_\Omega \mathcal{N}^{(j)}(\mathbf{x},\mathbf{y})d\mathbf{x} = 0, \quad \int_\Omega \mathbf{x} \times \mathcal{N}^{(j)}(\mathbf{x},\mathbf{y})d\mathbf{x} = 0.$$

The matrix function $\mathcal{N}$ is symmetric, i.e.

$$\mathcal{N}(\mathbf{x},\mathbf{y}) = \mathcal{N}^T(\mathbf{y},\mathbf{x}).$$

3.7.2 Asymptotics

To describe the asymptotics of Green's matrix $\mathcal{N}$ near the junction points we need one more coordinate system. That is, let the horizontal part of $\partial\Omega$, containing $\mathbf{a}^{(1)},\ldots,\mathbf{a}^{(K)}$, belong to the plane $x_3 = 0$, and let the axis Ox_3 be directed downwards.

LEMMA 3.4

The matrix-valued function $\mathbf{y} \to \mathcal{N}(\mathbf{x},\mathbf{y})$, $\mathbf{y} \in \Omega$, *can be extended by continuity to* $\mathbf{a}^{(j)}$ *for any* $\mathbf{x} \in \Omega$, *and the following asymptotic representation holds in a neighbourhood of* $\mathbf{a}^{(j)}$:

$$\mathcal{N}(\mathbf{x},\,\mathbf{a}^{(j)}) = \mathcal{B}(\mathbf{x},\mathbf{a}^{(j)}) + \mathbf{N}^{(j)}(\mathbf{x}), \tag{3.7.4}$$

$$\frac{1}{2}(\nabla_\mathbf{y} \times \mathcal{N}^T(\mathbf{x},\mathbf{y})|_{\mathbf{y}=\mathbf{a}^{(j)}})^T = \Big[-\partial_{x_2}\mathcal{B}^{(3)}(\mathbf{x}-\mathbf{a}^{(j)}), \partial_{x_1}\mathcal{B}^{(3)}(\mathbf{x}-\mathbf{a}^{(j)}),$$

$$\frac{1}{2}\Big(\partial_{x_2}\mathcal{B}^{(1)}(\mathbf{x}-\mathbf{a}^{(j)}) - \partial_{x_1}\mathcal{B}^{(2)}(\mathbf{x}-\mathbf{a}^{(j)})\Big)\Big] + \mathbf{D}^{(j)}(\mathbf{x}), \tag{3.7.5}$$

where $\mathcal{B}$ *is the Boussinesq tensor for the half-space* $\mathbb{R}^3_-$, *and the matrices* $\mathbf{N}^{(j)}$ *and* $\mathbf{D}^{(j)}$ *are smooth.*

Proof

Let $\mathbf{y} \in \Omega$ be located in a ball $B(\mathbf{a}^{(j)},\rho) = \{\|\mathbf{x}-\mathbf{a}^{(j)}\| < \rho\}$, where ρ is sufficiently small. We represent Green's matrix as

$$\mathcal{N}(\mathbf{x},\mathbf{y}) = \mathcal{M}(\mathbf{x},\mathbf{y})\zeta(\mathbf{x}) + \mathbf{M}(\mathbf{x},\mathbf{y}),$$

where ζ is a smooth cut-off function such that

$$\zeta(\mathbf{x}) = 1 \quad \text{as} \quad \|\mathbf{x}-\mathbf{a}^{(j)}\| < 2\rho$$

and

$$\zeta(\mathbf{x}) = 0 \quad \text{as} \quad \|\mathbf{x}-\mathbf{a}^{(j)}\| > 3\rho,$$

and $\mathcal{M}$ is Mindlin's solution.

The columns of $\mathbf{M}$ satisfy the Neumann boundary value problem in Ω with smooth right-hand sides. Thus the components of $\mathbf{M}$ are smooth functions in both $\mathbf{x}$ and $\mathbf{y}$ in the closure of $\Omega \cap B(\mathbf{a}^{(j)},\rho)$ (see Agmon, Douglis and Nirenberg (1959)). The result follows from Lemma 3.1 and formula (3.4.11). □

3.8 Korn's inequalities

We consider an isotropic homogeneous elastic body D and introduce the elastic energy functional:

$$\mathcal{E}(\mathbf{u};D) = \frac{1}{2} \sum_{i,j=1}^{3} \int_D \sigma_{ij}(\mathbf{u};\mathbf{x}) \varepsilon_{ij}(\mathbf{u};\mathbf{x}) d\mathbf{x}$$

$$= \frac{1}{2} \int_D \Big\{ 2\mu \sum_{i,j=1}^{3} \varepsilon_{ij}^2 + \lambda \Big(\sum_{i=1}^{3} \varepsilon_{ii} \Big)^2 \Big\} d\mathbf{x}. \tag{3.8.1}$$

This functional vanishes on the rigid-body translations and rotations only (see Section 3.1).

3.8.1 The case of bounded Lipschitz domains

LEMMA 3.5

Let D be a bounded domain with Lipschitz boundary, and $\mathbf{u} \in (H^1(D))^3$. Then:

(i) the displacement vector $\mathbf{u}$ satisfies the estimate

$$\|\mathbf{u}\|_{(H^1(D))^3}^2 \leq \mathrm{Const}\{\mathcal{E}(\mathbf{u};D) + \|\mathbf{u}\|_{(L_2(D))^3}^2\}. \tag{3.8.2}$$

(ii) Let V be a subspace of $(H^1(D))^3$ which does not include rigid-body translations and rotations $\mathbf{a} + \mathbf{b} \times \mathbf{x}$. Then, for any $\mathbf{u} \in V$,

$$\|\mathbf{u}\|_{(H^1(D))^3}^2 \leq \mathrm{Const}\,\mathcal{E}(\mathbf{u};D). \tag{3.8.3}$$

(iii) Let F_j, $j = 1, \ldots, 6$, be linear bounded functionals on $(H^1(D))^3$, which do not vanish simultaneously on rigid-body displacements. Then

$$\|\mathbf{u}\|_{(H^1(D))^3} \leq \mathrm{Const}\,\Big\{\mathcal{E}(\mathbf{u};D) + \sum_{j=1}^{6} |F_j(\mathbf{u})|^2\Big\}.$$

Proof

Assertions (i) and (ii) can be found, for example, in Oleinik, Shamaev and Yosifian (1992) (Theorems 2.4 and 2.5).

(iii) Let Z be the subspace of vector functions $\mathbf{v} \in (H^1(D))^3$ defined by
$$F_j(\mathbf{v}) = 0, \ j = 1, \ldots, 6.$$
We note that every $\mathbf{u} \in (H^1(D))^3$ can be uniquely represented as
$$\mathbf{u} = \mathbf{v} + \mathbf{a} + \mathbf{b} \times \mathbf{x},$$
where $\mathbf{v} \in Z$, and $\mathbf{a}, \mathbf{b} \in \mathbb{R}^3$. In fact, we can find the vectors $\mathbf{a}$ and $\mathbf{b}$ from the equations
$$F_j(\mathbf{a} + \mathbf{b} \times \mathbf{x}) = F_j(\mathbf{u}), \ j = 1, \ldots, 6.$$
These equations are solvable because six equalities $F_j(\mathbf{a} + \mathbf{b} \times \mathbf{x}) = 0$ imply $\mathbf{a} = 0$ and $\mathbf{b} = 0$.

By (ii)
$$\|\mathbf{v}\|^2_{(H^1(D))^3} \leq c\mathcal{E}(\mathbf{v}; D) = c\mathcal{E}(\mathbf{u}; D). \tag{3.8.4}$$
Since F_j, $j = 1, \ldots, 6$, do not vanish simultaneously on rigid-body displacements,
$$\|\mathbf{a}\| + \|\mathbf{b}\| \leq c \sum_{j=1}^{6} |F_j(\mathbf{a} + \mathbf{b} \times \mathbf{x})| = c \sum_{j=1}^{6} |F_j(\mathbf{u})|. \tag{3.8.5}$$

The result follows from (3.8.4) and (3.8.5). □

Further, we shall need two lemmas of the same nature.

LEMMA 3.6

Let D be a bounded domain with Lipschitz boundary. Also, let ω be a non-empty subdomain of D. The following inequality is valid for all $\mathbf{u} \in (H^1(D))^3$:
$$\int_D \|\mathbf{u}(\mathbf{x})\|^2 d\mathbf{x} \leq C\alpha_\omega \left(\mathcal{E}(\mathbf{u}; D) + \int_\omega \|\mathbf{u}\|^2 d\mathbf{x} \right) \tag{3.8.6}$$

with C independent of ω and
$$\alpha_\omega = \sup_{\mathbf{a},\mathbf{b}} \frac{\int_D \|\mathbf{a} + \mathbf{b} \times \mathbf{x}\|^2 d\mathbf{x}}{\int_\omega \|\mathbf{a} + \mathbf{b} \times \mathbf{x}\|^2 d\mathbf{x}}. \tag{3.8.7}$$

KORN'S INEQUALITIES

Here, the supremum is taken over all non-zero constant vectors $\mathbf{a}, \mathbf{b}$.

Proof

We introduce the space

$$\mathbf{V} = \{\mathbf{v} \in (H^1(D))^3 : \int_D \mathbf{v}(\mathbf{x}) \cdot (\mathbf{a} + \mathbf{b} \times \mathbf{x}) d\mathbf{x} = 0, \text{ for all } \mathbf{a}, \mathbf{b} \in \mathbb{R}^3\}.$$

Let $\mathbf{u} \in (H^1(D))^3$ be represented as $\mathbf{v} + \mathbf{w}$, where $\mathbf{w}$ is a rigid-body displacement and $\mathbf{v} \in \mathbf{V}$. By Lemma 3.5(ii), $\mathbf{v}$ satisfies the Korn inequality

$$\|\mathbf{v}\|^2_{(H^1(D))^3} \leq C\mathcal{E}(\mathbf{v}; D). \qquad (3.8.8)$$

Hence

$$\int_D \|\mathbf{u}\|^2 d\mathbf{x} = \int_D (\|\mathbf{v}\|^2 + \|\mathbf{w}\|^2) d\mathbf{x} \leq C\mathcal{E}(\mathbf{u}; D) + \alpha_\omega \int_\omega \|\mathbf{w}\|^2 d\mathbf{x}.$$

To complete the proof we use (3.8.8) and write the inequality

$$\int_\omega \|\mathbf{w}\|^2 d\mathbf{x} \leq 2 \int_\omega (\|\mathbf{u}\|^2 + \|\mathbf{v}\|^2) d\mathbf{x}$$

$$\leq 2 \int_\omega \|\mathbf{u}\|^2 d\mathbf{x} + 2C\mathcal{E}(\mathbf{u}; D). \qquad (3.8.9)$$

$\square$

LEMMA 3.7

Let D and ω be the same as in Lemma 3.6 and $\mathbf{u} \in H^1(D)$. Then

$$\int_D \|\nabla \mathbf{u}\|^2 d\mathbf{x} \leq C\beta_\omega \left(\mathcal{E}(\mathbf{u}; D) + \int_\omega \|\mathbf{u}\|^2 d\mathbf{x}\right), \qquad (3.8.10)$$

where

$$\beta_\omega = \sup_{\mathbf{a}, \mathbf{b}} \frac{\|\mathbf{b}\|^2}{\int_\omega \|\mathbf{a} + \mathbf{b} \times \mathbf{x}\|^2 d\mathbf{x}},$$

and the constant C is independent of ω and $\mathbf{u}$.

Proof

We use the same notation as in the proof of the previous lemma. By (3.8.8) and by the definition of β_ω we have

$$\int_D \|\nabla \mathbf{u}\|^2 d\mathbf{x} \leq 2\int_D (\|\nabla \mathbf{v}\|^2 + \|\nabla \mathbf{w}\|^2) d\mathbf{x}$$

$$\leq 2C\mathcal{E}(\mathbf{u}; D) + 2\beta_\omega \int_\omega \|\mathbf{w}\|^2 d\mathbf{x}.$$

To complete the proof, it remains to make use of (3.8.9). □

3.8.2 Half-space and a cylinder

We shall use the notation $E(\mathbb{R}^3_-)$ for the space of vector functions with finite norm

$$\|\mathbf{u}\| = \left(\mathcal{E}(\mathbf{u}; \mathbb{R}^3_-) + \int_{\mathbb{R}^3_-} \|\mathbf{u}\|^2 \frac{d\mathbf{x}}{\|\mathbf{x}\|^2}\right)^{1/2}.$$

LEMMA 3.8

The inequality

$$\|\nabla \mathbf{u}\|^2_{(L_2(\mathbb{R}^3_-))^3} \leq C\mathcal{E}(\mathbf{u}; \mathbb{R}^3_-) \qquad (3.8.11)$$

holds for all $\mathbf{u} \in E(\mathbb{R}^3_-)$.

Proof

Let $B_1 = \{\mathbf{x} : \|\mathbf{x}\| < 1\}$. Applying Lemma 3.5 (ii) to the domain $\mathbb{R}^3_- \cap B_1$ and to the subspace V_1 consisting of vector functions from $(H^1(\mathbb{R}^3_- \cap B_1))^3$ that vanish on $(\partial B_1) \cap \mathbb{R}^3_-$, we obtain (3.8.11) for such functions. Since a uniform scaling of variables does not affect the constant in (3.8.11), the inequality holds for any function with compact support. Since the set of such functions is dense in $E(\mathbb{R}^3_-)$ the result follows. □

The next statement gives Korn's inequality in a semi-infinite cylinder. In this and the next sections we use the notation

$$\Pi(T) = \{\mathbf{x} : (x_1, x_2) \in g, x_3 > T\},$$

where g is a bounded Lipschitz domain in $\mathbb{R}^2$. The notation $E(\Pi(T))$ stands for the space of vector functions $\mathbf{u}$ such that

$$\mathcal{E}(\mathbf{u}; \Pi(T)) + \int_g \|\mathbf{u}(\mathbf{x}', T)\|^2 d\mathbf{x}' < \infty,$$

where $\mathbf{x}' = (x_1, x_2)$.

LEMMA 3.9

The inequality below is valid for all $\mathbf{u} \in E(\Pi(1))$:

$$\int_{\Pi(1)} (|\nabla' u_3|^2 + |u_3|^2 + \|\partial_{x_3} \mathbf{u}'\|^2) \frac{d\mathbf{x}}{x_3^2} + \int_{\Pi(1)} (\|\nabla' \mathbf{u}'\|^2 + \|\mathbf{u}'\|^2) \frac{d\mathbf{x}}{x_3^4}$$

$$\leq C \Big(\mathcal{E}(\mathbf{u}, \Pi(1)) + \int_g \|\mathbf{u}(x', 1)\|^2 dx'\Big), \qquad (3.8.12)$$

where $\mathbf{u}' = (u_1, u_2)$ and $\nabla' = (\partial_{x_1}, \partial_{x_2})$.

Proof

Consider a bounded cylinder

$$\Pi_l(a) = \{\mathbf{x} : \mathbf{x}' \in g, a < x_3 < a+l\}, \ l \geq 1.$$

We shall prove the inequality

$$\frac{C_0}{l^2} \int_{\Pi_l(a)} (\|\nabla' u_3\|^2 + |u_3|^2) d\mathbf{x} + \frac{1}{2l} \int_g |u_3(x', a+l)|^2 dx'$$

$$\leq C_1 \sum_{i,j=1}^{3} \int_{\Pi_l(a)} |\varepsilon_{ij}(\mathbf{u})|^2 d\mathbf{x} + \frac{1}{l} \int_g |u_3(x', a)|^2 dx', \qquad (3.8.13)$$

where C_0 and C_1 do not depend of a, l and $\mathbf{u}$. Introducing new variables

$$x_3 = ly_3 + a, \ \mathbf{x}' = \mathbf{y}', \ U_3 = lu_3, \ \mathbf{U}' = \mathbf{u}',$$

we reduce the proof to the case $a = 0$, $l = 1$. The direct consequence of the Newton–Leibniz formula is

$$\int_g |U_3(\mathbf{y}',1)|^2 d\mathbf{y}' \leq C \int_{\Pi_1(0)} |\varepsilon_{33}(\mathbf{U})|^2 d\mathbf{y} + \frac{3}{2}\int_g |U_3(\mathbf{y}',0)|^2 d\mathbf{y}'. \quad (3.8.14)$$

By Lemma 3.5(iii) for all $\mathbf{U} \in (H^1(\Pi_1(0)))^3$

$$\|\mathbf{U}\|^2_{(H^1(\Pi_1(0)))^3} \leq C\Big(\mathcal{E}(\mathbf{U};\Pi_1(0)) + \sum_{j=1}^6 |F_j(\mathbf{U})|^2\Big),$$

where F_j are linear bounded functionals on $(H^1(\Pi_1(0)))^3$ which do not vanish simultaneously on the rigid-body displacements. Without loss of generality we can assume that equalities (3.5.1) are valid. Then, it is an easy matter to show that the functionals

$$F_j(\mathbf{U}) = \int_g U_3(\mathbf{y}',0)y_j d\mathbf{y}',\ j=1,2,$$

$$F_{6-j}(\mathbf{U}) = \int_g U_j(\mathbf{y}',0)d\mathbf{y}',\ j=1,2,3,$$

$$F_6(\mathbf{U}) = \int_g (U_1(\mathbf{y}',0)y_2 - U_2(\mathbf{y}',0)y_1)d\mathbf{y}'$$

are a good choice. Thus,

$$C_0 \int_{\Pi_1(0)} (\|\nabla' U_3\|^2 + |U_3|^2) d\mathbf{y} \leq \mathcal{E}(\mathbf{U};\Pi_1(0))$$

$$+ \frac{1}{4}\int_g |U_3(\mathbf{y}',0)|^2 d\mathbf{y}' + \sum_{j=4}^6 |F_j(\mathbf{U})|^2. \quad (3.8.15)$$

Instead of $\mathbf{U}$ one can substitute the vector

$$\mathbf{U} + C_1 \mathbf{e}^{(1)} + C_2 \mathbf{e}^{(2)} + C_3(\mathbf{y} \times \mathbf{e}^{(3)})$$

into (3.8.15). Then the left-hand side and the first two terms on the right do not change. The last term in the right-hand side will be zero provided the constants C_j satisfy

$$F_j(\mathbf{U}) + C_1 F_j(\mathbf{e}^{(1)}) + C_2 F_j(\mathbf{e}^{(2)}) + C_3 F_j(\mathbf{y} \times \mathbf{e}^{(3)}) = 0,$$

for $j = 4,5,6$. Since the matrix of this system is non-singular, the constants C_k are defined uniquely. Thus, the last term in the right-hand side of (3.8.15) may be omitted. Next, using (3.8.14) we obtain (3.8.13).

Let $a = l = 2^k$, $k = 0, 1, \ldots$. Then

$$C_0 2^{-2k} \int_{\Pi^{(k)}} (\|\nabla' u_3\|^2 + |u_3|^2) d\mathbf{x} + 2^{-k-1} \int_g |u_3(\mathbf{x}', 2^{k+1})|^2 d\mathbf{x}'$$

$$\leq C_1 \mathcal{E}(\mathbf{u}; \Pi^{(k)}) + 2^{-k} \int_g |u_3(\mathbf{x}', 2^k)|^2 d\mathbf{x}', \qquad (3.8.16)$$

where
$$\Pi^{(k)} = \{\mathbf{x} : \mathbf{x}' \in g, \ 2^k < x_3 < 2^{k+1}\}.$$

Summing (3.8.16) with respect to k we derive that

$$\int_{\Pi(1)} (\|\nabla' u_3\|^2 + |u_3|^2) \frac{d\mathbf{x}}{x_3^2} \leq C\Big(\mathcal{E}(\mathbf{u}; \Pi(1)) + \int_g |u_3(\mathbf{x}', 1)|^2 d\mathbf{x}'\Big). \quad (3.8.17)$$

Hence, by $\partial_{x_3} u_i(\mathbf{x}) = 2\varepsilon_{i3}(\mathbf{u}; \mathbf{x}) - \partial_{x_i} u_3(\mathbf{x})$, $i = 1, 2$, we obtain

$$\int_{\Pi(1)} \|\partial_{x_3} \mathbf{u}'\|^2 \frac{d\mathbf{x}}{x_3^2} \leq C\Big(\mathcal{E}(\mathbf{u}; \Pi(1)) + \int_g |u_3(\mathbf{x}', 1)|^2 d\mathbf{x}'\Big). \qquad (3.8.18)$$

Applying the Hardy inequality we arrive at

$$\int_{\Pi(1)} \|\mathbf{u}'\|^2 \frac{d\mathbf{x}}{x_3^4} \leq C\Big(\int_{\Pi(1)} \|\partial_{x_3} \mathbf{u}'\|^2 \frac{d\mathbf{x}}{x_3^2} + \int_g \|\mathbf{u}'(\mathbf{x}', 1)\|^2 d\mathbf{x}'\Big). \quad (3.8.19)$$

The inequalities (3.8.18) and (3.8.19) yield the estimate of the integral in the left-hand side of (3.8.19) by the expression in the right-hand side of (3.8.12).

Using the two-dimensional Korn inequality

$$\int_g \|\nabla' \mathbf{u}'(\mathbf{x}', x_3)\|^2 d\mathbf{x}'$$

$$\leq C\Big(\sum_{i,j=1}^{2} \int_g |\varepsilon_{ij}(\mathbf{u}; \mathbf{x}', x_3)|^2 d\mathbf{x}' + \int_g \|\mathbf{u}'(\mathbf{x}', x_3)\|^2 d\mathbf{x}'\Big), \qquad (3.8.20)$$

together with (3.8.19), we estimate the term, including $\nabla' \mathbf{u}'$ in (3.8.12), and complete the proof. □

LEMMA 3.10

Let $\Pi_l = \{\mathbf{x} : \mathbf{x}' \in g, 0 < x_3 < l\}$, $l \geq 1$. Then for all functions $\mathbf{u} \in (H^1(\Pi_l))^3$

$$l^{-2}\int_{\Pi_l}(\|\nabla'u_3\|^2 + |u_3|^2 + \|\partial_{x_3}\mathbf{u}'\|^2)d\mathbf{x} + l^{-4}\int_{\Pi_l}(\|\nabla'\mathbf{u}'\|^2 + \|\mathbf{u}'\|^2)d\mathbf{x}$$

$$\leq C\Big(\mathcal{E}(\mathbf{u};\Pi_l) + \int_g \|\mathbf{u}(\mathbf{x}',0)\|^2 d\mathbf{x}'\Big), \qquad (3.8.21)$$

where C does not depend of l and $\mathbf{u}$.

Proof

The estimate for the first two terms in the left-hand side follows from (3.8.13). Since $\partial_{x_3}u_i(\mathbf{x}) = 2\varepsilon_{i3}(\mathbf{u};\mathbf{x}) - \partial_{x_i}u_3(\mathbf{x})$, $i = 1,2$, the estimate of the third term on the left also follows from (3.8.13).

Using the Newton–Leibniz formula

$$\mathbf{u}'(\mathbf{x}',x_3) = \int_0^{x_3}\partial_{x_3}\mathbf{u}'(\mathbf{x}',x_3)dx_3 + \mathbf{u}'(\mathbf{x}',0)$$

we derive

$$\int_{\Pi_l}\|\mathbf{u}'\|^2 d\mathbf{x} \leq 2l\Big\{\int_g\Big(\int_0^l \|\partial_{x_3}\mathbf{u}'\|dx_3\Big)^2 d\mathbf{x}'$$

$$+ \int_g \|\mathbf{u}'(\mathbf{x}',0)\|^2 d\mathbf{x}'\Big\}$$

and by Cauchy's inequality applied to the first integral in the right-hand side

$$\int_{\Pi_l}\|\mathbf{u}'\|^2 d\mathbf{x} \leq 2l^2\int_{\Pi_l}\|\partial_{x_3}\mathbf{u}'\|^2 d\mathbf{x}$$

$$+ 2l\int_g \|\mathbf{u}'(\mathbf{x}',0)\|^2 d\mathbf{x}',$$

which gives the estimate for the last term on the left-hand side of (3.8.21). The estimate for the fourth term on the left-hand side of (3.8.21) follows from (3.8.20). □

3.9 Asymptotics at infinity for solutions to the traction problem for a half-cylinder

Let Π and $\Pi(T)$ be the same cylinders as in Sections 3.6 and 3.8. We introduce the notation

$$\Gamma(T) = \{\mathbf{X} : \mathbf{X} \in \partial\Pi,\ X_3 > T\}.$$

Let $\mathbf{u}$ solve the Lamé system in the half-cylinder

$$L(\partial_X)\mathbf{u}(\mathbf{X}) = 0, \quad \mathbf{X} \in \Pi(T), \tag{3.9.1}$$

where $T > 1$, complemented by the boundary condition on the lateral surface:

$$\sigma^{(n)}(\mathbf{u};\mathbf{X}) = 0, \quad \mathbf{X} \in \Gamma(T). \tag{3.9.2}$$

We assume that $\mathbf{u}$ satisfies

$$\mathcal{E}(\mathbf{u},\Pi(T)) + \int_g \|\mathbf{u}(\mathbf{X}',T)\|^2 d\mathbf{X}' < \infty. \tag{3.9.3}$$

The solution $\mathbf{u}$ is understood in the weak sense, i.e. the equality

$$\sum_{i,j=1}^{3} \int \sigma_{ij}(\mathbf{u})\varepsilon_{ij}(\mathbf{v})d\mathbf{X} = 0 \tag{3.9.4}$$

is valid for any function $\mathbf{v} \in (H^1(\Pi(T)))^3$ such that $\mathbf{v} = 0$ for $X_3 = T$. We note that, by Lemma 3.9, $\mathbf{u}$ satisfies (3.8.12) where Π is replaced by $\Pi(T)$ and $\mathbf{u}(\mathbf{x}',1)$ by $\mathbf{u}(\mathbf{x}',T)$.

THEOREM 3.3

Let $\mathbf{u}$ be a solution of (3.9.1), (3.9.2). Then there exist constant vectors $\mathbf{a}$ and $\mathbf{b}$ in $\mathbb{R}^3$ such that

$$\mathbf{u}(\mathbf{X}) = \mathbf{a} + \mathbf{b} \times \mathbf{X} + \mathbf{R}(\mathbf{X}),$$

where $\mathbf{R}$ satisfies

$$\|\partial_{X_3}^m \mathbf{R}(\cdot,X_3)\|_{(H^1(g))^3} \leq c_m e^{-\gamma X_3}, \tag{3.9.5}$$

for $X_3 \geq T+1$ and for $m = 0,1,\ldots$, with positive constants c_m and γ. Moreover, γ is independent of $\mathbf{u}$.

Proof

Let $\boldsymbol{\xi}^{(k)}$, $k = 1, \ldots, 6$, be the X_3- polynomial solutions constructed in Section 3.6.3.

Using Betti's formula (3.1.9) for the domain $\{\mathbf{X} : a < X_3 < b, \mathbf{X}' \in g\}$, with $\mathbf{u}$ and $\mathbf{v}$ replaced by $\boldsymbol{\xi}^{(k)}$ and $\mathbf{u}$, we obtain

$$\sum_{i=1}^{3} \int_g (u_i \sigma_{i3}(\boldsymbol{\xi}^{(k)}) - \sigma_{i3}(\mathbf{u})\xi_i^{(k)}) d\mathbf{X}'|_{X_3=b}$$

$$= \sum_{i=1}^{3} \int_g (u_i \sigma_{i3}(\boldsymbol{\xi}^{(k)}) - \sigma_{i3}(\mathbf{u})\xi_i^{(k)}) d\mathbf{X}'|_{X_3=a}$$

for every a, b, $T \leq a < b < \infty$. By C_k we denote the sum

$$\sum_{i=1}^{3} \int_g (u_i \sigma_{i3}(\boldsymbol{\xi}^{(k)}) - \sigma_{i3}(\mathbf{u})\xi_i^{(k)}) d\mathbf{X}',$$

which is independent of X_3. By Lemma 3.2, one can choose vectors $\mathbf{a}$ and $\mathbf{b}$ in such a way that

$$\int_g (\mathbf{a} + \mathbf{b} \times \mathbf{X}) \cdot \boldsymbol{\sigma}^{(3)}(\boldsymbol{\xi}^{(k)}) d\mathbf{X}' = C_k.$$

We put

$$\mathbf{R} = \mathbf{u} - \mathbf{a} - \mathbf{b} \times \mathbf{X}.$$

Then

$$\sum_{i=1}^{3} \int_g (R_i \sigma_{i3}(\boldsymbol{\xi}^{(k)}) - \sigma_{i3}(\mathbf{R})\xi_i^{(k)}) d\mathbf{X}' = 0, \ k = 1, \ldots, 6, \quad (3.9.6)$$

for all X_3, $X_3 \geq T$.

From Section 3.6.3 it follows that the vector functions $\boldsymbol{\sigma}^{(3)}(\boldsymbol{\xi}^{(k)})$, $k = 3, 4, 5, 6$, do not depend on X_3 and that for $j = 1, 2$

$$\boldsymbol{\sigma}^{(3)}(\boldsymbol{\xi}^{(4+j)}) = \left(0, 0, -\mu \frac{3\lambda + 2\mu}{\lambda + \mu} X_j\right).$$

We note that $\sigma_{13}(\boldsymbol{\xi}^{(k)})$ and $\sigma_{23}(\boldsymbol{\xi}^{(k)})$, $k = 1, 2$, are independent of X_3 whereas

$$\sigma_{33}(\boldsymbol{\xi}^{(k)}) = -\mu \frac{3\lambda + 2\mu}{\lambda + \mu} X_k X_3, \ k = 1, 2.$$

We introduce

$$\mathbf{p}^{(k)} = \sigma^{(3)}(\boldsymbol{\xi}^{(k)}) - X_3 \sigma^{(3)}(\boldsymbol{\xi}^{(4+k)}), \quad k = 1, 2;$$

$$\mathbf{p}^{(k)} = \sigma^{(3)}(\boldsymbol{\xi}^{(k)}), \quad k = 3, 4, 5, 6,$$

which do not depend on X_3. Let also

$$\boldsymbol{\zeta}^{(k)} = \boldsymbol{\xi}^{(k)} - X_3 \boldsymbol{\xi}^{(4+k)}, \quad k = 1, 2;$$

$$\boldsymbol{\zeta}^{(k)} = \boldsymbol{\xi}^{(k)}, \quad k = 3, 4, 5, 6.$$

Now (3.9.6) takes the form

$$\int_g \mathbf{R} \cdot \mathbf{p}^{(k)} d\mathbf{X}' = \int_g \sigma^{(3)}(\mathbf{R}) \cdot \boldsymbol{\zeta}^{(k)} d\mathbf{X}', \qquad (3.9.7)$$

with $k = 1, \ldots, 6$. Let $t > T$ and $\Pi_1(t) = \{\mathbf{X} \in \Pi : t < X_3 < t + 1\}$. It follows from (3.6.27) that

$$\int_{\Pi_1(t)} (\mathbf{a} + \mathbf{b} \times \mathbf{X}) \cdot \mathbf{p}^{(k)} d\mathbf{X} = 0, \quad k = 1, \ldots, 6,$$

implies $\mathbf{a} = \mathbf{b} = 0$. By Lemma 3.5 (iii) there exists $c > 0$ independent of t and $\mathbf{R}$ such that

$$\|\mathbf{R}\|^2_{(H^1(\Pi_1(t)))^3} \leq c \left\{ \mathcal{E}(\mathbf{R}; \Pi_1(t)) + \sum_{k=1}^6 \left(\int_{\Pi_1(t)} \mathbf{R} \cdot \mathbf{p}^{(k)} d\mathbf{X} \right)^2 \right\}.$$

Hence, by (3.9.7)

$$\|\mathbf{R}\|^2_{(H^1(\Pi_1(t)))^3} \leq ct^6 \mathcal{E}(\mathbf{R}; \Pi_1(t)). \qquad (3.9.8)$$

Now, let η be a smooth function such that $\eta(X_3) = 1$ for $X_3 > t + 1$ and $\eta(X_3) = 0$ for $X_3 < t$, where $t > T$. Then

$$\sum_{i,j=1}^3 \int_{\Pi(T)} \sigma_{ij}(\mathbf{R}_0) \varepsilon_{ij}(\eta^2 \mathbf{R}_0) d\mathbf{X} = 0,$$

where $\mathbf{R}_0 = \mathbf{R} - \boldsymbol{\alpha} - \boldsymbol{\beta} \times \mathbf{X}$, with constant vectors $\boldsymbol{\alpha}$ and $\boldsymbol{\beta}$ to be chosen later. This implies

$$\sum_{i,j=1}^3 \int_{\Pi(T)} \eta^2 \sigma_{ij}(\mathbf{R}_0) \varepsilon_{ij}(\mathbf{R}_0) d\mathbf{X} = - \sum_{i,j=1}^3 \int_{\Pi(T)} \eta \eta' \sigma_{ij}(\mathbf{R}_0)(R_{0i} + R_{0j}) d\mathbf{X}.$$

Using Cauchy's inequality we obtain

$$\sum_{i,j=1}^{3} \int_{\Pi(T)} \eta^2 \sigma_{ij}(\mathbf{R}_0) \varepsilon_{ij}(\mathbf{R}_0) d\mathbf{X} \leq c_1 \int_{\Pi(T)} \eta'^2 \|\mathbf{R}_0\|^2 d\mathbf{X}. \qquad (3.9.9)$$

Hence

$$\mathcal{E}(\mathbf{R};\Pi(t+1)) \leq c_2 \min_{\boldsymbol{\alpha},\boldsymbol{\beta}} \int_{\Pi_1(t)} \|\mathbf{R}(\mathbf{X}) - \boldsymbol{\alpha} - \boldsymbol{\beta} \times \mathbf{X}\|^2 d\mathbf{X}. \qquad (3.9.10)$$

Since the minimum is attained on $\boldsymbol{\alpha}_0$ and $\boldsymbol{\beta}_0$ such that $\mathbf{R} - \boldsymbol{\alpha}_0 - \boldsymbol{\beta}_0 \times \mathbf{X}$ is orthogonal in $(L_2(\Pi_1(t)))^3$ to rigid-body displacements, it follows from Lemma 3.5(ii) that the right-hand side in (3.9.10) is majorised by $c_3 \mathcal{E}(\mathbf{R};\Pi_1(t))$. Thus,

$$\mathcal{E}(\mathbf{R};\Pi(t+1)) \leq c_3(\mathcal{E}(\mathbf{R};\Pi(t)) - \mathcal{E}(\mathbf{R};\Pi(t+1))).$$

Therefore

$$\mathcal{E}(\mathbf{R};\Pi(t+1)) \leq \frac{c_3}{1+c_3} \mathcal{E}(\mathbf{R};\Pi(t)).$$

This implies

$$\mathcal{E}(\mathbf{R};\Pi(t)) \leq c_4 e^{-\sigma t},$$

with $\sigma = -\log c_4(1+c_4)^{-1}$. Hence and by (3.9.8)

$$\|\mathbf{R}\|_{H^1(\Pi_1(t))} \leq c e^{-\gamma t},$$

where $\gamma = \sigma/3$. Since $\partial_{X_3}^m \mathbf{R}$ also satisfies (3.9.1), (3.9.2), we obtain

$$\|\partial_{X_3}^m \mathbf{R}\|_{H^1(\Pi_1(t))} \leq c_m e^{-\gamma t}.$$

This implies (3.9.5). □

4
ELASTIC MULTI-STRUCTURE

In this chapter we analyse a mixed boundary value problem for the Lamé system in an elastic multi-structure which consists of a union of a three-dimensional domain and thin cylinders and has the same geometry as in Chapter 2. In addition, we assume that the number of thin cylinders exceeds two and that the junction points do not belong to the same straight line. Just for the sake of simplicity of presentation we assume that the cylinders have the same cross-sections.

Traction boundary conditions are prescribed on the surface of the elastic cap and on the lateral surface of the thin cylinders. The displacement components are given at the ends of thin rods.

We construct a multi-scaled asymptotic expansion for the displacement field in the multi-structure and give an estimate for the remainder.

The engineering approach to problems of this kind is described in Section 4.2.7, where the three-dimensional element is considered as a rigid body, and the thin rods are replaced by one-dimensional elastic piles clamped at junction points. In the engineering literature (see, for example, Asplund (1966)) this model is called the pile structure; the stress–strain state is usually described by the rigid-body displacements of the pile cap and by the reaction forces and moments at junction points (these are known as 'lock forces' and 'lock moments'). The basic relation in the pile structure model is a system of linear algebraic equations with respect to components of the rigid-body displacements of the pile cap. The 6×6 matrix of this system is called the stiffness matrix of the pile structure. Since the stiffness coefficients of elastic piles depend on the normalised thickness of the thin rods, the components of the stiffness matrix include ε. For the case where the piles are parallel to each other, the principal part of this matrix is singular, and a rigorous analysis is absent in the literature.

Here, we develop an algorithm for constructing the power asymptotic series in ε which is uniformly valid for elastic fields in the entire multi-structure. The leading order term of the displacement vector within the three-dimensional domain Ω is represented by rigid-body translations and rotations. In comparison with the engineering model of the pile structure we show, in particular, that the rigorous asymptotic analysis gives the same leading order displacement, provided that the transverse components of the principal force vector and third component of the principal moment do not vanish simultaneously. Otherwise, the formula for the displacement field is different (see Section 4.6).

In Section 4.1 we formulate the boundary value problem and describe the elastic multi-structure. Model boundary value problems are formulated and analysed in Section 4.2. Section 4.3 deals with a formal asymptotic expansion of the displacement field. In Section 4.4 we give estimates for the remainder term, and, therefore, justify the asymptotic expansion of the solution.

The coefficients in the asymptotic expansion of the solution involve solutions to the following 'model problems' in sets independent of ε:

(1) the Neumann problem for the Lamé system in the three-dimensional domain Ω (Fig. 2.2);

(2) the 'junction layer': the Neumann problem for the Lamé system in the scaled junction region (Fig. 2.3);

(3) the 'bottom layer': the Dirichlet–Neumann problem in the scaled bottom region of a thin cylinder (Fig. 2.4);

(4) the plane Neumann problems for the Lamé and the Laplace equations on the scaled cross-section of a thin rod (Fig. 2.5(a));

(5) the boundary value problems for ordinary differential equations on the axis of a thin rod (Fig. 2.5(b));

(6) the system of linear equations corresponding to the skeleton of the multi-structure (Fig. 2.6).

First five model problems are posed for elements of the multi-structure. In contrast, the algebraic system from (6) includes characteristics of all elements of the multi-structure and is closely related to the pile structure model.

The leading term of the displacement field and the principal parts of the lock forces and moments are analysed in Section 4.5, where the physical interpretation of the results is given.

4.1 Multi-structure and boundary value problem

Let Ω_ε be a region in $\mathbb{R}^3$ which depends on a small parameter $\varepsilon > 0$ and consists of a bounded domain Ω with the Lipschitz boundary and K

MODEL PROBLEMS

parallel thin cylinders $\Pi_\varepsilon^{(j)}$ of the finite length $l^{(j)}$, $j = 1, \ldots, K; K \geq 3$ (see Fig. 2.1). By 'thin' we mean that the diameter of the cross-section has the order ε. It is assumed that some open part of the boundary $\partial\Omega$ belongs to the plane $x_3 = 0$ and contains the junction points $\mathbf{a}^{(j)} = (a_1^{(j)}, a_2^{(j)}, 0), j = 1, \ldots, K$. In contrast with Chapter 2, we suppose that the junction points should not be located on the same straight line. The thin cylinder at $a^{(j)}$ is given by

$$\Pi_\varepsilon^{(j)} = g_j(\varepsilon) \times [0, l^{(j)}),$$

where

$$g_j(\varepsilon) = \{(x_1, x_2) : \varepsilon^{-1}(x_1 - a_1^{(j)}, x_2 - a_2^{(j)}) \in g\},$$

with g being a two-dimensional bounded Lipschitz domain independent of ε whose closure lies in the unit disk centred at the origin. We assume, for the sake of simplicity, that all thin rods have the same shape of cross-sections. That part of the boundary corresponding to the base region of the thin rod $\Pi_\varepsilon^{(j)}$ is denoted by $S_\varepsilon^{(j)}$, and

$$S_\varepsilon = \cup_{j=1}^K S_\varepsilon^{(j)}.$$

We consider a mixed Dirichlet–Neumann boundary value problem for the Lamé system in Ω_ε. The displacement vector $\mathbf{u}_\varepsilon(\mathbf{x})$ satisfies the system

$$-\mathbf{L}(\partial_\mathbf{x})\mathbf{u}(\mathbf{x}, \varepsilon) = \mathbf{F}(\mathbf{x}, \varepsilon), \ \mathbf{x} \in \Omega_\varepsilon, \qquad (4.1.1)$$

the traction boundary condition

$$\boldsymbol{\sigma}^{(n)}(\mathbf{u}; \mathbf{x}) = \mathbf{P}(\mathbf{x}, \varepsilon), \ x \in \partial\Omega_\varepsilon \setminus \overline{S}_\varepsilon, \qquad (4.1.2)$$

and the displacement boundary condition

$$\mathbf{u}(\mathbf{x}, \varepsilon) = \boldsymbol{\Phi}^{(j)}(\mathbf{x}, \varepsilon), \ \mathbf{x} \in S_\varepsilon^{(j)}, \ j = 1, \ldots, K, \qquad (4.1.3)$$

where $\mathbf{n}$ is the outward unit normal on $\partial\Omega_\varepsilon$; λ, μ are the Lamé constants and $\mathbf{L}$ and $\boldsymbol{\sigma}^{(n)}$ are defined by (3.1.5), (3.1.7). We suppose that the vector functions $\mathbf{F}$, $\mathbf{P}$ and $\boldsymbol{\Phi}^{(j)}$ belong to $(L_2(\Omega_\varepsilon))^3$, $(L_2(\partial\Omega \setminus S_\varepsilon))^3$ and $(H^{1/2}(S_\varepsilon^{(j)}))^3$, respectively. Solvability of (4.1.1)–(4.1.3) is well known (see Section 3 of Chapter 1 in Oleinik, Shamaev and Yosifian (1992)).

4.2 Model problems

Here we formulate basic results related to the solvability and asymptotic representation of solutions to model problems mentioned in the introduction to this chapter.

4.2.1 Limit domains

The model domains are the same as in Chapter 2 (see Section 2.2.1):
- *the bounded domain* Ω corresponding to the pile cap of the multi-structure;
- *the union* G of the half-space $\mathbb{R}^3_-$ and the semi-infinite cylinder

$$\Pi_+ = \{\mathbf{X} : X_3 \geq 0, (X_1, X_2) \in g\};$$

- *the semi-infinite cylinder*

$$\Pi_- = \{\mathbf{Y} : Y_3 < 0, (Y_1, Y_2) \in g\}$$

with the base region S;
- *the scaled cross-section* g of the thin cylinders;
- *the vertical segment* representing the axis of a thin cylinder;
- *the skeleton* of the multi-structure.

4.2.2 Model problem for the body Ω

In this section we consider the boundary value problem for the Lamé system in Ω with traction boundary conditions. The solution is allowed to be singular at the points $\mathbf{a}^{(j)}$, $j = 1, \ldots, K$.

We seek a vector-valued function $\mathbf{u}_\Omega$ subject to the Lamé system

$$-\mathbf{L}(\partial_\mathbf{x})\mathbf{u}_\Omega(\mathbf{x}) = \mathbf{f}(\mathbf{x}), \quad \mathbf{x} \in \Omega, \qquad (4.2.1)$$

and the traction boundary condition

$$\boldsymbol{\sigma}^{(n)}(\mathbf{u}_\Omega; \mathbf{x}) = \mathbf{p}(\mathbf{x}), \quad \mathbf{x} \in \partial\Omega \setminus \cup_j \mathbf{a}_j. \qquad (4.2.2)$$

We look for a solution of (4.2.1), (4.2.2) represented in the form

$$\mathbf{u}_\Omega(\mathbf{x}) = \mathbf{u}_r(\mathbf{x}) + \mathbf{u}_s(\mathbf{x}). \qquad (4.2.3)$$

The regular part $\mathbf{u}_r$ is a vector function from the Sobolev space $(H^1(\Omega))^3$, and the singular part is defined as

$$\mathbf{u}_s(\mathbf{x}) = \sum_{j=1}^{K} \Big\{ \mathcal{N}(\mathbf{x},\, \mathbf{a}^{(j)}) \mathfrak{F}^{(j)}$$

$$+ \frac{1}{2} (\nabla_y \times \mathcal{N}^t(\mathbf{x}, \mathbf{y})|_{\mathbf{y}=\mathbf{a}^{(j)}})^t \mathfrak{M}^{(j)} \Big\}. \qquad (4.2.4)$$

Here, $\mathfrak{F}^{(j)}$, $\mathfrak{M}^{(j)}$ are constant vectors; $\mathcal{N}(\mathbf{x},\mathbf{y})$ is Green's matrix for the Neumann boundary value problem for the Lamé operator in Ω (see Section 3.7). Representation (4.2.3) corresponds to the setting of the concentrated forces $\mathfrak{F}^{(j)}$ and moments $\mathfrak{M}^{(j)}$ at the junction points $\mathbf{a}^{(j)}$, $j = 1, \ldots, K$. By Lemma 3.4 and by definition of the Boussinesq-Cerruti solution (see Section 3.4.1),

$$-\mathbf{L}(\partial_\mathbf{x}) \mathbf{u}_\Omega(\mathbf{x}) = \mathbf{f}(\mathbf{x}), \quad \mathbf{x} \in \Omega,$$

$$\sigma^{(n)}(\mathbf{u}_\Omega; \mathbf{x}) = \mathbf{p}(\mathbf{x}) + \sum_{j=1}^{K} \Big\{ \mathfrak{F}^{(j)} \delta(\mathbf{x}' - \mathbf{a}'^{(j)})$$

$$- \frac{1}{2}(\nabla_\mathbf{x} \times (\delta(\mathbf{x}' - \mathbf{a}'^{(j)})\mathbf{I}) \mathfrak{M}^{(j)}) \Big\}, \quad \mathbf{x} \in \partial\Omega.$$

We note that the junction points $\mathbf{a}^{(j)}$ are located on the flat part of the boundary $\partial\Omega$ parallel to the (x_1,x_2)- plane, and the last term in the formula above does not include the derivatives with respect to x_3.

We shall use the following notations for the principal force and moment vectors:

$$\mathbb{F}(\mathbf{f},\mathbf{p},\Omega) = \int_\Omega \mathbf{f}(\mathbf{x}) d\mathbf{x} + \int_{\partial\Omega} \mathbf{p}(\mathbf{x}) ds, \qquad (4.2.5)$$

$$\mathbb{M}(\mathbf{f},\mathbf{p},\Omega) = \int_\Omega \mathbf{x} \times \mathbf{f}(\mathbf{x}) d\mathbf{x} + \int_{\partial\Omega} \mathbf{x} \times \mathbf{p}(\mathbf{x}) ds. \qquad (4.2.6)$$

THEOREM 4.1

Let $\mathbf{f} \in (L_2(\Omega))^3$, $\mathbf{p} \in (L_2(\partial\Omega))^3$, and let $\mathfrak{F}^{(j)}$, $\mathfrak{M}^{(j)}$, $j = 1, 2, \ldots, K$, be constant vectors such that

$$\mathbb{F}(\mathbf{f},\mathbf{p},\Omega) + \sum_{j=1}^{K} \mathfrak{F}^{(j)} = \mathbf{0}, \qquad (4.2.7)$$

$$\mathbb{M}(\mathbf{f},\mathbf{p},\Omega) + \sum_{j=1}^{K} \left\{ \mathfrak{M}^{(j)} + \mathbf{a}^{(j)} \times \mathfrak{F}^{(j)} \right\} = \mathbf{0}. \qquad (4.2.8)$$

Then, there exists a solution $\mathbf{u}$ *of problem* (4.2.1), (4.2.2), *which admits the representation* (4.2.3). *The solution* $\mathbf{u}$ *is unique provided its regular part* $\mathbf{u}_r$ *from* (4.2.3) *satisfies the orthogonality conditions*

$$\int_\Omega \mathbf{u}_r(\mathbf{x}) d\mathbf{x} = \mathbf{0}, \quad \int_\Omega \mathbf{x} \times \mathbf{u}_r(\mathbf{x}) d\mathbf{x} = \mathbf{0}. \qquad (4.2.9)$$

Proof

Since the statement is invariant with respect to translations and rotations of the coordinate system, we can use the coordinates where the axes coincide with the principal inertia axes of Ω. Substituting (4.2.3) in (4.2.1), (4.2.2) and using the definition of Green's matrix, we find that the vector $\mathbf{u}_r$ satisfies the boundary value problem

$$-\mathbf{L}(\partial_\mathbf{x})\mathbf{u}_r(\mathbf{x}) = \mathbf{f}(\mathbf{x}) + \mathfrak{C} + \mathfrak{A} \times \mathbf{x}, \quad \mathbf{x} \in \Omega, \qquad (4.2.10)$$
$$\sigma^{(n)}(\mathbf{u}_r;\mathbf{x}) = \mathbf{p}(\mathbf{x}), \quad \mathbf{x} \in \partial\Omega, \qquad (4.2.11)$$

where

$$\mathfrak{C} + \mathfrak{A} \times \mathbf{x} = \sum_{j=1}^{K} \Big\{ (\mathrm{mes}_3 \Omega)^{-1} \mathfrak{F}^{(j)}$$
$$+ J_j^{-1} \left(\mathbf{a}^{(j)} \times \mathfrak{F}^{(j)} + \mathfrak{M}^{(j)} \right) \times \mathbf{x} \Big\}.$$

Hence, applying (4.2.7), (4.2.8) one can see that the right-hand sides (4.2.10), (4.2.11) are orthogonal to the rigid-body translations and rotations. Thus, problem (4.2.10), (4.2.11) is solvable in $(H^1(\Omega))^3$ (see, for example, Oleinik, Shamaev and Yosifian (1992), Section 3 in Chapter 1). The orthogonality conditions (4.2.9) provide the uniqueness of $\mathbf{u}_r$. □

4.2.3 Junction layer

Let G denote the union of the half-space $\mathbb{R}^3_- = \{\mathbf{X} : X_3 < 0\}$ and the semi-infinite cylinder

$$\Pi_+ = \{\mathbf{X} : X_3 \geq 0, (X_1, X_2) \in g\},$$

located in $\mathbb{R}^3_+$, where g is a two-dimensional bounded domain with the Lipschitz boundary (see Fig. 2.3).

Function space $\mathbf{E}(G)$

A vector function $\mathbf{W}$ is said to belong to the space $\mathbf{E}(G)$ if $\mathbf{W} \in (\mathbf{H}^1(G \cap B_r))^3$ for all positive r, where $B_r = \{\mathbf{x} : \|\mathbf{x}\| < r\}$, and

$$\mathcal{E}(\mathbf{W}; G) + \int_{R_-^3} \|\mathbf{W}\|^2 \frac{d\mathbf{X}}{\|\mathbf{X}\|^2} < \infty.$$

LEMMA 4.1

The functional

$$\mathbf{W} \to (\mathcal{E}(\mathbf{W}; G))^{1/2}$$

defines a norm in the space $\mathbf{E}(G)$, and the following estimate holds:

$$\int_{\mathbb{R}_-^3} \|\mathbf{W}\|^2 \frac{d\mathbf{X}}{\|\mathbf{X}\|^2} + \int_{\mathbb{R}_-^3} \|\nabla \mathbf{W}\|^2 d\mathbf{X}$$

$$+ \int_{\Pi_+} (\|\nabla_{X'} W_3\|^2 + W_3^2 + \|\partial_{X_3} \mathbf{W}'\|^2) \frac{d\mathbf{X}}{(1+X_3)^2}$$

$$+ \int_{\Pi_+} (\|\nabla_{X'} \mathbf{W}'\|^2 + \|\mathbf{W}'\|^2) \frac{d\mathbf{X}}{(1+X_3)^4} \leq \text{Const } \mathcal{E}(\mathbf{W}; G), \quad (4.2.12)$$

where $\mathbf{W}' = (W_1, W_2)$.

Proof

The estimate of the Dirichlet integral over $\mathbb{R}_-^3$ follows from Lemma 3.8. The estimate of the first term in the left-hand side of (4.2.12) follows from (2.2.20). It remains to estimate the integral over Π_+.

We introduce a semi-infinite cylinder

$$\Pi_+^{(\alpha)} = \{X_3 > -\alpha, \ \mathbf{X}' \in g, \ \alpha \in (0, 1)\}.$$

Applying inequality (3.8.12) from Lemma 3.9 we obtain

$$\int_{\Pi_+^{(\alpha)}} (\|\nabla_{\mathbf{X}'} W_3\|^2 + W_3^2 + \|\partial_{X_3} \mathbf{W}'\|^2) \frac{d\mathbf{X}}{(2+X_3)^2}$$

$$+ \int_{\Pi_+^{(\alpha)}} (\|\nabla_{\mathbf{X}'} \mathbf{W}'\|^2 + \|\mathbf{W}'\|^2) \frac{d\mathbf{X}}{(2+X_3)^4}$$

$$\leq \text{Const} \left\{ \mathcal{E}(\mathbf{W}; \Pi_+^{(\alpha)}) + \int_g \|\mathbf{W}(\mathbf{X}', -\alpha)\|^2 d\mathbf{X}' \right\}.$$

Hence

$$\int_{\Pi_+} (\|\nabla_{\mathbf{X}'} W_3\|^2 + W_3^2 + \|\partial_{X_3} \mathbf{W}'\|^2) \frac{d\mathbf{X}}{(1+X_3)^2}$$

$$+ \int_{\Pi_+} (\|\nabla_{\mathbf{X}'} \mathbf{W}'\|^2 + \|\mathbf{W}'\|^2) \frac{d\mathbf{X}}{(1+X_3)^4}$$

$$\leq \text{Const} \left\{ \mathcal{E}(\mathbf{W}; G) + \int_g \int_{-1}^0 \|\mathbf{W}(\mathbf{X})\|^2 d\mathbf{X} \right\}.$$

Since the last integral can be estimated by the first term in the left-hand side of (4.2.12), the proof is complete. □

Boundary value problem

Consider the boundary value problem in G with prescribed tractions on ∂G:

$$-\mathbf{L}(\partial_{\mathbf{X}})\mathbf{W}(\mathbf{X}) = \mathbf{F}(\mathbf{X}), \quad \mathbf{X} \in G, \qquad (4.2.13)$$

$$\sigma^{(n)}(\mathbf{W}; \mathbf{X}) = \mathbf{P}(\mathbf{X}), \quad \mathbf{X} \in \partial G, \qquad (4.2.14)$$

where $\mathbf{F} \in (L_2(G))^3$ and $\mathbf{P} \in (L_2(\partial G))^3$ have bounded supports. By a solution of problem (4.2.13), (4.2.14) we mean a function $\mathbf{W} \in \mathbf{E}(G)$ which satisfies the relation

$$\sum_{i,j=1}^3 \int_G \sigma_{ij}(\mathbf{W}; \mathbf{X}) \varepsilon_{ij}(\mathcal{V}; \mathbf{X}) d\mathbf{X} = h(\mathcal{V}), \qquad (4.2.15)$$

for any $\mathcal{V} \in \mathbf{E}(G)$, where

$$h(\mathcal{V}) = \int_G \mathbf{F}(\mathbf{X}) \cdot \mathcal{V}(\mathbf{X}) d\mathbf{X} + \int_{\partial G} \mathbf{P}(\mathbf{X}) \cdot \mathcal{V}(\mathbf{X}) ds. \qquad (4.2.16)$$

Solvability in $\mathbf{E}(G)$ and asymptotics at infinity

THEOREM 4.2

Let $\mathbf{F}$ and $\mathbf{P}$ be vector-valued functions from $(\mathbf{L}_2(G))^3$ and $(\mathbf{L}_2(\partial G))^3$ with bounded supports.

MODEL PROBLEMS

(i) *Problem (4.2.13), (4.2.14) is uniquely solvable in* $\mathbf{E}(G)$.

(ii) *The solution* $\mathbf{W}$ *admits the asymptotic representation in the half-space* $\mathbb{R}^3_-$:

$$\mathbf{W}(\mathbf{X}) \sim \mathcal{B}(\mathbf{X})\mathfrak{F} + (\mathfrak{M}_1 \partial_{X_1} - \mathfrak{M}_2 \partial_{X_2})\mathcal{B}^{(3)}(\mathbf{X})$$
$$+ \frac{1}{2}\mathfrak{M}_3(-\partial_{X_1}\mathcal{B}^{(2)}(\mathbf{X}) + \partial_{X_2}\mathcal{B}^{(1)}(\mathbf{X})) + \sum_{i=1}^{2} b_i \partial_{X_i} \mathcal{B}^{(i)}(\mathbf{X})$$
$$+ \frac{1}{2}b_3(\partial_{X_1}\mathcal{B}^{(2)}(\mathbf{X}) + \partial_{X_2}\mathcal{B}^{(1)}(\mathbf{X})) + \sum_{|\alpha|\geq 2} \partial^{\alpha}_{\mathbf{X}'} \mathcal{B}(\mathbf{X})\mathbf{B}_{\alpha},$$
$$\mathbf{X} \in \mathbb{R}^3_-,\ \|\mathbf{X}\| \to \infty, \qquad (4.2.17)$$

where $\mathcal{B}$ is the Boussinesq–Cerruti solution (see Section 3.4.1), b_j are constants, $\mathbf{B}_\alpha$ are constant vectors, and $\mathfrak{F}$ and $\mathfrak{M}$ are given by

$$\mathfrak{F} = \mathbb{F}(\mathbf{F},\mathbf{P},G) \quad and \quad \mathfrak{M} = \mathbb{M}(\mathbf{F},\mathbf{P},G), \qquad (4.2.18)$$

where the functionals $\mathbb{F}$, $\mathbb{M}$ were defined in (4.2.5), (4.2.6). The asymptotic expansion (4.2.17) should be understood in the following sense: the difference of $\mathbf{W}(\mathbf{X})$ and the sum of terms, homogeneous of order m, $m \geq -M$, from the right hand side is $O(\|\mathbf{X}\|^{-M-1})$ and the above asymptotic expansion can be differentiated.

(iii) *In the cylindrical extension* Π_+ *the vector* $\mathbf{W}$ *has the asymptotics*

$$\mathbf{W}(\mathbf{X}) = \mathbf{a} + \mathbf{b} \times \mathbf{X} + \mathbf{R}(\mathbf{X}), \qquad (4.2.19)$$

where $\mathbf{a}$, $\mathbf{b}$ are constant vectors, and

$$\|\partial^j_{X_3}\mathbf{R}(\cdot,X_3)\|_{(H^1(g))^3} \leq C_j e^{-\gamma X_3}, \ \gamma > 0, \qquad (4.2.20)$$

for large positive X_3 and some constant C_j, $j = 0,1,\ldots$.

Proof

(i) Owing to the boundedness of the supports of $\mathbf{F}$ and $\mathbf{P}$ and the estimate (4.2.12), formula (4.2.16) defines a linear continuous functional on $\mathbf{E}(G)$. From the Riesz theorem we obtain the existence and uniqueness for the variational problem (4.2.15).

(ii) Let r be a positive constant such that the ball B_r contains the supports of $\mathbf{F}$ and $\mathbf{P}$. We introduce a smooth function $\theta = \theta(\mathbf{Y})$ in such a way that $\theta(\mathbf{Y}) = 0$ in $B_r \cup \Pi_+$ and $\theta(\mathbf{Y}) = 1$ in $\mathbb{R}^3 \setminus B_{2r}$. Since the supports of the vector functions $\mathbf{L}\theta\mathbf{W}$ and $\sigma^{(n)}(\theta\mathbf{W}; \mathbf{X})$ are bounded and are situated in $\overline{\mathbb{R}}^3_-$, one can represent $\mathbf{W}$ by means of Mindlin's solution $\mathcal{M}$ (see Section 3.4.2):

$$\mathbf{W}(\mathbf{X}) = -\int_{\mathbb{R}^3_-} \mathcal{M}(\mathbf{X}' - \mathbf{Y}', X_3, Y_3)\mathbf{L}(\partial_\mathbf{Y})(\theta(\mathbf{Y})\mathbf{W}(\mathbf{Y}))d\mathbf{Y}$$

$$+ \int_{X_3=0} \mathcal{M}(\mathbf{X}' - \mathbf{Y}', X_3, Y_3)\sigma^{(3)}(\theta\mathbf{W}; \mathbf{Y})dS,$$

for $\mathbf{X} \in \overline{\mathbb{R}}^3_- \setminus B_{2r}$. We note that

$$\mathcal{M}(\mathbf{X}' - \mathbf{Y}', X_3, Y_3) = \|\mathbf{X}\|^{-1}\mathcal{M}\left(\frac{\mathbf{X}' - \mathbf{Y}'}{\|\mathbf{X}\|}, \frac{X_3}{\|\mathbf{X}\|}, \frac{Y_3}{\|\mathbf{X}\|}\right).$$

Expanding the last matrix function in Taylor's series with respect to $\mathbf{Y}'/\|\mathbf{X}\|$ and $Y_3/\|\mathbf{X}\|$ in a neighbourhood of the origin and using Lemma 3.1 we arrive at (4.2.17).

(iii) Let the supports of $\mathbf{F}$ and $\mathbf{P}$ be contained in a ball of radius T centred at the origin. Then $\mathbf{u}$ satisfies (3.9.3), (3.9.4). Hence, the result follows from Theorem 3.3. □

Junction layer

Theorem 4.2 gives the solution $\mathbf{W}$ which does not decay in the cylindrical extension Π_+ and vanishes like $O(\|\mathbf{X}\|^{-1})$ at infinity in the half-space $\mathbb{R}^3_-$. We change the boundary value problem by introducing 12 constants in its formulation in such a way that the solution decays in both directions (compare with Section 2.2.4).

Using Theorem 4.2 we shall construct a vector function $\mathbf{W}^{(1)}$ in G satisfying (as well as $\mathbf{W}$) (4.2.13), (4.2.14) and subject to (4.2.17) with $\mathfrak{F}_j = \mathfrak{M}_j = 0$, $j = 1, 2, 3$. Besides, $\mathbf{W}^{(1)}$ admits the following asymptotic representation in the cylindrical extension:

$$\mathbf{W}^{(1)}(\mathbf{X}) = \sum_{j=1}^{6}(d_j^{(I)}\boldsymbol{\xi}^{(j)}(\mathbf{X}) + d_j^{(II)}\boldsymbol{\psi}^{(j)}(\mathbf{X})) + \mathbf{R}^{(1)}(\mathbf{X}),$$

where $\mathbf{R}^{(1)}$ decays exponentially as $X_3 \to +\infty$. Here, $\boldsymbol{\xi}^{(j)}$ and $\boldsymbol{\psi}^{(j)}$ are solutions in the cylinder defined in Section 3.6.3.

MODEL PROBLEMS

We shall look for $\mathbf{W}^{(1)}$ in the form

$$\mathbf{W}^{(1)}(\mathbf{X}) = \eta(X_3) \sum_{j=1}^{6} d_j^{(I)} \boldsymbol{\xi}^{(j)}(\mathbf{X}) + \mathbf{w}(\mathbf{X}),$$

where $\mathbf{w} \in \mathbf{E}(G)$ and η is a smooth function such that $\eta(X_3) = 0$ for $X_3 < 1$ and $\eta(X) = 1$ for $X_3 > 2$. Then

$$-\mathbf{L}(\partial_\mathbf{X})\mathbf{w}(\mathbf{X}) = \mathbf{F}(\mathbf{X}) + \sum_{j=1}^{6} d_j^{(I)} \mathbf{L}(\partial_\mathbf{X})(\eta(X_3)\boldsymbol{\xi}^{(j)}(\mathbf{X})), \quad \mathbf{X} \in G,$$

$$\sigma^{(n)}(\mathbf{w};\mathbf{X}) = \mathbf{P}(\mathbf{X}) - \sum_{j=1}^{6} d_j^{(I)} \sigma^{(n)}(\eta \boldsymbol{\xi}^{(j)};\mathbf{X}), \quad \mathbf{X} \in \partial G.$$

By Theorem 4.2, this problem has a unique solution in $\mathbf{E}(G)$ and this solution admits the asymptotic representations (4.2.17) and (4.2.19). By (4.2.18), the coefficients $\mathfrak{F}$ and $\mathfrak{M}$ in (4.2.17) are zero if and only if

$$d_q^{(I)} = D_q^{-1} \mathbb{F}_q(\mathbf{F},\mathbf{P},G), \quad q = 1, 2, \tag{4.2.21}$$

$$d_3^{(I)} = -D_3^{-1} \mathbb{F}_3(\mathbf{F},\mathbf{P},G), \tag{4.2.22}$$

$$d_4^{(I)} = -D_4^{-1} \mathbb{M}_3(\mathbf{F},\mathbf{P},G), \tag{4.2.23}$$

$$d_5^{(I)} = -D_1^{-1} \mathbb{M}_2(\mathbf{F},\mathbf{P},G), \tag{4.2.24}$$

$$d_6^{(I)} = D_2^{-1} \mathbb{M}_1(\mathbf{F},\mathbf{P},G). \tag{4.2.25}$$

(To obtain this one should use the biorthogonality relation (3.6.29) when calculating the principal force and moment of the right-hand sides in the problem for $\mathbf{w}$.) With this choice of $d_j^{(I)}$, $j = 1,\ldots,6$, the vector function $\mathbf{W}^{(1)}$ has been constructed. This function is unique since the difference of two such solutions is in $\mathbf{E}(G)$, satisfies the homogeneous problem (4.2.13), (4.2.13) and equals zero by Theorem 4.2.

Now, we introduce the vector function

$$\boldsymbol{\mathcal{W}}(\mathbf{X}) = \mathbf{W}^{(1)}(\mathbf{X}) - \eta(X_3) \sum_{j=1}^{6} (d_j^{(I)} \boldsymbol{\xi}_j(\mathbf{X}) + d_j^{(II)} \boldsymbol{\psi}_j(\mathbf{X})).$$

Since

$$\boldsymbol{\mathcal{W}}(\mathbf{X}) = \mathbf{R}^{(1)}(\mathbf{X}) = \mathbf{w}(\mathbf{X}) - \sum_{j=1}^{6} d_j^{(II)} \boldsymbol{\psi}_j(\mathbf{X}), \quad X_3 > 2,$$

it follows from Theorem 4.2(iii) applied to the vector function **w** that

$$\|\partial^j_{X_3}\boldsymbol{\mathcal{W}}(\cdot,X_3)\|_{(H^1(g))^3} \le C_j e^{-\gamma X_3} \quad (4.2.26)$$

for large positive X_3. Furthermore, the coefficients $\mathfrak{F}_j$ and $\mathfrak{M}_j$, $j = 1, 2, 3$, in (4.2.17) vanish. Clearly, $\boldsymbol{\mathcal{W}}$ solves the boundary value problem

$$-\mathbf{L}(\partial_\mathbf{X})\boldsymbol{\mathcal{W}}(\mathbf{X}) = \mathbf{F}(\mathbf{X}) + \sum_{j=1}^{6} \Big\{ d_j^{(I)} \mathbf{L}(\partial_\mathbf{X})\eta(X_3)\boldsymbol{\xi}^{(j)}(\mathbf{X})$$
$$+ d_j^{(II)} \mathbf{L}(\partial_\mathbf{X})\eta(X_3)\boldsymbol{\psi}^{(j)}(\mathbf{X}) \Big\}, \quad \mathbf{X} \in G, \quad (4.2.27)$$

$$\sigma^{(n)}(\boldsymbol{\mathcal{W}};\mathbf{X}) = \mathbf{P}(\mathbf{X}) - \sum_{j=1}^{6} \Big\{ d_j^{(I)} \sigma^{(n)}(\eta\boldsymbol{\xi}^{(j)};\mathbf{X})$$
$$+ d_j^{(II)} \sigma^{(n)}(\eta\boldsymbol{\psi}^{(j)};\mathbf{X}) \Big\}, \quad \mathbf{X} \in \partial G. \quad (4.2.28)$$

One can consider (4.2.27), (4.2.28) as a problem of finding the collection

$$\{\boldsymbol{\mathcal{W}}, d_1^{(I)}, \ldots, d_6^{(I)}, d_1^{(II)}, \ldots, d_6^{(II)}\}, \quad (4.2.29)$$

where $\boldsymbol{\mathcal{W}} \in \mathbf{E}(G)$, the estimate (4.2.26) holds and the coefficients $\mathfrak{F}_j$ and $\mathfrak{M}_j$, $j = 1, 2, 3$, are equal to zero in the asymptotic representation (4.2.17) for $\boldsymbol{\mathcal{W}}$.

The function $\boldsymbol{\mathcal{W}}$, which plays an important role in the subsequent asymptotic analysis, will be called the *junction layer*.

LEMMA 4.2

The collection (4.2.29) is defined uniquely.

Proof

Suppose $\mathbf{F} = 0$ and $\mathbf{P} = 0$ in (4.2.27), (4.2.28).

We begin with the constants $(d_1^{(I)}, \ldots, d_6^{(I)}, d_1^{(II)}, \ldots, d_6^{(II)})$ and construct the solution $\mathbf{W}^{(1)}$ of problem (4.2.27), (4.2.28). Modulo exponentially vanishing terms, $\mathbf{W}^{(1)}$ is equal to the sum

$$\sum_{j=1}^{6} (D_j^{(I)} \boldsymbol{\xi}^{(j)}(\mathbf{X}) + D_j^{(II)} \boldsymbol{\psi}^{(j)})$$

with constant coefficients $D_j^{(I)}, D_j^{(II)}$, for large positive X_3. Thus, we have defined the linear mapping

$$(d_1^{(I)}, \ldots, d_6^{(I)}, d_1^{(II)}, \ldots, d_6^{(II)}) \to (D_1^{(I)}, \ldots, D_6^{(I)},$$
$$D_1^{(II)}, \ldots, D_6^{(II)}). \quad (4.2.30)$$

Let $D_j^{(I)}, D_j^{(II)}$, $j = 1, \ldots, 6$, be arbitrary constants. Then

$$\eta(\mathbf{X}) \sum_{j=1}^{6} (D_j^{(I)} \xi^{(j)}(\mathbf{X}) + D_j^{(II)} \psi^{(j)}(\mathbf{X}))$$

solves problem (4.2.27), (4.2.28) with $d_j^{(I)} = D_j^{(I)}$ and $d_j^{(II)} = D_j^{(II)}$. Hence the mapping (4.2.30) is surjective, and therefore it is a one-to-one mapping.

Let problem (4.2.27), (4.2.28) have a junction layer solution $\mathcal{W}$ which enters the collection (4.2.29). Then $\mathcal{W} = \mathbf{W}^{(1)}$ and therefore $D_j^{(I)} = D_j^{(II)} = 0$, $j = 1, \ldots, 6$. This implies $d_j^{(I)} = d_j^{(II)} = 0$, $j = 1, \ldots, 6$. By Theorem 4.2, $\mathcal{W} = \mathbf{0}$. □

Thus, we have arrived at the main result of the present section.

THEOREM 4.3

Let $\mathbf{F} \in (\mathbf{L}_2(G))^3$, $\mathbf{P} \in (\mathbf{L}_2(\partial G))^3$ have bounded supports. There exists a unique set

$$\{\mathcal{W}; d_1^{(I)}, \ldots, d_6^{(I)}; d_1^{(II)}, \ldots, d_6^{(II)}\}, \quad (4.2.31)$$

which satisfies the boundary value problem (4.2.27), (4.2.28), where $\mathcal{W} \in \mathbf{E}(G)$ admits the asymptotic expansion (4.2.17) in $\mathbb{R}^3_-$ with $\mathfrak{F} = 0$, $\mathfrak{M} = 0$, and decays exponentially at infinity in Π_+. The constants $d_j^{(I)}$ are given by (4.2.21)–(4.2.25).

4.2.4 Model problem for the bottom layer

Let Π_- be the semi-cylinder $\{\mathbf{Y} : Y_3 < 0, (Y_1, Y_2) \in g\}$ (see Fig. 2.3), with g being the same as in the previous section. We denote by S the base of Π_-.

Function spaces

A vector-valued function $\mathbf{V}$ is said to belong to the space $\mathbf{E}(\Pi_-)$ if $\mathbf{V} \in (H^1(\Pi_- \cap B_r))^3$ for any positive r, and $\mathcal{E}(\mathbf{V}; \Pi_-) < \infty$. In the space $\mathbf{E}(\Pi_-)$ we define the norm

$$\|\mathbf{V}\|_{\mathbf{E}(\Pi_-)} = \left(\mathcal{E}(\mathbf{V}, \Pi_-) + \int_g \|\mathbf{V}(\mathbf{Y}', 0)\|^2 d\mathbf{Y}'\right)^{1/2}.$$

Clearly, $\mathbf{E}(\Pi_-)$ includes the rigid-body displacements. By Lemma 3.9,

$$\int_{\Pi_-} (|\nabla' V_3|^2 + V_3^2 + \|\partial_{Y_3} \mathbf{V}'\|^2) \frac{d\mathbf{Y}}{1+Y_3^2}$$

$$+ \int_{\Pi_-} (\|\nabla' \mathbf{V}\|^2 + \|\mathbf{V}'\|^2) \frac{d\mathbf{Y}}{(1+Y_3^2)^2}$$

$$\leq \text{Const } \|\mathbf{V}\|^2_{\mathbf{E}(\Pi_-)}. \quad (4.2.32)$$

We introduce the subspace

$$\mathbf{E}^0(\Pi_-; S) = \{\mathbf{V} \in \mathbf{E}(\Pi_-) : \mathbf{V} = 0 \text{ on } S\}$$

of $\mathbf{E}(\Pi_-)$. We note that $(\mathcal{E}(\mathbf{V}; \Pi_-))^{1/2}$ is the norm in $\mathbf{E}^0(\Pi_-; S)$.

Boundary value problem

Consider the mixed boundary value problem: the displacement vector $\mathbf{V}$ satisfies the Lamé system

$$-\mathbf{L}(\partial_\mathbf{Y})\mathbf{V}(\mathbf{Y}) = \mathcal{G}(\mathbf{Y}), \quad \mathbf{Y} \in \Pi_-, \quad (4.2.33)$$

the traction boundary condition

$$\sigma^{(n)}(\mathbf{V}; \mathbf{Y}) = \mathbf{H}(\mathbf{Y}), \quad \mathbf{Y} \in \partial \Pi_- \setminus S, \quad (4.2.34)$$

and the displacement boundary condition

$$\mathbf{V}(\mathbf{Y}) = \mathbf{\Phi}(\mathbf{Y}), \quad \mathbf{Y} \in S, \quad (4.2.35)$$

where $\mathcal{G} \in (L_2(\Pi_-))^3$, $\mathbf{H} \in (L_2(\partial \Pi_- \setminus S))^3$ and $\mathbf{\Phi} \in (H^{1/2}(S))^3$. We assume that $\mathcal{G}$ and $\mathbf{H}$ have bounded supports.

MODEL PROBLEMS

By a solution of (4.2.33)–(4.2.35) we understand a vector function $\mathbf{V} \in \mathbf{E}(\Pi_-)$ satisfying the equation

$$\sum_{i,j=1}^{3} \int_{\Pi_-} \sigma_{ij}(\mathbf{V};\mathbf{Y})\varepsilon_{ij}(\mathcal{U};\mathbf{Y})d\mathbf{Y} = h(\mathcal{U}), \qquad (4.2.36)$$

for all $\mathcal{U} \in \mathbf{E}^0(\Pi_-; S)$ and the boundary condition (4.2.35). The right-hand side $h(\mathcal{U})$ is given by

$$h(\mathcal{U}) = \int_{\Pi_-} \mathcal{G}(\mathbf{Y}) \cdot \mathcal{U}(\mathbf{Y})d\mathbf{Y} + \int_{\partial\Pi_-\backslash S} \mathbf{H}(\mathbf{Y}) \cdot \mathcal{U}(\mathbf{Y})ds.$$

Solvability in $\mathbf{E}(\Pi_-)$ and asymptotics

THEOREM 4.4

The problem (4.2.35)–(4.2.34), where $\mathcal{G} \in (L_2(\Pi_-))^3$ and $\mathbf{H} \in (L_2(\partial\Pi_-\backslash S))^3$ have bounded supports, is uniquely solvable in $\mathbf{E}(\Pi_-)$ and its solution admits the asymptotic representation

$$\mathbf{V}(\mathbf{Y}) = \mathbf{a} + \mathbf{b} \times \mathbf{Y} + \mathbf{R}(\mathbf{Y}), \qquad (4.2.37)$$

where $\mathbf{a}$, $\mathbf{b}$ are constant vectors, and

$$\|\partial_{Y_3}^j \mathbf{R}(\cdot, Y_3)\|_{(H^1(g))^3} \leq C_j e^{\gamma Y_3}, \ \gamma > 0, \qquad (4.2.38)$$

for large $|Y_3|$ and some constant C_j, $j = 0, 1, \ldots$.

Proof

There exists a function $\mathbf{V}_0 \in (H^1(\Pi_-))^3$ and hence $\mathbf{V}_0 \in \mathbf{E}(\Pi_-)$ such that $\mathbf{V}_0 = \mathbf{\Phi}$ on S and

$$\|\mathbf{V}_0\|_{\mathbf{E}(\Pi_-)} \leq \text{Const } \|\mathbf{\Phi}\|_{(H^{1/2}(S))^3}.$$

We seek a solution in the form $\mathbf{V} = \mathbf{V}_0 + \mathbf{V}_1$. Then, the vector function $\mathbf{V}_1$ satisfies the variational problem (4.2.36) with h given by

$$h(\mathcal{U}) = -2\mathcal{E}(\mathbf{V}_0, \mathcal{U}) + \left\{ \int_{\Pi_-} \mathcal{G} \cdot \mathcal{U}d\mathbf{Y} + \int_{\partial\Pi_-\backslash S} \mathbf{H} \cdot \mathcal{U}ds \right\},$$

and $\mathbf{V}_1 = 0$ on S. It is clear that h is a linear bounded functional on $\mathbf{E}^0(\Pi_-; S)$. Using the Riesz theorem we obtain the existence and uniqueness of $\mathbf{V}_1$.

For sufficiently large $|T|$ (T is negative), the function $\mathbf{V}$ satisfies homogeneous problem (4.2.33), (4.2.34), where Π_- and $\partial\Pi_- \setminus S$ are replaced by

$$\{\mathbf{Y} : Y_3 < T, (Y_1, Y_2) \in g\} \text{ and } \{\mathbf{Y} : Y_3 < T, (Y_1, Y_2) \in \partial g\}.$$

Hence, (4.2.37) and (4.2.38) follow from Theorem 3.3. □

REMARK 4.1

Formula (4.2.37) can be rewritten in the form

$$\mathbf{V}(\mathbf{Y}) = \sum_{j=1}^{6} c_j \psi^{(j)}(\mathbf{Y}) + \mathbf{R}(\mathbf{Y}), \qquad (4.2.39)$$

where $\psi^{(j)}$ are the rigid-body displacements specified in Section 3.6.3. Clearly,

$$c_i = a_i, \; i = 1, 2, 3; \; c_4 = b_3, \; c_5 = b_2, \; c_6 = -b_1.$$

Solvability in the space of functions vanishing at infinity

Similar to the junction layer problem (4.2.27), (4.2.28) we introduce six constants and change the boundary conditions at S in such a way that the solution of the new problem decays at infinity.

Consider the problem (4.2.33)–(4.2.35). By Theorem 4.4 and Remark 4.1 this problem has a solution $\mathbf{V} \in \mathbf{E}(\Pi_-)$ with asymptotics (4.2.39). Then, the components of the vector function

$$\mathcal{V}(\mathbf{Y}) = \mathbf{V}(\mathbf{Y}) - \sum_{j=1}^{6} c_j \psi^{(j)}(\mathbf{Y}) \qquad (4.2.40)$$

decay exponentially as $Y_3 \to -\infty$. The vector function $\mathcal{V}$ satisfies system (4.2.33), the traction boundary condition (4.2.34) and the displacement condition on the base region S:

$$\mathcal{V}(\mathbf{Y}) = \Phi(\mathbf{Y}) - \sum_{j=1}^{6} c_j \psi^{(j)}(Y_1, Y_2, 0), \; \mathbf{Y} \in S. \qquad (4.2.41)$$

The function $\mathcal{V}$ will be called *the bottom layer*.

THEOREM 4.5

Let $\mathcal{G} \in (\mathbf{L}_2(\Pi_-))^3$ and $\mathbf{H} \in (\mathbf{L}_2(\partial\Pi_- \setminus S))^3$ have bounded supports and $\boldsymbol{\Phi} \in (H^{1/2}(S))^3$. There exists a unique set

$$\{\mathcal{V}, c_1, \ldots, c_6\} \in \mathbf{E}(\Pi_-) \times \mathbb{R}^6, \qquad (4.2.42)$$

which satisfies the boundary value problem (4.2.33), (4.2.34), (4.2.41), where $\mathcal{V}$ decays exponentially at infinity:

$$\|\partial_{Y_3}^j \mathcal{V}(\cdot, Y_3)\|_{(H^1(g))^3} \leq C_j e^{\gamma Y_3}, \qquad (4.2.43)$$

with $\gamma > 0$ and C_j being constants.

Proof

Let $\mathbf{V}$ be a solution of (4.2.33)–(4.2.35) given by Theorem 4.4. Then, the constants c_j in (4.2.42) coincide with constants from (4.2.39) and the bottom layer $\mathcal{V}$ is given by (4.2.40). The estimate (4.2.43) follows from (4.2.38) of Theorem 4.4.

In order to prove the uniqueness we suppose

$$\mathbf{G} = 0, \ \mathbf{H} = 0 \ \text{and} \ \boldsymbol{\Phi} = 0.$$

Then, assuming that the coefficients c_j in (4.2.41) are given, we construct a unique solution of (4.2.33), (4.2.34) and (4.2.41) from $\mathbf{E}(\Pi_-)$. By Theorem 4.4, the solution is

$$-\sum_{j=1}^{6} c_j \psi^{(j)}(\mathbf{Y}).$$

Since this solution must decay at infinity, all the coefficients c_j are equal to zero. Hence, by Theorem 4.4, $\mathcal{V} = 0$. □

4.2.5 Model problems for a bounded two-dimensional domain

Here we collect information on a model boundary value problem for the scaled cross-section of a thin elastic rod to be used in the sequel.

Let $\mathbf{U} = (U_1, U_2, U_3)^T$ satisfy

$$-\mathbf{L}_0(\partial_{\mathbf{X}'})\mathbf{U}(\mathbf{X}') = \mathcal{F}(\mathbf{X}'), \quad \mathbf{X}' = (X_1, X_2) \in g, \qquad (4.2.44)$$

$$\mathbf{T}_0(\mathbf{n},\partial_{\mathbf{X}'})\mathbf{U}(\mathbf{X}') = \mathcal{P}(\mathbf{X}'), \quad \mathbf{X}' \in \partial g, \qquad (4.2.45)$$

where $\mathbf{T}_0$ are specified in (3.6.5) and (3.6.8), and $\mathcal{F} \in (L_2(g))^3, \mathcal{P} \in (L_2(\partial g))^3$. This problem can be split into the plane strain and anti-plane shear problems introduced in Section 3.2.

Problem (4.2.44), (4.2.45) is solvable in $(H^1(g))^3$ if and only if the right-hand sides satisfy the orthogonality conditions:

$$\int_g \mathcal{F}(\mathbf{X}')d\mathbf{X}' + \int_{\partial g} \mathcal{P}(\mathbf{X}')ds = 0, \qquad (4.2.46)$$

$$\int_g (\mathcal{F}_1(\mathbf{X}')X_2 - \mathcal{F}_2(\mathbf{X}')X_1)d\mathbf{X}'$$

$$+ \int_{\partial g} (\mathcal{P}_1(\mathbf{X}')X_2 - \mathcal{P}_2(\mathbf{X}')X_1)ds = 0. \qquad (4.2.47)$$

To provide the uniqueness, we assume that the displacement vector is subject to

$$\int_g \mathbf{U}(\mathbf{X}')d\mathbf{X}' = 0, \quad \int_g (U_1(\mathbf{X}')X_2 - U_2(\mathbf{X}')X_1)d\mathbf{X}' = 0. \qquad (4.2.48)$$

Otherwise, one would be allowed to add any rigid-body displacement.

4.2.6 Model problems on the axis of an elastic rod

It will be shown further that the leading order terms of the longitudinal, transverse and torsional displacements within a one-dimensional elastic rod (see Fig. 2.5(b)) satisfy the ordinary differential equations of the form

$$\frac{d^4 v}{dz^4}(z) = R(z), \quad z \in (0,l), \qquad (4.2.49)$$

$$\frac{d^2 w}{dz^2}(z) = r(z), \quad z \in (0,l), \qquad (4.2.50)$$

and the boundary conditions at the ends of the interval $(0,l)$:

$$v(\tfrac{1}{2}(l \pm l)) = B_0^\pm, \qquad (4.2.51)$$
$$v'(\tfrac{1}{2}(l \pm l)) = B_1^\pm, \qquad (4.2.52)$$

MODEL PROBLEMS

$$w(\tfrac{1}{2}(l \pm l)) = b^{\pm}. \tag{4.2.53}$$

Here, we include the elastic constants characterising the rod into the right-hand sides R and r. The solutions of these problems can be represented as

$$v(z) = \frac{1}{2}\left\{\int_0^z \left[lt^2\phi_{II}\left(\frac{z}{l}\right) - \frac{t^3}{3}\phi_I\left(\frac{z}{l}\right)\right]R(t)dt\right.$$

$$\left. - \int_z^l \left[\frac{1}{3}(l-t)^3\psi_I\left(\frac{z}{l}\right) + l(l-t)^2\psi_{II}\left(\frac{z}{l}\right)\right]R(t)dt\right\}$$

$$+ \phi_I\left(\frac{z}{l}\right)B_0^- + l\phi_{II}\left(\frac{z}{l}\right)B_1^- + \psi_I\left(\frac{z}{l}\right)B_0^+ + l\psi_{II}\left(\frac{z}{l}\right)B_1^+, \tag{4.2.54}$$

where

$$\phi_I(z) = (z-1)^2(2z+1), \quad \phi_{II}(z) = (z-1)^2 z,$$
$$\psi_I(z) = z^2(3-2z), \quad \psi_{II}(z) = z^2(z-1),$$

and

$$w(z) = \left(\frac{z}{l} - 1\right)\int_0^z r(t)t\,dt + \frac{z}{l}\int_z^l (t-l)r(t)dt$$

$$+ \left(1 - \frac{z}{l}\right)b^- + \frac{z}{l}b^+. \tag{4.2.55}$$

Also, in further sections we shall need the following linear functionals on the right-hand sides of (4.2.49)–(4.2.53):

$$\mathbb{Q}^{(I)}(r, b^+, b^-) := \left.\frac{dw}{dz}\right|_{z=0} = \int_0^l \left(\frac{t}{l} - 1\right)r(t)dt + \frac{1}{l}(b^+ - b^-), \tag{4.2.56}$$

$$\mathbb{Q}^{(II)}(R, B_0^+, B_0^-, B_1^+, B_1^-) := \left.\frac{d^2v}{dz^2}\right|_{z=0}$$

$$= \int_0^l t\left(1 - \frac{t}{l}\right)^2 R(t)dt + \frac{6}{l^2}(B_0^+ - B_0^-) - \frac{2}{l}(2B_1^- + B_1^+), \tag{4.2.57}$$

$$\mathbb{Q}^{(III)}(R, B_0^+, B_0^-, B_1^+, B_1^-) := \left.\frac{d^3v}{dz^3}\right|_{z=0}$$

$$= -\int_0^l \left(1 - \frac{t}{l}\right)^2\left(1 + \frac{2t}{l}\right)(t)dt + \frac{12}{l^3}(B_0^- - B_0^+) + \frac{6}{l^2}(B_1^- + B_1^+). \tag{4.2.58}$$

4.2.7 Model matrices and the pile structure

In the present section we introduce a set of matrices independent of ε which play an important role for the asymptotic evaluation of the solution of the problem (4.1.1)–(4.1.3). The components of these matrices include information about all elements of the multi–structure. The matrices mentioned can be interpreted as the normalised stiffness and coupling matrices for the pile structure related to the skeleton of the multi-structure.

Model matrices

The further asymptotic algorithm requires certain model matrices:

$$\mathcal{A} = \sum_{k=1}^{K} \frac{12}{l_k^3} \begin{pmatrix} D_1 & 0 & -D_1 a_2^{(k)} \\ 0 & D_2 & D_2 a_1^{(k)} \\ -D_1 a_2^{(k)} & D_2 a_1^{(k)} & (a_1^{(k)})^2 D_2 + (a_2^{(k)})^2 D_1 \\ & & + l_k^2 (D_4/12) \end{pmatrix}, \quad (4.2.59)$$

$$\mathcal{B} = \sum_{k=1}^{K} \frac{D_3}{l_k} \begin{pmatrix} 1 & a_2^{(k)} & -a_1^{(k)} \\ a_2^{(k)} & (a_2^{(k)})^2 & -a_1^{(k)} a_2^{(k)} \\ -a_1^{(k)} & -a_1^{(k)} a_2^{(k)} & (a_1^{(k)})^2 \end{pmatrix}, \quad (4.2.60)$$

$$\mathcal{D} = \sum_{k=1}^{K} \frac{6}{l_k^2} \begin{pmatrix} 0 & 0 & 0 \\ 0 & -D_2 & -D_2 a_1^{(k)} \\ D_1 & 0 & -D_1 a_2^{(k)} \end{pmatrix}, \quad (4.2.61)$$

and

$$\mathcal{E} = \sum_{k=1}^{K} \frac{4}{l_k} \begin{pmatrix} 0 & 0 & 0 \\ 0 & D_2 & 0 \\ 0 & 0 & D_1 \end{pmatrix}, \quad (4.2.62)$$

where D_i, $i = 1, \ldots, 4$, are the normalised stiffness coefficients introduced in Section 3.6.4. The quantities $a_j^{(k)}$ are coordinates of the junction points calculated in the coordinate system such that the plane $Ox_1 x_2$ contains the junction points, the Ox_3 axis is directed downwards, and the Ox_1, Ox_2 axes are parallel to the principal inertia axes of the cross-sections of thin rods.

Since for any non-zero constant vector $X \in \mathbb{R}^3$

$$X^t \mathcal{A} X = 12 \sum_k \frac{1}{l_k^3} \Big\{ D_1 (X_1 - X_3 a_2^{(k)})^2$$

MODEL PROBLEMS   175

$$+ D_2(X_2 + X_3 a_1^{(k)})^2 + X_3^2 l_k^2 D_4/12 \bigg\} > 0, \qquad (4.2.63)$$

the matrix $\mathcal{A}$ is non-singular. Similarly,

$$X^t \mathcal{B} X = D_3 \sum_k \frac{1}{l_k}(X_1 + a_2^{(k)} X_2 - a_1^{(k)} X_3)^2 > 0, \qquad (4.2.64)$$

rovided the junction points $\mathbf{a}^{(k)}$, $k = 1,\ldots,K$, are not located on the same straight line.

LEMMA 4.3

The inequality

$$X^t \mathcal{A} X + 2 Y^t \mathcal{D} X + Y^t \mathcal{E} Y \geq \frac{1}{4} X^t \mathcal{A} X \qquad (4.2.65)$$

holds for any $X, Y \in \mathbb{R}^3$.

Proof

We denote by $\mathcal{A}_k$, $\mathcal{D}_k$ and $\mathcal{E}_k$ the terms with the number k from the sums (4.2.59), (4.2.61), (4.2.62), respectively. Let

$$\mathcal{E}_k^{(1)} = \frac{l_k}{4}\mathrm{diag}\{0, D_2^{-1}, D_1^{-1}\}.$$

Then

$$|Y^t \mathcal{D}_k X| = |Y^t \mathcal{E}_k \mathcal{E}_k^{(1)} \mathcal{D}_k X| \leq \left\{(Y^t \mathcal{E}_k Y)(X^t \mathcal{D}_k^t \mathcal{E}_k^{(1)} \mathcal{D}_k X)\right\}^{1/2}.$$

One can verify directly that

$$\mathcal{D}_k^t \mathcal{E}_k^{(1)} \mathcal{D}_k = \frac{3}{4}\bigg\{\mathcal{A}_k - \mathrm{diag}\{0, 0, l_k^{-1} D_4\}\bigg\}.$$

Therefore,

$$|Y^t \mathcal{D}_k X| \leq \left\{\frac{3}{4}(Y^t \mathcal{E}_k Y)(X^t A_k X)\right\}^{1/2} \leq \frac{1}{2}(Y^t \mathcal{E}_k Y + \frac{3}{4} X^t \mathcal{A}_k X).$$

Taking the summation with respect to k we arrive at (4.2.65).   □

The matrices $\mathcal{A}$, $\mathcal{B}$, $\mathcal{D}$ and $\mathcal{E}$ occur naturally in the pile structure model described in the next section.

The pile structure

The classical pile structure model describes a 3D–1D multi-structure as a set of elastic piles of different orientations (i.e. the vectors directed along the piles span $\mathbb{R}^3$) encased at their heads in the rigid pile cap. The main relation is the 6×6 linear system which provides the correspondence between the rigid-body displacements of the pile cap and the principal forces and moments of the external load (see Asplund (1966), pp. 291–297).

We fix the same Cartesian coordinate system as above. The elastic rod r_k, corresponding to the junction point $\mathbf{x} = \mathbf{a}^{(k)} = (a_1^{(k)}, a_2^{(k)}, 0)$, is assumed to have the length l_k, and it is characterised by the stiffness coefficients C_1, C_2, C_3, C_4 which are supposed to be positive (see Section 3.3). The first two coefficients C_1 and C_2 are related to transverse displacements in the direction of the Ox_1 and Ox_2 axes. The coefficients C_3 and C_4 correspond to the longitudinal displacement and the axial rotation, respectively. Our multi-structure is related to the pile structure by

$$C_i = \varepsilon^4 D_i,\ i = 1,2,4;\ C_3 = \varepsilon^2 D_3, \qquad (4.2.66)$$

where D_k, $k = 1,\ldots,4$, are the same constants as in Section 3.6.4, and ε is the non-dimensional small parameter (see Section 4.1). The region Ω is assumed to be rigid, and the displacements of the ends of the rods at the junction points and the rotations are equal to the displacements and the rotations of Ω at the same points. The dead ends of elastic rods are assumed to be clamped. The interaction between the rigid-body Ω and elastic rods is characterised by lock forces $\mathfrak{F}^{(k)}$ and lock moments $\mathfrak{M}^{(k)}$ which are applied at the junction points. Assume that the external force $\mathfrak{F}$ and the moment $\mathfrak{M}$ are applied to Ω.

The equilibrium equations have the form

$$\sum_{k=1}^{K} \mathfrak{F}^{(k)} + \mathfrak{F} = 0, \qquad (4.2.67)$$

$$\sum_{k=1}^{K} (\mathfrak{M}^{(k)} + \mathbf{a}^{(k)} \times \mathfrak{F}^{(k)}) + \mathfrak{M} = 0. \qquad (4.2.68)$$

The objective is to derive a system of linear equations with respect to components of rigid-body translations and rotations of Ω and a representation for components of lock forces and moments at the junction

MODEL PROBLEMS

points. The displacements of points of the one-dimensional elastic rod r_k are characterised by functions $v_1^{(k)}$, $v_2^{(k)}$, $v_3^{(k)}$, $v_4^{(k)}$ which correspond to the transverse displacements, the longitudinal displacement and the axial rotation, respectively.

Mathematical model for the pile structure

Now we describe the mathematical model. The rigid-body displacements of Ω are specified by

$$\boldsymbol{\alpha} + \boldsymbol{\beta} \times \mathbf{x}, \tag{4.2.69}$$

where $\boldsymbol{\alpha}$ and $\boldsymbol{\beta}$ are constant vectors.

The functions $v_i^{(k)}$, $i = 1, 2, 3, 4$, satisfy the ordinary differential equations

$$\frac{d^4 v_i^{(k)}}{dx_3^4}(x_3) = 0, \quad x_3 \in (0, l_k), \ i = 1, 2, \tag{4.2.70}$$

$$\frac{d^2 v_i^{(k)}}{dx_3^2}(x_3) = 0, \quad x_3 \in (0, l_k), \ i = 3, 4. \tag{4.2.71}$$

The clamping conditions at the dead ends of the rods r_k are

$$v_i^{(k)}(l_k) = 0, \ i = 1, 2, 3, 4, \tag{4.2.72}$$

$$\frac{dv_i^{(k)}}{dx_3}(l_k) = 0, \ i = 1, 2. \tag{4.2.73}$$

The continuity conditions of rotations and translations at the junction points $\mathbf{a}^{(k)}$, $k = 1, \ldots, K$, can be written as

$$v_i^{(k)}(0) = \alpha_i + (\boldsymbol{\beta} \times \mathbf{a}^{(k)})_i, \ i = 1, 2, 3, \tag{4.2.74}$$

$$v_4^{(k)}(0) = \beta_3, \ \frac{dv_1^{(k)}}{dx_3}(0) = \beta_2, \ \frac{dv_2^{(k)}}{dx_3}(0) = -\beta_1. \tag{4.2.75}$$

Forces and moments at junction points, exerted by elastic rods on the body Ω, are specified by

$$\mathfrak{F}_i^{(k)} = -C_i \frac{d^3 v_i^{(k)}}{dx_3^3}(0), \ i = 1, 2, \tag{4.2.76}$$

$$\mathfrak{F}_3^{(k)} = C_3 \frac{dv_3^{(k)}}{dx_3}(0), \tag{4.2.77}$$

$$\mathfrak{M}_{3-i}^{(k)} = (-1)^{i+1} C_i \frac{d^2 v_i^{(k)}}{dx_3^2}(0), \ i = 1, 2, \tag{4.2.78}$$

$$\mathfrak{M}_3^{(k)} = C_4 \frac{dv_4^{(k)}}{dx_3}(0). \qquad (4.2.79)$$

The solution of the pile structure equations

By formulae (4.2.54), (4.2.55), the solutions of (4.2.70)–(4.2.75) have the form

$$v_1^{(k)} = (\alpha_1 - a_2^{(k)}\beta_3)\phi_I\left(\frac{x_3}{l_k}\right) + \beta_2 l_k \phi_{II}\left(\frac{x_3}{l_k}\right), \qquad (4.2.80)$$

$$v_2^{(k)} = (\alpha_2 + a_1^{(k)}\beta_3)\phi_I\left(\frac{x_3}{l_k}\right) - \beta_1 l_k \phi_{II}\left(\frac{x_3}{l_k}\right), \qquad (4.2.81)$$

$$v_3^{(k)} = (\alpha_3 + \beta_1 a_2^{(k)} - a_1^{(k)}\beta_2)\left(1 - \frac{x_3}{l_k}\right), \qquad (4.2.82)$$

$$v_4^{(k)} = \beta_3\left(1 - \frac{x_3}{l_k}\right), \qquad (4.2.83)$$

where ϕ_I, ϕ_{II} are given in Section 4.2.6.

Evaluating the right-hand sides of (4.2.76)–(4.2.79) we obtain

$$\mathfrak{F}_1^{(k)} = -\frac{6C_1}{l_k^3}\left\{2(\alpha_1 - \beta_3 a_2^{(k)}) + l_k\beta_2\right\}, \qquad (4.2.84)$$

$$\mathfrak{F}_2^{(k)} = -\frac{6C_2}{l_k^3}\left\{2(\alpha_2 + \beta_3 a_2^{(k)}) - l_k\beta_1\right\}, \qquad (4.2.85)$$

$$\mathfrak{F}_3^{(k)} = -\frac{C_3}{l_k}\left\{\alpha_3 + \beta_1 a_2^{(k)} - a_1^{(k)}\beta_2\right\}, \qquad (4.2.86)$$

$$\mathfrak{M}_1^{(k)} = \frac{2C_2}{l_k^2}\left\{3(\alpha_2 + \beta_3 a_1^{(k)}) - 2l_k\beta_1\right\}, \qquad (4.2.87)$$

$$\mathfrak{M}_2^{(k)} = -\frac{2C_1}{l_k^2}\left\{3(\alpha_1 - \beta_3 a_2^{(k)}) + 2l_k\beta_2\right\}, \qquad (4.2.88)$$

$$\mathfrak{M}_3^{(k)} = -\frac{C_4\beta_3}{l_k}. \qquad (4.2.89)$$

Substituting these expressions into the balance relations (4.2.67), (4.2.68) we arrive at the following system of linear equations:

$$\begin{pmatrix} \varepsilon^2 \mathcal{A} & \varepsilon^2 \mathcal{D}^t \\ \varepsilon^2 \mathcal{D} & \mathcal{B} + \varepsilon^2 \mathcal{E} \end{pmatrix}(\alpha_1, \alpha_2, \beta_3, \alpha_3, \beta_1, \beta_2)^t$$

$$= \varepsilon^{-2}(\mathfrak{F}_1, \mathfrak{F}_2, \mathfrak{M}_3, \mathfrak{F}_3, \mathfrak{M}_1, \mathfrak{M}_2)^t, \qquad (4.2.90)$$

where the matrices $\mathcal{A}$, $\mathcal{B}$, $\mathcal{D}$ and $\mathcal{E}$ are given by (4.2.59)–(4.2.62).

From (4.2.63)–(4.2.65), the matrix of system (4.2.90) is positive definite, and, hence, there exists a unique set of the coefficients α_j, β_j, $j = 1, 2, 3$, solving this system. The components of the lock forces and moments are evaluated by (4.2.84)–(4.2.89).

The system (4.2.90) is the basic relation in the described model. The diagonal 3×3 blocks $\mathcal{A}$, $\mathcal{B}$ and $\mathcal{E}$ can be interpreted as the normalised stiffness matrices for transverse and longitudinal displacements. The matrix $\mathcal{D}$ is the coupling matrix.

4.2.8 Special cases

The matrix of system (4.2.90) degenerates for $\varepsilon = 0$. This causes displacement components of the body Ω to have different orders of magnitude for transverse and longitudinal loads.

The case $|\mathfrak{F}_1| + |\mathfrak{F}_2| + |\mathfrak{M}_3| \neq 0$

Using (4.2.90) we obtain

$$\mathcal{A}(\alpha_1, \alpha_2, \beta_3)^t = \varepsilon^{-4}(\mathfrak{F}_1, \mathfrak{F}_2, \mathfrak{M}_3)^t + O(\varepsilon^{-2})$$

and

$$\mathcal{B}(\alpha_3, \beta_1, \beta_2)^t + \varepsilon^2 \mathcal{D}(\alpha_1, \alpha_2, \beta_3)^t = \varepsilon^{-2}(\mathfrak{F}_3, \mathfrak{M}_1, \mathfrak{M}_2)^t + O(1).$$

Although the quantities α_3, β_1, β_2 are small in comparison with α_1, α_2, β_3, we need them for the evaluation of the longitudinal components of the lock forces (see (4.2.86)). The components of the lock forces and moments are determined by (4.2.84)–(4.2.89). The last terms in curly brackets in (4.2.84), (4.2.85), (4.2.87)–(4.2.89) are much smaller than the remaining terms (their ratio is of order $O(\varepsilon^2)$).

The above analysis shows that one can use a simplified model of the pile structure where the local rotations with respect to Ox_1, Ox_2 are neglected, i.e. the right-hand sides of the two last relations (4.2.75) are zero.

The case $\mathfrak{F}_1 = \mathfrak{F}_2 = \mathfrak{M}_3 = 0$

Again, neglecting terms of order ε^2 one has

$$\mathcal{B}(\alpha_3, \beta_1, \beta_2)^t = \varepsilon^{-2}(\mathfrak{F}_3, \mathfrak{M}_1, \mathfrak{M}_2)^t + O(1), \qquad (4.2.91)$$

$$\mathcal{A}(\alpha_1, \alpha_2, \beta_3)^t = -\mathcal{D}^t(\alpha_3, \beta_1, \beta_2)^t. \tag{4.2.92}$$

It follows from (4.2.84)–(4.2.89) that all components of the lock forces and moments, except $\mathfrak{F}_3^{(k)}$, are $O(\varepsilon^2)$, and

$$\mathfrak{F}_3^{(k)} = -\frac{\varepsilon^2 D_3}{l_k}\{\alpha_3 + \beta_1 a_2^{(k)} - a_1^{(k)}\beta_2\}.$$

We note that the components of the rigid-body translations and rotations have the same order of magnitude.

4.3 Asymptotic expansion of the solution

Multi-scaled asymptotic representations of the right-hand sides are specified in Section 4.3.1. In Section 4.3.2 we describe the asymptotic expansion of the displacement field in Ω_ε. The general term of this asymptotic series is constructed in Sections 4.3.4–4.3.8. In Section 4.3.9 we present the recurrent sequence for construction of all the terms of the asymptotic series.

4.3.1 Asymptotic representation of the right-hand sides

We use the notations introduced in Section 2.3.1. That is, the global system of coordinates $\mathbf{x} = (x_1, x_2, x_3)$ is chosen in such a way that the junction points $\mathbf{a}^{(j)}$, $j = 1, \ldots, K$, are located on the plane $x_3 = 0$.

In the vicinity of the junction point $\mathbf{a}^{(j)}$ we shall use the coordinates $\mathbf{X}^{(j)}$ defined by (2.3.1). The coordinate system $(\zeta_1^{(j)}, \zeta_2^{(j)}, z)$ defined in (2.3.2) is introduced to describe fields in thin cylinders $\Pi_\varepsilon^{(j)}$. Scaled variables $\mathbf{Y}^{(j)}$ introduced in (2.3.3) are used to describe the bottom layer in the vicinity of the end of $\Pi_\varepsilon^{(j)}$.

Since the cross-sections of the thin cylinders are the same, the limit domains introduced in Section 2.3.1 are independent of the index j associated with thin cylinders.

In the text below we shall use the notations G for the junction model domain, Π_- for the bottom limit domain and S for the base of Π_-. Also, we shall use the cut-off functions $\mathfrak{X}$, χ, η and Ξ, the same as those introduced in Section 2.3.2.

The right-hand sides of the boundary value problem (4.1.1)–(4.1.3) are represented as asymptotic sums of four types of terms which correspond to the region Ω, the junction regions, thin rods and the bottom region. Namely,

$$\mathbf{F}(\mathbf{x},\varepsilon) \sim \sum_{k=0}^{\infty} \varepsilon^k \Big\{ \mathfrak{X}(\mathbf{x},\varepsilon)\mathbf{f}^{(k)}(\mathbf{x}) + \sum_{j=1}^{K} [\varepsilon^{-4}\mathbf{F}^{(k,j)}(\mathbf{X}^{(j)})$$
$$+ \varepsilon^{-2}\boldsymbol{\mathcal{F}}^{(k,j)}(\zeta^{(j)},z)\eta(z/\varepsilon)$$
$$+ \varepsilon^{-4}\boldsymbol{\mathcal{G}}^{(k,j)}(\mathbf{Y}^{(j)})]\Big\}, \quad \mathbf{x} \in \Omega_\varepsilon, \qquad (4.3.1)$$

where
$$\mathbf{f}^{(k)} \in (L_2(\Omega))^3, \ \mathbf{F}^{(k,j)} \in (L_2(G))^3,$$

and
$$\boldsymbol{\mathcal{F}}^{(k,j)} \in C^{\infty}([0,l_j];(L_2(g))^3), \ \boldsymbol{\mathcal{G}}^{(k,j)} \in (L_2(\Pi_-))^3.$$

It is assumed that the term $\mathbf{f}^{(k)}$ is smooth in fixed neighbourhoods of the junction points, and the vector functions $\mathbf{F}^{(k,j)}$ and $\boldsymbol{\mathcal{G}}^{(k,j)}$ are supported by fixed balls centred at the origin of the corresponding coordinate system. The notation $\sim$ in (4.3.1) means that the $(L_2(\Omega_\varepsilon))^3$ norm of the difference between $\mathbf{F}$ and the sum of the first $N+1$ terms of the series (4.3.1) ($0 \leq k \leq N$) is estimated by Const $\varepsilon^{N-3/2}$.

The right-hand side in (4.1.2) is defined as

$$\mathbf{P}(\mathbf{x},\varepsilon) \sim \sum_{k=0}^{\infty} \varepsilon^k \Big\{ \mathfrak{X}(\mathbf{x},\varepsilon)\mathbf{p}^{(k)}(\mathbf{x}) + \sum_{j=1}^{K} [\varepsilon^{-3}\mathbf{P}^{(k,j)}(\mathbf{X}^{(j)})$$
$$+ \varepsilon^{-1}\boldsymbol{\mathcal{P}}^{(k,j)}(\zeta^{(j)},z)\eta(z/\varepsilon)$$
$$+ \varepsilon^{-3}\mathbf{H}^{(k,j)}(\mathbf{Y}^{(j)})]\Big\}, \quad \mathbf{x} \in \partial\Omega_\varepsilon \setminus \bar{S}_\varepsilon, \qquad (4.3.2)$$

where
$$\mathbf{p}^{(k)} \in (L_2(\partial\Omega))^3, \ \mathbf{P}^{(k,j)} \in (L_2(\partial G))^3,$$

and
$$\boldsymbol{\mathcal{P}}^{(k,j)} \in C^{\infty}([0,l_j];(L_2(\partial g))^3), \ \mathbf{H}^{(k,j)} \in (L_2(\partial\Pi_- \setminus S))^3.$$

The vector functions $\mathbf{p}^{(k)}$ are smooth in fixed neighbourhoods of the junction points, and the vector functions $\mathbf{P}^{(k,j)}$ and $\mathbf{H}^{(k,j)}$ are supported by fixed balls centred at the origin. The asymptotic relation (4.3.2) is understood in the sense that the $(L_2(\partial\Omega_\varepsilon \setminus S_\varepsilon))^3$ norm of the difference between $\mathbf{P}$ and the sum of the first $N+1$ terms of the series (4.3.2) ($0 \leq k \leq N$) is estimated by Const ε^{N-1}.

The right-hand side in (4.1.3) is represented in the form

$$\boldsymbol{\Phi}^{(j)}(\mathbf{x},\varepsilon) \sim \sum_{k=0}^{\infty} \varepsilon^{k-2} \boldsymbol{\Phi}^{(k,j)}(Y_1^{(j)}, Y_2^{(j)}), \ \mathbf{x} \in S_\varepsilon, \qquad (4.3.3)$$

where $\boldsymbol{\Phi}^{(k,j)} \in (H^{1/2}(g))^3$. The asymptotic relation (4.3.3) means that the $(H^{1/2}(S_\varepsilon^{(j)}))^3$ norm of the difference between $\boldsymbol{\Phi}^{(j)}$ and the first $N+1$ terms of the series (4.3.3) is estimated by Const $\varepsilon^{N-1/2}$.

We also impose the balance condition

$$\int_G \mathbf{F}^{(0,j)} d\mathbf{X} + \int_{\partial G} \mathbf{P}^{(0,j)} ds = 0. \qquad (4.3.4)$$

Since this restriction is applied to the terms $\mathbf{F}^{(k,j)}, \mathbf{P}^{(k,j)}$ with $k = 0$ only, but not to the other terms in (4.3.1), (4.3.2), it does not cause any loss of generality.

4.3.2 Description of the asymptotic series for the solution

The displacement field $\mathbf{u}(\mathbf{x},\varepsilon)$ will be sought in the form of the asymptotic series

$$\mathbf{u}(\mathbf{x},\varepsilon) \sim \sum_{k=-1}^{\infty} \varepsilon^k \mathbf{u}^{(k)}(\mathbf{x},\varepsilon) \qquad (4.3.5)$$

with

$$\mathbf{u}^{(k)}(\mathbf{x},\varepsilon) = \varepsilon^{-4} \Bigg\{ \sum_{i=1}^{2} \alpha_i^{(k)} \mathbf{e}^{(i)} + \varepsilon^2 \alpha_3^{(k)} \mathbf{e}^{(3)}$$

$$+ \left(\beta_3^{(k)} \mathbf{e}^{(3)} + \varepsilon^2 \sum_{i=1}^{2} \beta_i^{(k)} \mathbf{e}^{(i)} \right) \times \mathbf{x} \Bigg\}$$

$$+ \mathfrak{X}(\mathbf{x},\varepsilon) \mathbf{u}_\Omega^{(k)}(\mathbf{x}) + \sum_{j=1}^{K} \Big\{ \eta(z/\varepsilon) \mathfrak{U}^{(k,j)}(\zeta, z, \varepsilon) \qquad (4.3.6)$$

$$+ \varepsilon^{-2} \Xi(\mathbf{x}-\mathbf{a}^{(j)}) \boldsymbol{\mathcal{W}}^{(k,j)}(\mathbf{X}^{(j)}) + \varepsilon^{-2} \eta(z/\varepsilon) \boldsymbol{\mathcal{V}}^{(k,j)}(\mathbf{Y}^{(j)}) \Big\}.$$

Here $\alpha_i^{(k)}$, $\beta_i^{(k)}$, $i = 1, 2, 3$, are constants, and the terms in the first curly brackets in (4.3.6) represent rigid-body displacements; these coefficients are defined by the system of linear equations of the form (4.2.90). The vector-valued function $\mathbf{u}_\Omega^{(k)}$ solves the problem of type (4.2.1), (4.2.2), and can be represented as

$$\mathbf{u}_\Omega^{(k)}(\mathbf{x}) = \mathbf{u}_r^{(k)}(\mathbf{x}) + \mathbf{u}_s^{(k)}(\mathbf{x}). \tag{4.3.7}$$

The field $\mathbf{u}_r^{(k)} \in (H^1(\Omega))^3$ is a solution of the boundary value problem (4.2.10), (4.2.11) with self-balanced right-hand sides from L_2 which are smooth in the vicinity of the junction points; hence it is also smooth in neighbourhoods of the junction points. The vector function $\mathbf{u}_s^{(k)}$ is defined by

$$\mathbf{u}_s^{(k)}(\mathbf{x}) = \sum_{j=1}^{K} \left\{ \mathcal{N}(\mathbf{x}, \mathbf{a}^{(i)}) \mathfrak{F}^{(k,j)} \right.$$

$$\left. + \frac{1}{2} (\nabla_{\mathbf{x}^*} \times \mathcal{N}^t(\mathbf{x}, \mathbf{x}^*)|_{\mathbf{x}^*=\mathbf{a}^{(i)}})^t \mathfrak{M}^{(k,j)} \right\}, \tag{4.3.8}$$

where $\mathcal{N}$ is Green's matrix from Section 4.2.2; $\mathfrak{F}^{(k,j)}$, $\mathfrak{M}^{(k,i)}$ are constant vectors, which satisfy the following balance relations:

$$\mathbb{F}(\mathbf{f}^{(k)}, \mathbf{p}^{(k)}, \Omega) + \sum_{j=1}^{K} \mathfrak{F}^{(k,j)} = 0, \tag{4.3.9}$$

$$\mathbb{M}(\mathbf{f}^{(k)}, \mathbf{p}^{(k)}, \Omega) + \sum_{j=1}^{K} \left\{ \mathfrak{M}^{(k,j)} + \mathbf{a}^{(j)} \times \mathfrak{F}^{(k,j)} \right\} = 0, \tag{4.3.10}$$

with $\mathbb{F}$ and $\mathbb{M}$ being introduced in (4.2.5) and (4.2.6).

The product

$$\mathfrak{X}(\mathbf{x}, \varepsilon) \mathbf{u}_\Omega^{(k)}$$

is extended as zero into thin cylinders.

The vector function $\mathfrak{U}^{(k,j)}$ corresponds to the displacement in a thin cylinder $\Pi_\varepsilon^{(j)}$, and it is given by

$$\mathfrak{U}^{(k,j)}(\zeta, z, \varepsilon) = \varepsilon^{-4} \sum_{i=1}^{2} \sum_{q=0}^{3} \varepsilon^q \frac{d^q}{dz^q} v_i^{(k,j)}(z) \phi^{(i,q)}(\zeta)$$

$$+ \varepsilon^{-2} \sum_{i=3}^{4} \sum_{q=0}^{1} \varepsilon^q \frac{d^q}{dz^q} v_i^{(k,j)}(z) \phi^{(i,q)}(\zeta) + \mathbf{U}^{(k,j)}(\zeta, z), \tag{4.3.11}$$

where the functions $v_i^{(k,j)}$ depend on the longitudinal variable and are subject to ordinary differential equations of fourth and second orders of the type (4.2.49), (4.2.50); the vector function $\mathbf{U}^{(k,j)}$ belongs to

$C^\infty([0, l_j]; (H^1(g))^3)$ and for every $z \in [0, l_j]$ it satisfies a boundary value problem on the cross-section g of the form (4.2.44), (4.2.45), and the orthogonality conditions (4.2.48).

The junction layer $\mathcal{W}^{(k,j)}$ and the constants $d_i^{(I,k,j)}$, $d_i^{(II,k,j)}$, $i = 1, 2, \ldots, 6$, give a solution of the boundary value problem of the type (4.2.27), (4.2.28) in G, and $\mathcal{W}^{(k,j)}$ decays at infinity (see Theorem 4.3).

The bottom layer $\mathcal{V}^{(k,j)}$ and constants $c_i^{(k,j)}$, $i = 1, 2, \ldots, 6$, are specified as a solution of (4.2.33)–(4.2.41), where $\mathcal{V}^{(k,j)}$ decays at infinity (see Theorem 4.5). The constants $d_i^{(I,k,j)}$, $d_i^{(II,k,j)}$, $c_i^{(k,j)}$ will be included in the boundary conditions for the functions $v_i^{(k,j)}$.

In the following sections we describe an algorithm for calculation of the term $\mathbf{u}^{(m)}$ in (4.3.5). Looking at the expansion (4.3.6) one can deduce that the first group of terms in curly brackets leaves a discrepancy in the Dirichlet boundary condition on S_ε; the term $\mathfrak{X}\mathbf{u}_\Omega^{(k)}$ will produce an error in the vicinity of the junction points; the term $\eta \mathfrak{U}^{(k,j)}$ yields a discrepancy in the vicinity of the junction point as well as at S_ε. Since $\mathcal{W}^{(k,j)}$ decays exponentially in the cylindrical extension, we just have to take into account an error in the three-dimensional region Ω. The last term does not give any discrepancy in our asymptotic expansion. In the asymptotic procedure the right-hand sides absorb discrepancy terms from the previous steps. The modified right-hand sides will be denoted by the same letters with '$\sim$'; the structure of the modified right-hand sides will be described in each case.

The idea of the construction of the asymptotic expansion is described as follows.

We re-expand the right-hand sides and all terms of the expansion for $\mathbf{u}$ in every element of the multi-structure (the domain Ω, the junction regions, the thin cylinders, the bottom regions) using appropriate coordinates. Comparing the terms of equal orders we derive a system of recurrent relations for coefficients of the series for $\mathbf{u}$ via a set of boundary value problems posed in model domains. Finally, we show the sequence of steps required to find all the coefficients.

As in Chapter 2 when constructing asymptotic expansions, we shall use the notations $\mathbf{X}, \mathbf{Y}, \zeta$ without an upper index.

4.3.3 Auxiliary solutions of the Lamé system in a thin elastic rod

Consider a thin cylinder $\Pi_\varepsilon = \{\mathbf{x} : 0 < x_3 < l, \varepsilon^{-1}(x_1, x_2) \in g \subset \mathbb{R}^2\}$. Here $\varepsilon > 0$ is a non-dimensional small parameter, g is a bounded domain with the Lipschitz boundary. In this section we introduce $\zeta = \varepsilon^{-1}(x_1, x_2)$

ASYMPTOTIC EXPANSION OF THE SOLUTION

and $z = x_3$. The notations used follow Section 3.6.

Here, we find certain asymptotic solutions of the Lamé system in Π_ε to be used in the next section for the asymptotic analysis of $\mathbf{u}(\mathbf{x}; \varepsilon)$ in thin cylinders.

THEOREM 4.6

Let $\mathcal{F} \in C^\infty([0, l]; (L_2(g))^3)$ and $\mathcal{P} \in C^\infty([0, l]; (L_2(\partial g))^3)$. Then the vector-valued function

$$\mathfrak{U}(\zeta, z, \varepsilon) = \varepsilon^{-4} \sum_{k=1}^{2} \sum_{j=0}^{3} \varepsilon^j \frac{d^j}{dz^j} v_k(z) \phi^{(k,j)}(\zeta)$$

$$+ \varepsilon^{-2} \sum_{k=3}^{4} \sum_{j=0}^{1} \varepsilon^j \frac{d^j}{dz^j} v_k(z) \phi^{(k,j)}(\zeta) + \mathbf{U}(\zeta, z) \quad (4.3.12)$$

satisfies the equations

$$-\mathbf{L}(\partial_\mathbf{x})\mathfrak{U} - \varepsilon^{-2} \mathcal{F}(\zeta, z) = \varepsilon^{-1} \mathcal{F}^{(0)}(\zeta, z) + \mathcal{F}^{(1)}(\zeta, z), \quad (4.3.13)$$

for $\mathbf{x} \in \Pi_\varepsilon$, and the boundary condition

$$\sigma^{(n)}(\mathfrak{U}; \mathbf{x}) - \varepsilon^{-1} \mathcal{P}(\zeta, z) = \mathcal{P}^{(0)}(\zeta, z), \ (\zeta, z) \in \partial g \times [0, l], \quad (4.3.14)$$

with

$$\mathcal{F}^{(j)} \in C^\infty([0, l]; (L_2(g))^3), \ j = 0, 1,$$

and

$$\mathcal{P}^{(0)} \in C^\infty([0, l]; (L_2(\partial g))^3),$$

provided

$$\frac{d^4}{dz^4} v_k(z) = -D_k^{-1}\left\{ \int_g \mathcal{F}_k(\zeta, z) d\zeta + \int_{\partial g} \mathcal{P}_k(\zeta, z) dl_\zeta \right\},$$
$$\text{for } k = 1, 2, \quad (4.3.15)$$

$$\frac{d^2}{dz^2} v_3(z) = D_3^{-1}\left\{ \int_g \mathcal{F}_3(\zeta, z) d\zeta + \int_{\partial g} \mathcal{P}_3(\zeta, z) dl_\zeta \right\}, \quad (4.3.16)$$

$$\frac{d^2}{dz^2} v_4(z) = D_4^{-1}\left\{ \int_g (\zeta_1 \mathcal{F}_2(\zeta, z) - \zeta_2 \mathcal{F}_1(\zeta, z)) d\zeta \right.$$
$$\left. + \int_{\partial g} (\zeta_1 \mathcal{P}_2(\zeta, z) - \zeta_2 \mathcal{P}_1(\zeta, z)) dl_\zeta \right\}, \quad (4.3.17)$$

and the vector $\mathbf{U} \in C^\infty([0,l];(H^1(g))^3)$ satisfies the two-dimensional system

$$-\mathbf{L}_0 \mathbf{U} = \mathcal{F}(\zeta, z) \qquad (4.3.18)$$
$$+ \mathbf{L}_1 \left(\frac{d^4 v_1}{dz^4} \phi^{(1,3)} + \frac{d^4 v_2}{dz^4} \phi^{(2,3)} + \frac{d^2 v_4}{dz^2} \phi^{(4,1)} + \frac{d^2 v_3}{dz^2} \phi^{(3,1)} \right)$$
$$+ \mathbf{L}_2 \left(\frac{d^4 v_1}{dz^4} \phi^{(1,2)} + \frac{d^4 v_2}{dz^4} \phi^{(2,2)} + \frac{d^2 v_3}{dz^2} \phi^{(3,0)} + \frac{d^2 v_4}{dz^2} \phi^{(4,0)} \right)$$

for $\zeta \in g$, and boundary conditions

$$\mathbf{T}_0 \mathbf{U} = \mathcal{P}(\zeta, z) - \mathbf{T}_1 \left(\frac{d^4 v_1}{dz^4} \phi^{(1,3)} + \frac{d^4 v_2}{dz^4} \phi^{(2,3)} \right.$$
$$\left. + \frac{d^2 v_3}{dz^2} \phi^{(3,1)} + \frac{d^2 v_4}{dz^2} \phi^{(4,1)} \right), \quad \zeta \in \partial g, \qquad (4.3.19)$$

which is solvable for every $z \in [0,l]$. Here $\mathbf{L}_1$, $\mathbf{L}_2$ and $\mathbf{T}_0$, $\mathbf{T}_1$ are matrix operators defined by (3.6.7)–(3.6.9). The functions $\phi^{(k,j)}$ are defined in Section 3.6.2.

Proof

Applying the operators

$$\mathbf{L}(\partial_\mathbf{x}) = \varepsilon^{-2} \mathbf{L}_0(\partial_\zeta) + \varepsilon^{-1} \mathbf{L}_1(\partial_\zeta) \frac{d}{dz} + \mathbf{L}_2 \frac{d^2}{dz^2}, \qquad (4.3.20)$$

$$\boldsymbol{\sigma}^{(n)}(\cdot; \mathbf{x}) = \varepsilon^{-1} \mathbf{T}_0(\mathbf{n}, \partial_\zeta) + \mathbf{T}_1(\mathbf{n}) \frac{d}{dz}, \qquad (4.3.21)$$

(compare with (3.6.3), (3.6.4)) to the vector $\mathfrak{U}$ and using equalities (3.6.15), (3.6.16) we derive

$$\mathbf{L}(\partial_\mathbf{x})\mathfrak{U} = \varepsilon^{-2} \sum_{k=1}^{2} \left\{ \frac{d^4 v_k}{dz^4} (\mathbf{L}_2 \phi^{(k,2)} + \mathbf{L}_1(\partial_\zeta)\phi^{(k,3)}) \right.$$
$$\left. + \frac{d^2 v_{2+k}}{dz^2} (\mathbf{L}_2 \phi^{(2+k,0)} + \mathbf{L}_1(\partial_\zeta)\phi^{(2+k,1)}) \right\}$$
$$+ \varepsilon^{-2} \mathbf{L}_0 \mathbf{U} + \varepsilon^{-1} \mathcal{F}^{(0)} + \mathcal{F}^{(1)}, \qquad (4.3.22)$$

where

$$\mathcal{F}^{(0)} = \sum_{k=1}^{2} \left\{ \frac{d^5 v_k}{dz^5} \mathbf{L}_2 \phi^{(k,3)} + \frac{d^3 v_{k+2}}{dz^3} \mathbf{L}_2 \phi^{(k+2,1)} \right\} + \mathbf{L}_1 \mathbf{U},$$

and
$$\mathcal{F}^{(1)} = \mathbf{L}_2 \mathbf{U}.$$

Similarly,

$$\sigma^{(n)}(\mathfrak{U}; \mathbf{x}) = \varepsilon^{-1} \mathbf{T}_1(\mathbf{n}) \sum_{k=1}^{2} \left\{ \frac{d^4 v_k}{dz^4} \phi^{(k,3)} + \frac{d^2 v_{k+2}}{dz^2} \phi^{(k+2,1)} \right\}$$
$$+ \varepsilon^{-1} \mathbf{T}_0(\mathbf{n}, \partial_\zeta) \mathbf{U} + \mathcal{P}^{(0)}, \qquad (4.3.23)$$

where
$$\mathcal{P}^{(0)} = \mathbf{T}_1(\mathbf{n}) \mathbf{U}.$$

Substituting (4.3.22), (4.3.23) into the system (4.3.13) and the traction boundary condition (4.3.14) and equating the coefficients near ε^{-2}, ε^{-1}, respectively, we arrive at the two-dimensional boundary value problem (4.3.18), (4.3.19) which must be satisfied for every $z \in [0,l]$. Problem (4.3.18), (4.3.19) is solvable if and only if (4.2.46) and (4.2.47) are valid. This is equivalent to the orthogonality of the right-hand sides and the vectors $\psi^{(j)}$, $j = 1, 2, 3, 4$, in $(L_2(g))^3 \times (L_2(\partial g))^3$, for every $z \in [0,l]$. By the biorthogonality condition (3.6.28), this is equivalent to (4.3.15)–(4.3.17). □

In order to provide the uniqueness of $\mathbf{U}$, we assume

$$\int_g \mathbf{U}(\zeta, z) d\zeta = 0, \quad \int_g \left(\zeta_1 U_2^{(0)}(\zeta, z) - \zeta_2 U_1^{(0)}(\zeta, z) \right) d\zeta = 0. \quad (4.3.24)$$

REMARK 4.2

It is appropriate to mention that the right-hand sides (4.3.13)–(4.3.19) are specified uniquely by $\mathcal{F}$ and $\mathcal{P}$. The vector function $\mathbf{U}$ is defined uniquely by $\mathcal{F}$ and $\mathcal{P}$, whereas for functions v_k, $k = 1, 2, 3, 4$, we still need the boundary conditions at the ends of the segment $[0, l]$.

4.3.4 Expansions for displacement in a thin rod

The field $\mathfrak{U}^{(m,j)}$ is a special solution of the elasticity problem in a thin cylinder described in Theorem 4.6; the vector functions $\mathcal{F}$, $\mathcal{P}$

in (4.3.13) and (4.3.14) should be replaced by $\tilde{\mathcal{F}}^{(m,j)}$, $\tilde{\mathcal{P}}^{(m,j)}$, absorbing $\mathcal{F}^{(m,j)}$, $\mathcal{P}^{(m,j)}$ and discrepancy terms in the right-hand sides (4.3.13), (4.3.14) which occur on the previous steps of the asymptotic algorithm. Therefore, from Remark 4.1, $\tilde{\mathcal{F}}^{(m,j)}$, $\tilde{\mathcal{P}}^{(m,j)}$ depend on $\mathcal{F}^{(k,j)}$, $\mathcal{P}^{(k,j)}$, $k \le m$, only.

The functions $v_k^{(m,j)}$ in the expansion (4.3.12) for the field $\mathfrak{U}^{(m,j)}$ satisfy the ordinary differential equations (4.3.15)–(4.3.17) on $(0, l_j)$, and the vector function $\mathbf{U}^{(k,j)}$ can be found as a solution of the boundary value problem (4.3.18), (4.3.19). In all these equations the vector functions $\mathcal{F}, \mathcal{P}$ should be replaced by $\tilde{\mathcal{F}}^{(m,j)}$, $\tilde{\mathcal{P}}^{(m,j)}$.

The boundary conditions for $v_k^{(m,j)}$ will be derived in the sections related to the analysis of the junction and the bottom layers. To this end we shall need the following expansion of $\mathfrak{U}^{(m,j)}$ in the vicinity of the end point $z = 0$:

$$\mathfrak{U}^{(m,j)}(\zeta, z, \varepsilon) = \sum_{k=1}^{4} \left\{ \varepsilon^{-m_k} v_k^{(m,j)}(0) \psi^{(k)}(\mathbf{X}) \right.$$

$$\left. + \varepsilon^{-1} \frac{d^{m_k - 1} v_k^{(m,j)}}{dz^{m_k - 1}}(0) \boldsymbol{\xi}^{(k)}(\mathbf{X}) \right\}$$

$$+ \sum_{k=1}^{2} \left\{ \varepsilon^{-3} \frac{dv_k^{(m,j)}}{dz}(0) \psi^{(4+k)}(\mathbf{X}) + \varepsilon^{-3} \frac{d^2 v_k^{(m,j)}}{dz^2}(0) \boldsymbol{\xi}^{(4+k)}(\mathbf{X}) \right\}$$

$$+ \tilde{\mathfrak{U}}^{(m,j)}(\zeta, X_3, \varepsilon), \qquad (4.3.25)$$

which follows from (4.3.11). Here $\psi^{(k)}$, $\boldsymbol{\xi}^{(k)}$ and m_k are the same as in Section 3.6; $\tilde{\mathfrak{U}}^{(m,j)}(\zeta, X_3, \varepsilon)$ is smooth with respect to X_3, ε, and belongs to $(H^1(g))^3$ as a function of ζ.

Near the end point $z = l_j$ of a thin rod $\Pi_\varepsilon^{(j)}$ one can use the expansion of $\mathfrak{U}^{(m,j)}$ in scaled coordinates $\mathbf{Y}$. This expansion is similar to (4.3.25) and has the form

$$\mathfrak{U}^{(m,j)}(\zeta, z, \varepsilon) = \sum_{k=1}^{4} \left\{ \varepsilon^{-m_k} v_k^{(m,j)}(l_j) \psi^{(k)}(\mathbf{Y}) \right.$$

$$\left. + \varepsilon^{-1} \frac{d^{m_k - 1} v_k^{(m,j)}}{dz^{m_k - 1}}(l_j) \boldsymbol{\xi}^{(k)}(\mathbf{Y}) \right\}$$

$$+ \sum_{k=1}^{2} \left\{ \varepsilon^{-3} \frac{dv_k^{(m,j)}}{dz}(l_j) \psi^{(4+k)}(\mathbf{Y}) + \varepsilon^{-3} \frac{d^2 v_k^{(m,j)}}{dz^2}(l_j) \boldsymbol{\xi}^{(4+k)}(\mathbf{Y}) \right\}$$

$$+ \tilde{\mathfrak{V}}^{(m,j)}(\zeta, Y_3, \varepsilon), \qquad (4.3.26)$$

where $\tilde{\mathfrak{V}}^{(m,j)}$ is smooth with respect to Y_3, ε, and belongs to $(H^1(g))^3$ as a function of ζ. The vector functions $\tilde{\mathfrak{U}}^{(m,j)}$, $\tilde{\mathfrak{V}}^{(m,j)}$ include high-order derivatives of the functions $v_k^{(m,j)}$, $k = 1, 2, 3, 4$, $\mathbf{U}^{(m,j)}$, and, therefore, depend on the $\mathcal{F}^{(k,j)}$, $\mathcal{P}^{(k,j)}$, $k \leq m$.

4.3.5 Junction layer

We shall use the notation $\mathbf{T}(\mathbf{x}, \partial_\mathbf{x})$ for the first-order boundary differential operator specified by

$$\mathbf{T}(\mathbf{x}, \partial_\mathbf{x})\mathbf{v} = \sigma^{(n)}(\mathbf{v}; \mathbf{x}),$$

and let $[A, B] = AB - BA$ denote the commutator of operators A and B.

The junction layer $\mathcal{W}^{(m,j)}$ should compensate the right-hand sides $\mathbf{F}^{(m,j)}$, $\mathbf{P}^{(m,j)}$ and the discrepancy terms caused by the fields in Ω and in thin cylinders.

First, we compensate the error terms brought by the vector functions $\tilde{\mathfrak{U}}^{(k,j)}$, $k \leq m$, in (4.3.25), by $\mathbf{u}_s^{(m)}$ and by the bounded (near the junction points) part of $\mathbf{u}_\Omega^{(k)}$ with $k < m$ (see (4.3.8), (4.3.7)). This can be done by solution of the junction layer problem

$$\begin{pmatrix} -\mathbf{L}(\partial_\mathbf{x}) \\ \mathbf{T}(\mathbf{X}, \partial_\mathbf{X}) \end{pmatrix} \mathcal{W}^{(m,j)}(\mathbf{X}) = \begin{pmatrix} \tilde{\mathbf{F}}^{(m,j)}(\mathbf{X}) \\ \tilde{\mathbf{P}}^{(m,j)}(\mathbf{X}) \end{pmatrix} \qquad (4.3.27)$$

$$+ \begin{pmatrix} [\mathbf{L}(\partial_\mathbf{x}), \chi(\mathbf{X})] \\ -[\mathbf{T}(\mathbf{X}, \partial_\mathbf{X}), \chi(\mathbf{X})] \end{pmatrix} \Big\{ \mathcal{M}(\mathbf{X}, 0) \mathfrak{F}^{(m-1,j)}$$

$$+ \frac{1}{2} (\nabla_{\mathbf{X}^*} \times \mathcal{M}^t(\mathbf{X}, \mathbf{X}^*)|_{\mathbf{X}^* = 0})^t \mathfrak{M}^{(m,j)} \Big\}$$

$$+ \begin{pmatrix} [\mathbf{L}(\partial_\mathbf{x}), \eta(X_3)] \\ -[\mathbf{T}(\mathbf{X}, \partial_\mathbf{X}), \eta(X_3)] \end{pmatrix} \sum_{k=1}^{6} \Big\{ d_k^{(I,m,j)} \xi^{(k)}(\mathbf{X}) + d_k^{(II,m,j)} \psi^{(k)}(\mathbf{X}) \Big\},$$

on $G \times \partial G$.

The vector-valued functions $\tilde{\mathbf{F}}^{(m,j)}$, $\tilde{\mathbf{P}}^{(m,j)}$ absorb $\mathbf{F}^{(m,j)}$, $\mathbf{P}^{(m,j)}$ and two kinds of discrepancy terms: first, terms that come from the expansion (4.3.25) and contain $\tilde{\mathfrak{U}}^{(k,j)}$, $k \leq m - 2$, and their derivatives, and, consequently, $\mathcal{F}^{(k,j)}$, $\mathcal{P}^{(k,j)}$ for $k \leq m - 2$; second, terms that come from the expansion of the bounded part of $\mathbf{u}_\Omega^{(k)}$, $k \leq m - 2$, at junction points.

Using Theorem 4.3 we find that there exists a unique set

$$\{\mathcal{W}^{(m,j)}; d_1^{(I,m,j)}, \ldots, d_6^{(I,m,j)}; d_1^{(II,m,j)}, \ldots, d_6^{(II,m,j)}\}, \qquad (4.3.28)$$

such that the vector function $\mathcal{W}^{(m,j)}$ satisfies the conditions of decay from Theorem 4.3.

The last term in (4.3.27) should be compensated by the discrepancy that comes from the first two sums of the expansion (4.3.25) multiplied by the cut-off function η. Thus, we relate the values $v_k^{(s,j)}$ and their derivatives to the coefficients $d_k^{(i,m,j)}$, $k = 1, \ldots, 6$; $i = I, II$. Using (4.2.21)–(4.2.25) we deduce that

$$v_q^{(m+m_q-2,j)}(0) = d_q^{(II,m,j)}, \quad q = 1, 2, 3, 4; \qquad (4.3.29)$$

$$\frac{dv_q^{(m+1,j)}}{dz}(0) = d_{4+q}^{(II,m,j)}, \quad q = 1, 2; \qquad (4.3.30)$$

$$\frac{d^3 v_q^{(m-1,j)}}{dz^3}(0) = d_q^{(I,m,j)} = D_q^{-1}\Big\{-\mathfrak{F}_q^{(m-1,j)} + \mathbb{F}_q(\tilde{\mathbf{F}}^{(m,j)}, \tilde{\mathbf{P}}^{(m,j)}, G)\Big\}, \quad \text{for } q = 1, 2, \qquad (4.3.31)$$

$$\frac{dv_3^{(m-1,j)}}{dz}(0) = d_3^{(I,m,j)}$$
$$= -D_3^{-1}\Big\{-\mathfrak{F}_3^{(m-1,j)} + \mathbb{F}_3(\tilde{\mathbf{F}}^{(m,j)}, \tilde{\mathbf{P}}^{(m,j)}, G)\Big\}, (4.3.32)$$

$$\frac{dv_4^{(m-1,j)}}{dz}(0) = d_4^{(I,m,j)}$$
$$= -D_4^{-1}\Big\{-\mathfrak{M}_3^{(m,j)} + \mathbb{M}_3(\tilde{\mathbf{F}}^{(m,j)}, \tilde{\mathbf{P}}^{(m,j)}, G)\Big\}, \qquad (4.3.33)$$

$$\frac{d^2 v_q^{(m,j)}}{dz^2}(0) = d_{4+q}^{(I,m,j)}$$
$$= (-1)^q D_q^{-1}\Big\{-\mathfrak{M}_{3-q}^{(m,j)} + \mathbb{M}_{3-q}(\tilde{\mathbf{F}}^{(m,j)}, \tilde{\mathbf{P}}^{(m,j)}, G)\Big\},$$
$$\text{for } q = 1, 2, \qquad (4.3.34)$$

where the functionals $\mathbb{F}$, $\mathbb{M}$ were defined in (4.2.5) and (4.2.6).

4.3.6 Displacement in Ω

The vector function $\mathbf{u}_\Omega^{(m)}$ from (4.3.7) has to compensate the right-hand sides $\mathbf{f}^{(m)}$, $\mathbf{p}^{(m)}$ and the discrepancies caused by the junction layer fields. It satisfies the boundary value problem

$$\begin{pmatrix} -\mathbf{L}(\partial_\mathbf{x}) \\ \mathbf{T}(\mathbf{x},\partial_\mathbf{x}) \end{pmatrix} \mathbf{u}_\Omega^{(m)}(\mathbf{x}) = \begin{pmatrix} \tilde{\mathbf{f}}^{(m)}(\mathbf{x}) \\ \tilde{\mathbf{p}}^{(m)}(\mathbf{x}) \end{pmatrix} \qquad (4.3.35)$$

$$+ \sum_{j=1}^{K} \begin{pmatrix} [\mathbf{L}(\partial_\mathbf{x}), \Xi(\mathbf{x} - \mathbf{a}^{(j)})] \\ -[\mathbf{T}(\mathbf{x},\partial_\mathbf{x}), \Xi(\mathbf{x} - \mathbf{a}^{(j)})] \end{pmatrix} \Big\{ \sum_{q=1}^{2} b_q^{(m,j)} \partial_{x_q} \boldsymbol{\mathcal{B}}^{(q)}(\mathbf{x} - \mathbf{a}^{(j)})$$

$$+ \frac{1}{2} b_3^{(m,j)} (\partial_{x_1} \boldsymbol{\mathcal{B}}^{(2)}(\mathbf{x} - \mathbf{a}^{(j)}) + \partial_{x_2} \boldsymbol{\mathcal{B}}^{(1)}(\mathbf{x} - \mathbf{a}^{(j)})) \Big\} \text{ in } \begin{pmatrix} \Omega \\ \partial\Omega \setminus \cup_j \mathbf{a}^{(j)} \end{pmatrix},$$

where $b_q^{(m,j)}$ are the coefficients from the expansion (4.2.17) for the function $\mathcal{W}^{(m,j)}$; the principal force and moment vectors for the last sum in (4.3.35) are equal to zero. The vectors $\tilde{\mathbf{f}}^{(m)}$, $\tilde{\mathbf{p}}^{(m)}$ are given by the sum of $\mathbf{f}^{(m)}$, $\mathbf{p}^{(m)}$ and the self-balanced junction layer discrepancies from the previous steps of the asymptotic algorithm. These error terms are brought by the commutators of the cut-off functions and differential operators of the Lamé system and traction boundary conditions applied to corresponding terms of order $O(\|\mathbf{X}\|^{-k})$, $k \geq 3$, of the asymptotic expansion of the junction layer $\mathcal{W}^{(k,j)}$, $k < m$, in a half-space at infinity. Therefore, the principal force and moment vectors for the right-hand side of (4.3.35) are the same as for $\mathbf{f}^{(m)}$, $\mathbf{p}^{(m)}$.

By Theorem 4.1, for the constant vectors $\mathfrak{F}^{(m,j)}$, $\mathfrak{M}^{(m,j)}$, satisfying the balance relations (4.3.9), (4.3.10), there exists a unique solution of the problem (4.3.35) which admits the representation (4.3.7) where the regular part is orthogonal to the rigid-body rotations and translations in $(L_2(\Omega))^3$. Since the right-hand side (4.3.35) is smooth in the vicinity of the junction points, the regular part $\mathbf{u}_r^{(m)}$ is also smooth there (see (4.2.10), (4.2.11)).

4.3.7 Bottom layer

The special solution $\mathfrak{U}^{(m,j)}$ of the elasticity problem in a thin cylinder and the rigid-body translations and rotations from the expansion (4.3.6) give the error terms in the Dirichlet boundary condition on $S_\varepsilon^{(j)}$. Hence, using (4.3.26) we find that the boundary value problem for the bottom layer has the form

$$-\mathbf{L}(\partial_\mathbf{Y})\mathcal{V}^{(m,j)}(\mathbf{Y}) = \mathcal{G}^{(m,j)}(\mathbf{Y}), \; \mathbf{Y} \in \Pi_-, \qquad (4.3.36)$$

$$\mathbf{T}(\mathcal{V}^{(m,j)}; \mathbf{Y}) = \mathbf{H}^{(m,j)}(\mathbf{Y}), \; \mathbf{Y} \in \partial\Pi_- \setminus S, \qquad (4.3.37)$$

$$\mathcal{V}^{(m,j)}(\mathbf{Y}) = \tilde{\boldsymbol{\Phi}}^{(m,j)}(\mathbf{Y}) - \sum_{k=1}^{2} v_k^{(m+2,j)}(l_j)\boldsymbol{\psi}^{(k)} - v_3^{(m,j)}(l_j)\boldsymbol{\psi}^{(3)}$$

$$- v_4^{(m,j)}(l_j)\psi^{(4)}(\mathbf{Y}) - \sum_{k=1}^{2}\left\{\frac{dv_k^{(m+1,j)}}{dz}(l_j)\psi^{(4+k)}(\mathbf{Y}) + \alpha_k^{(m+2)}\mathbf{e}^{(k)}\right\}$$

$$- \alpha_3^{(m)}\mathbf{e}^{(3)} - \left(\sum_{k=1}^{2}\beta_k^{(m)}\mathbf{e}^{(k)} + \beta_3^{(m+2)}\mathbf{e}^{(3)}\right) \times (l_j\mathbf{e}^{(3)} + a_1^{(j)}\mathbf{e}^{(1)} + a_2^{(j)}\mathbf{e}^{(2)})$$

$$- \left(\sum_{k=1}^{2}\beta_k^{(m-1)}\mathbf{e}^{(k)} + \beta_3^{(m+1)}\mathbf{e}^{(3)}\right) \times \mathbf{Y}, \ \mathbf{Y} \in S, \qquad (4.3.38)$$

where $\tilde{\boldsymbol{\Phi}}^{(m,j)}$ involves $\boldsymbol{\Phi}^{(m,j)}$ and traces on S_ε of the functions $v_1^{(k,j)}$, $v_2^{(k,j)}$, $k \leq m$, $v_3^{(k,j)}, v_4^{(k,j)}$, $k \leq m-1$, and the vector functions $\mathfrak{V}^{(k,j)}$, $k \leq m-2$, and their derivatives which correspond to the terms in (4.3.26), different from the rigid-body displacements (multipliers of $\psi^{(j)}$, $j = 1,\ldots,6$).

By Theorem 4.5, there exists a unique set

$$\left\{\boldsymbol{\mathcal{V}}^{(m,j)};\ v_k^{(m+2,j)}(l_j),\ \frac{dv_k^{(m+1,j)}}{dz}(l_j), v_{k+2}^{(m,j)}(l_j),\ k=1,2\right\}, \qquad (4.3.39)$$

which satisfies the problem (4.3.36)–(4.3.38), and $\boldsymbol{\mathcal{V}}^{(m,j)}$ decays exponentially at infinity.

One can represent the constant components from (4.3.39) in the form

$$v_1^{(m+2,j)}(l_j) = -(\alpha_1^{(m+2)} + l_j\beta_2^{(m)}$$
$$- \beta_3^{(m+2)}a_2^{(j)}) + c_1^{(m,j)}, \qquad (4.3.40)$$
$$v_2^{(m+2,j)}(l_j) = -(\alpha_2^{(m+2)} + \beta_3^{(m+2)}a_1^{(j)}$$
$$- \beta_1^{(m)}l_j) + c_2^{(m,j)}, \qquad (4.3.41)$$
$$v_3^{(m,j)}(l_j) = -(\alpha_3^{(m)} + \beta_1^{(m)}a_2^{(j)}$$
$$- \beta_2^{(m)}a_1^{(j)}) + c_3^{(m,j)}, \qquad (4.3.42)$$
$$v_4^{(m,j)}(l_j) = -\beta_3^{(m+1)} + c_4^{(m,j)}, \qquad (4.3.43)$$
$$\frac{dv_1^{(m+1,j)}}{dz}(l_j) = -\beta_2^{(m-1)} + c_5^{(m,j)}, \qquad (4.3.44)$$
$$\frac{dv_2^{(m+1,j)}}{dz}(l_j) = \beta_1^{(m-1)} + c_6^{(m,j)}. \qquad (4.3.45)$$

Then the set

$$\{\boldsymbol{\mathcal{V}}^{(m,j)}; c_q^{(m,j)},\ q = 1,2,\ldots,6\} \qquad (4.3.46)$$

is a unique solution of system (4.3.36) with the traction boundary condition (4.3.37) and the following displacement boundary condition:

$$\mathcal{V}^{(m,j)}(\mathbf{Y}) = \tilde{\Phi}^{(m,j)}(Y_1, Y_2) - \sum_{q=1}^{6} c_q^{(m,j)} \psi^{(q)}(\mathbf{Y}), \ \mathbf{Y} \in S. \quad (4.3.47)$$

Therefore, the set (4.3.46) depends on $\mathcal{G}^{(m,j)}$, $\mathbf{H}^{(m,j)}$, $\Phi^{(m,j)}$, $\mathcal{F}^{(k,j)}$, $\mathcal{P}^{(k,j)}$ with $k \leq m-2$, the functions $v_1^{(k,j)}$, $v_2^{(k,j)}$, $k \leq m$, and $v_3^{(k,j)}$, $v_4^{(k,j)}$, $k \leq m-1$.

4.3.8 Functions $v_k^{(m,j)}$

Let $Q_i^{(m,j)}$ denote the right-hand sides of the problems (4.3.15)–(4.3.17) corresponding to $v_i^{(m,j)}$, with $\mathcal{F}_i$, $\mathcal{P}_i$ replaced by $\tilde{\mathcal{F}}_i^{(m,j)}$, $\tilde{\mathcal{P}}_i^{(m,j)}$. Equations (4.3.15) and the boundary conditions (4.3.29), (4.3.30) for $z = 0$ and (4.3.40)–(4.3.45) for $z = l_j$ enable us to find $v_i^{(m,j)}$, $i = 1, 2$, in the form

$$v_1^{(m,j)}(z) = -(\alpha_1^{(m)} + l_j \beta_2^{(m-2)} - a_2^{(j)} \beta_3^{(m)}) \psi_I\left(\frac{z}{l_j}\right)$$
$$- \beta_2^{(m-2)} l_j \psi_{II}\left(\frac{z}{l_j}\right) + w_1^{(m,j)}(z), \quad (4.3.48)$$

$$v_2^{(m,j)}(z) = -(\alpha_2^{(m)} + a_1^{(j)} \beta_3^{(m)} - \beta_1^{(m-2)} l_j) \psi_I\left(\frac{z}{l_j}\right)$$
$$+ \beta_1^{(m-2)} l_j \psi_{II}\left(\frac{z}{l_j}\right) + w_2^{(m,j)}(z), \quad (4.3.49)$$

with ψ_I, ψ_{II} being the same as in Section 4.2.6; the functions $w_i^{(m,j)}$, $i = 1, 2$, are given as the solutions of the boundary value problems

$$\frac{d^4 w_i^{(m,j)}}{dz^4}(z) = Q_i^{(m,j)}(z), \ z \in (0, l_j), \quad (4.3.50)$$

$$w_i^{(m,j)}(0) = d_i^{(II, m-2, j)}, \quad (4.3.51)$$

$$\frac{dw_i^{(m,j)}}{dz}(0) = d_{i+4}^{(II, m-1, j)}, \quad (4.3.52)$$

$$w_i^{(m,j)}(l_j) = c_i^{(m-2,j)}, \quad (4.3.53)$$

$$\frac{dw_i^{(m,j)}}{dz}(l_j) = c_{4+i}^{(m-1,j)}, \ i = 1, 2. \quad (4.3.54)$$

Equations (4.3.16), (4.3.17) and the boundary conditions (4.3.29), (4.3.42), (4.3.43) lead to the representations

$$v_3^{(m,j)}(z) = -(\alpha_3^{(m)} + \beta_1^{(m)} a_2^{(j)} - \beta_2^{(m)} a_1^{(j)})\frac{z}{l_j} + w_3^{(m,j)}, \quad (4.3.55)$$

$$v_4^{(m,j)}(z) = -\beta_3^{(m+1)}\frac{z}{l_j} + w_4^{(m,j)}, \quad (4.3.56)$$

where $w_i^{(m,j)}$, $i = 3, 4$, satisfy the boundary value problems

$$\frac{d^2 w_i^{(m,j)}}{dz^2}(z) = Q_i^{(m)}(z), \ z \in (0, l_j), \quad (4.3.57)$$

$$w_i^{(m,j)}(0) = d_i^{(II,m,j)}, \quad (4.3.58)$$

$$w_i^{(m,j)}(l_j) = c_i^{(m,j)}, \ i = 3, 4. \quad (4.3.59)$$

Calculation of the coefficients $\mathfrak{F}^{(k)}$, $\mathfrak{M}^{(k)}$

Formulae (4.3.31)–(4.3.34) yield the relations between the components of forces and moments and the constants $\alpha_j^{(k)}, \beta_j^{(k)}$:

$$\mathfrak{F}_1^{(m,j)} = \mathbb{F}_1(\tilde{\mathbf{F}}^{(m+1,j)}, \tilde{\mathbf{P}}^{(m+1,j)}; G)$$
$$- D_1 \left\{ \frac{12}{l_j^3}(\alpha_1^{(m)} - a_2^{(j)}\beta_3^{(m)}) + \frac{6\beta_2^{(m-2)}}{l_j^2} \right. \quad (4.3.60)$$

$$\left. + \mathbb{Q}^{(III)}(Q_1^{(m,j)}, d_1^{(II,m-2,j)}, c_1^{(m-2,j)}, d_5^{(II,m-1,j)}, c_5^{(m-1,j)}) \right\},$$

$$\mathfrak{F}_2^{(m,j)} = \mathbb{F}_2(\tilde{\mathbf{F}}^{(m+1,j)}, \tilde{\mathbf{P}}^{(m+1,j)}; G)$$
$$- D_2 \left\{ \frac{12}{l_j^3}(\alpha_2^{(m)} + a_1^{(j)}\beta_3^{(m)}) - \frac{6\beta_1^{(m-2)}}{l_j^2} \right. \quad (4.3.61)$$

$$\left. + \mathbb{Q}^{(III)}(Q_2^{(m,j)}, d_2^{(II,m-2,j)}, c_2^{(m-2,j)}, d_6^{(II,m-1,j)}, c_6^{(m-1,j)}) \right\},$$

$$\mathfrak{M}_3^{(m,j)} = \mathbb{M}_3(\tilde{\mathbf{F}}^{(m,j)}, \tilde{\mathbf{P}}^{(m,j)}; G) + D_4 \left\{ -\frac{\beta_3^{(m)}}{l_j} \right.$$
$$\left. + \mathbb{Q}^{(I)}(Q_4^{(m-1,j)}, d_4^{(II,m-1,j)}, c_4^{(m-1,j)}) \right\}, \quad (4.3.62)$$

$$\mathfrak{F}_3^{(m,j)} = \mathbb{F}_3(\tilde{\mathbf{F}}^{(m+1,j)}, \tilde{\mathbf{P}}^{(m+1,j)}; G)$$

ASYMPTOTIC EXPANSION OF THE SOLUTION

$$+ D_3 \Big\{ -(\alpha_3^{(m)} + a_2^{(j)}\beta_1^{(m)} - a_1^{(j)}\beta_2^{(m)})/l_j$$
$$+ \mathbb{Q}^{(I)}(Q_3^{(m,j)}, d_3^{(II,m,j)}, c_3^{(m,j)}) \Big\}, \qquad (4.3.63)$$

$$\mathfrak{M}_1^{(m,j)} = \mathbb{M}_1(\tilde{\mathbf{F}}^{(m,j)}, \tilde{\mathbf{P}}^{(m,j)}; G)$$
$$- D_2 \Big\{ -\frac{6}{l_j^2}(\alpha_2^{(m)} + a_1^{(j)}\beta_3^{(m)}) + \frac{4\beta_1^{(m-2)}}{l_j}$$
$$+ \mathbb{Q}^{(II)}(Q_2^{(m,j)}, d_2^{(II,m-2,j)}, c_2^{(m-2,j)},$$
$$d_6^{(II,m-1,j)}, c_6^{(m-1,j)}) \Big\}, \qquad (4.3.64)$$

$$\mathfrak{M}_2^{(m,j)} = \mathbb{M}_2(\tilde{\mathbf{F}}^{(m,j)}, \tilde{\mathbf{P}}^{(m,j)}; G)$$
$$+ D_1 \Big\{ -\frac{6}{l_j^2}(\alpha_1^{(m)} - a_2^{(j)}\beta_3^{(m)}) - \frac{4\beta_2^{(m-2)}}{l_j}$$
$$+ \mathbb{Q}^{(II)}(Q_1^{(m,j)}, d_1^{(II,m-2,j)}, c_1^{(m-2,j)},$$
$$d_5^{(II,m-1,j)}, c_5^{(m-1,j)}) \Big\}. \qquad (4.3.65)$$

The functionals $\mathbb{Q}^{(I)}$, $\mathbb{Q}^{(II)}$, $\mathbb{Q}^{(III)}$ were defined in Section 4.2.6 where the length l of the interval on the Oz axis should be replaced by l_j (we hope that this notation will not cause any confusion).

A system of algebraic equations for $\boldsymbol{\alpha}^{(m)}$, $\boldsymbol{\beta}^{(m)}$

Using the balance equations (4.3.9), (4.3.10) we arrive at the system of recurrent relations with respect to components of $\boldsymbol{\alpha}^{(m)}$, $\boldsymbol{\beta}^{(m)}$:

$$\mathcal{A} \begin{pmatrix} \alpha_1^{(m)} \\ \alpha_2^{(m)} \\ \beta_3^{(m)} \end{pmatrix} + \mathcal{D}^t \begin{pmatrix} \alpha_3^{(m-2)} \\ \beta_1^{(m-2)} \\ \beta_2^{(m-2)} \end{pmatrix} = \begin{pmatrix} h_1^{(m)} \\ h_2^{(m)} \\ h_3^{(m)} \end{pmatrix} \qquad (4.3.66)$$

$$\mathcal{D} \begin{pmatrix} \alpha_1^{(m)} \\ \alpha_2^{(m)} \\ \beta_3^{(m)} \end{pmatrix} + \mathcal{B} \begin{pmatrix} \alpha_3^{(m)} \\ \beta_1^{(m)} \\ \beta_2^{(m)} \end{pmatrix} + \mathcal{E} \begin{pmatrix} \alpha_3^{(m-2)} \\ \beta_1^{(m-2)} \\ \beta_2^{(m-2)} \end{pmatrix} = \begin{pmatrix} h_4^{(m)} \\ h_5^{(m)} \\ h_6^{(m)} \end{pmatrix} \qquad (4.3.67)$$

where $\mathcal{A}$, $\mathcal{B}$, $\mathcal{D}$ and $\mathcal{E}$ are matrices from (4.2.59)–(4.2.62). The quantities $h_i^{(m)}$, $i = 1, \ldots, 6$, are

$$h_q^{(m)} = \mathbb{F}_q(\mathbf{f}^{(m)}, \mathbf{p}^{(m)}; \Omega) + \sum_j \Big\{ \mathbb{F}_q(\tilde{\mathbf{F}}^{(m+1,j)}, \tilde{\mathbf{P}}^{(m+1,j)}, G) \quad (4.3.68)$$
$$- D_q \mathbb{Q}^{(III)}(Q_q^{(m,j)}, d_q^{(II,m-2,j)}, c_q^{(m-2,j)}, d_{q+4}^{(II,m-1,j)}, c_{q+4}^{(m-1,j)}) \Big\},$$
for $q = 1, 2$,
$$h_3^{(m)} = \mathbb{M}_3(\mathbf{f}^{(m)}, \mathbf{p}^{(m)}; \Omega) + \sum_j \Big\{ \mathbb{M}_3(\tilde{\mathbf{F}}^{(m,j)}, \tilde{\mathbf{P}}^{(m,j)}; G) \quad (4.3.69)$$
$$+ a_1^{(j)} \mathbb{F}_2(\tilde{\mathbf{F}}^{(m+1,j)}, \tilde{\mathbf{P}}^{(m+1,j)}, G) - a_2^{(j)} \mathbb{F}_1(\tilde{\mathbf{F}}^{(m+1,j)}, \tilde{\mathbf{P}}^{(m+1,j)}, G)$$
$$+ D_4 \mathbb{Q}^{(I)}(Q_4^{(m-1,j)}, d_4^{(II,m-1,j)}, c_4^{(m-1,j)})$$
$$- a_1^{(j)} D_2 \mathbb{Q}^{(III)}(Q_2^{(m,j)}, d_2^{(II,m-2,j)}, c_2^{(m-2,j)}, d_6^{(II,m-1,j)}, c_6^{(m-1,j)})$$
$$+ a_2^{(j)} D_1 \mathbb{Q}^{(III)}(Q_1^{(m,j)}, d_1^{(II,m-2,j)}, c_1^{(m-2,j)}, d_5^{(II,m-1,j)}, c_5^{(m-1,j)}) \Big\},$$
$$h_4^{(m)} = \mathbb{F}_3(\mathbf{f}^{(m)}, \mathbf{p}^{(m)}, \Omega) + \sum_j \Big\{ \mathbb{F}_3(\tilde{\mathbf{F}}^{(m+1,j)}, \tilde{\mathbf{P}}^{(m+1,j)}, G) \quad (4.3.70)$$
$$+ D_3 \mathbb{Q}^{(I)}(Q_3^{(m,j)}, d_3^{(II,m,j)}, c_3^{(m,j)}) \Big\},$$
$$h_5^{(m)} = \mathbb{M}_1(\mathbf{f}^{(m)}, \mathbf{p}^{(m)}; \Omega) + \sum_j \Big\{ \mathbb{M}_1(\tilde{\mathbf{F}}^{(m,j)}, \tilde{\mathbf{P}}^{(m,j)}; G) \quad (4.3.71)$$
$$+ a_2^{(j)} \Big[\mathbb{F}_3(\tilde{\mathbf{F}}^{(m+1,j)}, \tilde{\mathbf{P}}^{(m+1,j)}, G) + D_3 \mathbb{Q}^{(I)}(Q_3^{(m,j)}, d_3^{(II,m,j)}, c_3^{(m,j)}) \Big]$$
$$- D_2 \mathbb{Q}^{(II)}(Q_2^{(m,j)}, d_2^{(II,m-2,j)}, c_2^{(m-2,j)}, d_6^{(II,m-1,j)}, c_6^{(m-1,j)}) \Big\},$$
$$h_6^{(m)} = \mathbb{M}_2(\mathbf{f}^{(m)}, \mathbf{p}^{(m)}; \Omega) + \sum_j \Big\{ \mathbb{M}_2(\tilde{\mathbf{F}}^{(m,j)}, \tilde{\mathbf{P}}^{(m,j)}; G) \quad (4.3.72)$$
$$- a_1^{(j)} \Big[D_3 \mathbb{F}_3(\tilde{\mathbf{F}}^{(m+1,j)}, \tilde{\mathbf{P}}^{(m+1,j)}, G) + \mathbb{Q}^{(I)}(Q_3^{(m,j)}, d_3^{(II,m,j)}, c_3^{(m,j)}) \Big]$$
$$+ D_1 \mathbb{Q}^{(II)}(Q_1^{(m,j)}, d_1^{(II,m-2,j)}, c_1^{(m-2,j)}, d_5^{(II,m-1,j)}, c_5^{(m-1,j)}) \Big\}.$$

We remind the reader that, by assumption, $a_3^{(j)} = 0$, $j = 1, \ldots, K$.

The coefficients $\boldsymbol{\alpha}^{(k)}$, $\boldsymbol{\beta}^{(k)}$ with negative indices are assumed to be equal to zero. The recurrent relations (4.3.67), (4.3.68) together with the equalities (4.3.69)–(4.3.73) for $h_q^{(m)}$ yield the values of the components of $\boldsymbol{\alpha}^{(m)}$, $\boldsymbol{\beta}^{(m)}$ and the components of the lock forces $\mathfrak{F}^{(m,j)}$ and moments $\mathfrak{M}^{(m,j)}$, $j = 1, 2, \ldots, K$.

4.3.9 The recurrent procedure for the asymptotic expansion

To conclude, we shall describe the order of the steps of the asymptotic algorithm.

First, we solve the boundary value problem (4.3.50)–(4.3.54) for $w_k^{(m,j)}$, $k = 1,2$. The knowledge of these functions enables us to find $h_i^{(m)}$, $i = 1, 2, 3$, from the right-hand side of (4.3.67) and, therefore, to obtain the coefficients $\alpha_1^{(m)}$, $\alpha_2^{(m)}$, $\beta_3^{(m)}$ near the rigid-body displacements of Ω.

Using the formulae (4.3.48), (4.3.49) we obtain $v_1^{(m,j)}$, $v_2^{(m,j)}$.

Formulae (4.3.60), (4.3.61) and (4.3.65), (4.3.66) give the transverse components $\mathfrak{F}_1^{(m,j)}, \mathfrak{F}_2^{(m,j)}$ of forces and components $\mathfrak{M}_k^{(m,j)}$, $k = 1, 2, 3$, of moments at junction points.

Next, we obtain the solution

$$\left\{ \mathcal{W}^{(m,j)}, d_1^{(I,m,j)}, \ldots, d_6^{(I,m,j)}; d_1^{(II,m,j)}, \ldots, d_6^{(II,m,j)} \right\}$$

of the junction layer problem (4.3.27).

The bottom layer problem (4.3.36), (4.3.37) and (4.3.47) gives the set

$$\left\{ \mathcal{V}^{(m,j)}; c_q^{(m,j)}, q = 1, 2, \ldots, 6 \right\}.$$

Then, the boundary value problems (4.3.57)–(4.3.59) yield the functions $w_3^{(m,j)}$, $w_4^{(m,j)}$.

Consequently, one can calculate $h_4^{(m)}$, $h_5^{(m)}$, $h_6^{(m)}$ from the right-hand side of (4.3.68) and determine the remaining coefficients $\alpha_3^{(m)}, \beta_1^{(m)}, \beta_2^{(m)}$ near the rigid-body displacements of Ω.

The longitudinal components $\mathfrak{F}_3^{(m,j)}$ of forces at junction points are calculated in accordance with (4.3.64). Knowing the components of forces and moments at junction points one can find the singular part $\mathbf{u}_s^{(m)}$ of the displacement in Ω (see (4.3.8), $k = m$).

In accordance with Section 4.3.6, we obtain the regular term of the field $\mathbf{u}_\Omega^{(m)}$ in Ω, and, consequently, complete the evaluation of $\mathbf{u}_\Omega^{(m)}$.

Following the formulae (4.3.55), (4.3.56) we determine the functions $v_3^{(m,j)}$ completely, and $v_4^{(m,j)}$ up to a linear term $\beta_3^{(m,j)} z/l_j$. The boundary value problem (4.2.7), (4.2.8) with $\mathbf{U}$, $v_k, \mathcal{F}, \mathcal{P}$ being replaced by $\mathbf{U}^{(m,j)}, v_k^{(m,j)}, \tilde{\mathcal{F}}^{(m,j)}, \tilde{\mathcal{P}}^{(m,j)}$, respectively, gives the vector function $\mathbf{U}^{(m,j)}$. Thus, the displacement field $\mathfrak{U}^{(m,j)}$ in a thin cylinder has been determined up to the constant $\beta_3^{(m+1)}$. Repeating the first two steps of the iterative procedure with m replaced by $m+1$ we obtain the constants $\alpha_1^{(m+1)}, \alpha_2^{(m+1)}, \beta_3^{(m+1)}$ and complete the construction of the coefficient $\mathbf{u}^{(m)}$ in the asymptotic expansion (4.3.5).

LEMMA 4.4

Substitution of

$$s_m = \sum_{j=-1}^{m} \varepsilon^j \mathbf{u}^{(j)}(\mathbf{x}, \varepsilon) \qquad (4.3.73)$$

into the equations (4.1.1)–(4.1.3) *gives the right-hand sides*

$$\mathbf{F} - \mathbf{F}_m, \quad \mathbf{P} - \mathbf{P}_m, \quad \mathbf{\Phi}^{(k)} - \mathbf{\Phi}_m^{(k)},$$

where the terms $\mathbf{F}_m, \mathbf{P}_m, \mathbf{\Phi}_m^{(k)}$ *admit the asymptotic representation of the form* (4.3.1)–(4.3.3) *with* $k \geq m+1$.

Proof

Since the coefficients $\mathbf{u}^{(k)}$ have the form (4.3.6), the differential operators of the problem (4.1.1)–(4.1.3) applied to the sum s_m give the right-hand sides of the form (4.3.1)–(4.3.3). From the recurrent procedure just described we find that the first m terms of the expansions of the right-hand sides are the same as in (4.3.1)–(4.3.3). □

4.4 Justification of the asymptotic expansion

4.4.1 Korn's inequality in Ω_ε

THEOREM 4.7

For any $\mathbf{u} \in (H^1(\Omega_\varepsilon))^3$ *the following estimate*

$$\|\mathbf{u}\|_{(H^1(\Omega_\varepsilon))^3}^2 \leq \text{Const } \varepsilon^{-4} \Big\{ \mathcal{E}(\mathbf{u}; \Omega_\varepsilon)$$

$$+ \varepsilon^{-1} \sum_{j=1}^{K} \int_{S_\varepsilon^{(j)}} \|\mathbf{u}(\mathbf{x}', l^{(j)})\|^2 d\mathbf{x}' \Big\} \qquad (4.4.1)$$

holds.

Proof

Let $\pi_\varepsilon = \{\mathbf{x} : \varepsilon^{-1}\mathbf{x}' \in g,\ 0 < x_3 < 1\}$. First, we prove the inequality

$$\int_{\pi_\varepsilon} [|u_3|^2 + \varepsilon^2(\|\nabla' u_3\|^2 + \|\partial_{x_3}\mathbf{u}'\|^2 + \|\mathbf{u}'\|^2) + \varepsilon^4 \|\nabla'\mathbf{u}'\|^2] d\mathbf{x}$$

$$\leq C\Big(\mathcal{E}(\mathbf{u};\pi_\varepsilon) + \varepsilon^{-1} \int_{\varepsilon g} \|\mathbf{u}(\mathbf{x}',0)\|^2 d\mathbf{x}'\Big), \qquad (4.4.2)$$

where C does not depend on ε and $\mathbf{u}$.

We shall make use of Lemma 3.10. Taking l to be ε^{-1} and substituting $\varepsilon^{-1}\mathbf{x}$ instead of $\mathbf{x}$ we derive inequality (4.4.2).

Assume that $\Pi_{\varepsilon,\delta}^{(j)}$ are the same thin cylinders as in the proof of Lemma 2.1. By (4.4.2)

$$\|\mathbf{u}\|^2_{(H^1(\Pi_{\varepsilon,\delta}^{(j)}))^3} + \varepsilon^{-2}\|\mathbf{u}\|^2_{(L_2(\Pi_{\varepsilon,\delta}^{(j)}))^3}$$

$$\leq \text{Const } \varepsilon^{-4} \Big\{ \mathcal{E}(\mathbf{u}; \Pi_{\varepsilon,\delta}^{(j)}) + \varepsilon^{-1} \int_{S_\varepsilon^{(j)}} \|\mathbf{u}(\mathbf{x}', l^{(j)})\|^2 d\mathbf{x}' \Big\}. \qquad (4.4.3)$$

We adopt the notation $\omega = \Omega \cap (\cup_j \Pi_{\varepsilon,\delta}^{(j)})$. Since the junction points $\mathbf{a}^{(j)}$, $j = 1,\ldots,K$, do not belong to the same straight line, one can estimate the constants α_ω and β_ω in (3.8.7) and (3.8.10), with $D = \Omega$, by Const ε^{-2}. Applying Lemmas 3.6 and 3.7 we obtain

$$\|\mathbf{u}\|_{(H^1(\Omega))^3} \leq \text{Const } \varepsilon^{-2} \Big\{ \mathcal{E}(\mathbf{u}; \Omega) + \int_\omega \|\mathbf{u}\|^2 d\mathbf{x} \Big\}. \qquad (4.4.4)$$

Estimate (4.4.1) follows from (4.4.3) and (4.4.4). □

4.4.2 An estimate for the solution

THEOREM 4.8

Let $\mathbf{F} \in (L_2(\Omega_\varepsilon))^3$; $\mathbf{P} \in (L_2(\partial\Omega \setminus S_\varepsilon))^3$ and $\mathbf{\Phi}^{(j)} \in (H^{1/2}(S_\varepsilon^{(j)}))^3$, $j = 1,\ldots,K$. Then the problem (4.1.1)–(4.1.3) has a unique solution in the class $(H^1(\Omega_\varepsilon))^3$, and

$$\|\mathbf{u}\|_{(H^1(\Omega_\varepsilon))^3} \leq \text{Const } \varepsilon^{-4} \Big\{ \|\mathbf{F}\|_{(L_2(\Omega_\varepsilon))^3}$$

$$+ \varepsilon^{-1} \|\mathbf{P}\|_{(L_2(\partial\Omega_\varepsilon \setminus S_\varepsilon))^3} + \varepsilon^2 \sum_j \|\mathbf{\Phi}^{(j)}\|_{(H^{1/2}(S_\varepsilon))^3} \Big\}, \qquad (4.4.5)$$

where the constant does not depend on ε.

Proof

The uniqueness of the solution follows from the Betti formula (3.1.9) (when $\mathbf{u} = \mathbf{v}$). Next we prove the solvability and the estimate (4.4.5). Let $\mathbf{u}$ be represented by

$$\mathbf{u} = \mathbf{u}^{(1)} + \mathbf{u}^{(2)},$$

where $\mathbf{u}^{(1)}$ solves the problem (4.1.1)–(4.1.3) with $\mathbf{\Phi}^{(j)} = 0$, $j = 1, \ldots, K$, and $\mathbf{u}^{(2)}$ is a solution of (4.1.1)–(4.1.3) where $\mathbf{F} = \mathbf{P} = 0$.

Existence and estimate of $\mathbf{u}^{(1)}$. Let $\mathcal{H}$ denote the subspace of $(H^1(\Omega_\varepsilon))^3$ whose elements vanish at S_ε. Then, by Lemma 3.5(ii), the functional

$$\mathbf{v} \to (\mathcal{E}(\mathbf{v}; \Omega_\varepsilon))^{1/2}$$

is an equivalent norm on $\mathcal{H}$. By (3.1.9)

$$\sum_{i,j=1}^3 \int_{\Omega_\varepsilon} \sigma_{ij}(\mathbf{u}^{(1)}) \varepsilon_{ij}(\mathbf{v}) dx = - \int_{\Omega_\varepsilon} \mathbf{F} \cdot \mathbf{v} dx + \int_{\partial\Omega_\varepsilon \setminus S_\varepsilon} \mathbf{P} \cdot \mathbf{v} ds \qquad (4.4.6)$$

for any $\mathbf{v} \in \mathcal{H}$. The right-hand side in (4.4.6) is a continuous functional on $\mathcal{H}$. Thus, by the Riesz theorem there exists a solution $\mathbf{u}^{(1)}$ of (4.4.6). Taking $\mathbf{v} = \mathbf{u}^{(1)}$ and using Cauchy's inequality we derive that

$$\mathcal{E}(\mathbf{u}^{(1)}; \Omega_\varepsilon) \le \text{Const} \Big\{ \|\mathbf{F}\|_{(L_2(\Omega_\varepsilon))^3} \|\mathbf{u}^{(1)}\|_{(L_2(\Omega_\varepsilon))^3}$$

$$+ \|\mathbf{P}\|_{(L_2(\partial\Omega_\varepsilon \setminus S_\varepsilon))^3} \|\mathbf{u}^{(1)}\|_{(L_2(\partial\Omega_\varepsilon \setminus S_\varepsilon))^3} \Big\}. \qquad (4.4.7)$$

On the cross-section of a thin cylinder we have the following estimate:

$$\int_{\partial g_\varepsilon} \|\mathbf{u}^{(1)}\|^2 ds \le \text{Const} \Big\{ \varepsilon \int_{g_\varepsilon} \|\nabla' \mathbf{u}^{(1)}\|^2 dx + \varepsilon^{-1} \int_{g_\varepsilon} \|\mathbf{u}^{(1)}\|^2 dx \Big\}$$

which follows from the trace theorem for $\varepsilon = 1$ and scaling of the variables of integration. Integrating this inequality along the cylinder axis and using the trace theorem for Ω we obtain

JUSTIFICATION OF THE EXPANSION

$$\|\mathbf{u}^{(1)}\|_{(L_2(\partial\Omega_\varepsilon\setminus S_\varepsilon))^3} \leq \text{Const } \varepsilon^{-1}\|\mathbf{u}^{(1)}\|_{(H^1(\Omega_\varepsilon))^3}.$$

We also note that this inequality is a direct consequence of Lemma 2.7. This inequality and Korn's inequality (4.4.1) together with (4.4.7) yield

$$\|\mathbf{u}\|_{(H^1(\Omega_\varepsilon))^3} \leq \text{Const } \varepsilon^{-4}\left\{\|\mathbf{F}\|_{(L_2(\Omega_\varepsilon))^3} + \varepsilon^{-1}\|\mathbf{P}\|_{(L_2(\partial\Omega_\varepsilon\setminus S_\varepsilon))^3}\right\}. \quad (4.4.8)$$

Existence and estimate of $\mathbf{u}^{(2)}$. There exists a vector function $\mathbf{H} \in (H^1(\Omega_\varepsilon))^3$ with the support in thin cylinders $\Pi_\varepsilon^{(i)}, i = 1,\ldots,K$, such that

$$\mathbf{H}|_{S_\varepsilon^{(i)}} = \mathbf{\Phi}^{(i)},$$

and

$$\|\mathbf{H}\|_{H^1(\Omega_\varepsilon)} \leq \text{Const} \sum_{i=1}^{K} \|\mathbf{\Phi}^{(i)}\|_{H^{1/2}(S_\varepsilon^{(i)})} \quad (4.4.9)$$

(compare with (2.6.16)). We seek $\mathbf{u}^{(2)}$ in the form

$$\mathbf{u}^{(2)} = \mathbf{H} + \mathbf{W}.$$

Then,

$$\sum_{i,j=1}^{3}\int_{\Omega_\varepsilon}\sigma_{ij}(\mathbf{W})\varepsilon_{ij}(\mathbf{V})d\mathbf{x} = -\sum_{i,j=1}^{3}\int_{\Omega_\varepsilon}\sigma_{ij}(\mathbf{H})\varepsilon_{ij}(\mathbf{V})d\mathbf{x}, \quad (4.4.10)$$

for all $\mathbf{V} \in \mathcal{H}$; moreover, $\mathbf{W}$ should vanish on S_ε. The left-hand side is a linear bounded functional on $\mathcal{H}$. By the Riesz theorem there exists a solution $\mathbf{W} \in \mathcal{H}$ of the above variational problem. Taking $\mathbf{V} = \mathbf{W}$ in (4.4.10) and using Cauchy's inequality for the integral on the right-hand side we obtain

$$\mathcal{E}(\mathbf{W};\Omega_\varepsilon) \leq \text{Const } \|\nabla\mathbf{H}\|^2_{L_2(\Omega_\varepsilon)}.$$

By (4.4.9) and Korn's inequality (4.4.1)

$$\|\mathbf{W}\|_{(H^1(\Omega_\varepsilon))^3} \leq \text{Const } \varepsilon^{-2}\sum_{i=1}^{K}\|\mathbf{\Phi}^{(i)}\|_{H^{1/2}(S_\varepsilon^{(i)})}. \quad (4.4.11)$$

The estimate (4.4.11) together with (4.4.9) and (4.4.8) gives (4.4.5).

□

The remainder estimate

THEOREM 4.9

The solution $\mathbf{u}(\mathbf{x}, \varepsilon)$ *of the problem* (4.1.1)–(4.1.3) *with right-hand sides of the form* (4.3.1)–(4.3.3) *has the asymptotic representation* (4.3.5) *in the following sense:*

$$\left\|\mathbf{u} - \sum_{j=-1}^{N} \varepsilon^j \mathbf{u}^{(j)}(\mathbf{x}, \varepsilon)\right\|_{(H^1(\Omega_\varepsilon))^3} \leq \text{Const } \varepsilon^{N-3} \qquad (4.4.12)$$

where the constant does not depend on ε.

Proof

For large n we write the inequality

$$\|\mathbf{u} - \mathbf{s}_N\|^2_{(H^1(\Omega_\varepsilon))^3} \leq 2(\|\mathbf{u} - \mathbf{s}_n\|^2_{(H^1(\Omega_\varepsilon))^3} + \|\mathbf{s}_n - \mathbf{s}_N\|^2_{(H^1(\Omega_\varepsilon))^3}), \quad (4.4.13)$$

where s_m denotes the sum (4.3.74).

Let us denote by $\mathbf{F}_n, \mathbf{P}_n, \mathbf{\Phi}_n^{(j)}$ the right-hand sides of the problem (1.2.1)–(1.2.3) for $\mathbf{u} - \mathbf{s}_n$. Then, by Lemma 4.4, the vector functions $\mathbf{F}_n, \mathbf{P}_n$ and $\mathbf{\Phi}_n^{(j)}$ have the asymptotic form (4.3.1)–(4.3.3) with $k \geq n+1$. Therefore,

$$\|\mathbf{F}_n\|_{(L_2(\Omega_\varepsilon))^3} + \varepsilon^{-1}\|\mathbf{P}_n\|_{(L_2(\partial\Omega_\varepsilon \setminus S_\varepsilon))^3}$$

$$+ \varepsilon^2 \sum_j \|\mathbf{\Phi}_n^{(j)}\|_{(H^{1/2}(S_\varepsilon^{(j)}))^3} \leq \text{Const } \varepsilon^{n-2}.$$

Using
Theorem 4.8 we obtain

$$\|\mathbf{u} - \mathbf{s}_n\|_{(H^1(\Omega_\varepsilon))^3} \leq \text{Const } \varepsilon^{n-6}.$$

It is verified directly that the last term from (4.4.13) satisfies the estimate

$$\|\mathbf{s}_n - \mathbf{s}_N\|^2_{(H^1(\Omega_\varepsilon))^3} \leq \text{Const } \varepsilon^{N-3}.$$

Choosing n such that $n \geq N+3$ we complete the proof. $\square$

4.5 The leading order approximation

Here, we construct two terms of orders ε^{-1} and ε^0 of the asymptotic expansion (4.3.5). We proceed in accordance with the sequence of steps described in Section 4.3.9.

4.5.1 The term $\mathbf{u}^{(-1)}$

First, consider the term with $k = -1$ in the expansion (4.3.5). Since $w_1^{(-1,j)} = w_2^{(-1,j)} = 0$, and, therefore, the right-hand sides of (4.3.67) and (4.3.68), $m = -1$, are equal to zero, one has

$$\alpha_i^{(-1)} = \beta_i^{(-1)} = 0, \ i = 1, 2, 3. \qquad (4.5.1)$$

It follows from (4.3.48), (4.3.49) that $v_1^{(-1,j)} = v_2^{(-1,j)} = 0$. Thus, by (4.3.60)–(4.3.66) and (4.3.4), the components $\mathfrak{F}_1^{(-1,j)}, \mathfrak{F}_2^{(-1,j)}$ and $\mathfrak{M}_k^{(-1,j)}$, $k = 1, 2, 3$, are also zero. From (4.3.27), $m = -1$, we obtain that the set

$$\left\{ \mathcal{W}^{(-1,j)}, d_1^{(I,-1,j)}, \ldots, d_6^{(I,-1,j)}, d_1^{(II,-1,j)}, \ldots, d_6^{(II,-1,j)} \right\}$$

is trivial.

The right-hand sides of (4.3.36), (4.3.37), (4.3.47) with $m = -1$ equal zero, and, consequently,

$$\mathcal{V}^{(-1,j)} = 0, \ c_q^{(-1,j)} = 0, \ q = 1, 2, \ldots, 6.$$

The homogeneous boundary value problems (4.3.57)–(4.3.59), $m = -1$, give $w_3^{(-1,j)} = w_4^{(-1,j)} = 0$.

From (4.3.71)–(4.3.73) we derive that $h_4^{(-1)} = h_5^{(-1)} = h_6^{(-1)} = 0$, and, therefore, $\alpha_3^{(-1)} = \beta_1^{(-1)} = \beta_2^{(-1)} = 0$.

In accordance with (4.3.4), (4.3.64) the longitudinal components $\mathfrak{F}_3^{(-1,j)}$ of forces at junction points are zero. Thus, $\mathbf{u}_s^{(-1)} = 0$. Since $\mathbf{f}^{(-1)} = 0$, $\mathbf{p}^{(-1)} = 0$, the regular term $\mathbf{u}_r^{(-1)}$ from (4.3.7) also vanishes, and, therefore, $\mathbf{u}_\Omega^{(-1)} = 0$.

Formulae (4.3.55), (4.3.56) yield $v_3^{(-1,j)} = 0$, and $v_4^{(-1,j)}(z) = \beta_3^{(0)} z / l_j$, and, consequently (see (4.3.18), (4.3.19), (4.3.24)), the vector function $\mathbf{U}^{(-1,j)}$ is zero.

Thus, we have shown that

$$\mathcal{U}^{(-1,j)}(\zeta,z,\varepsilon) = \varepsilon^{-2} v_4^{(-1,j)}(z)\phi^{(4,0)}(\zeta) + \varepsilon^{-1}\frac{d}{dz}v_4^{(-1,j)}(z)\phi^{(4,1)}(\zeta),$$

where $\phi^{(4,0)}$, $\phi^{(4,1)}$ were defined by (3.6.14), (3.6.18), and, consequently, we obtain

$$\mathbf{u}^{(-1)}(\mathbf{x},\varepsilon) = \varepsilon^{-1}\beta_3^{(0)}\sum_{j=1}^{K}\eta(z/\varepsilon)\frac{1}{l_j}\Big\{z((a_2-x_2^{(j)})\mathbf{e}^{(1)} + (x_1-a_1^{(j)})\mathbf{e}^{(2)})$$

$$+ \varphi\Big(\frac{x_1-a_1^{(j)}}{\varepsilon},\frac{x_2-a_2^{(j)}}{\varepsilon}\Big)\mathbf{e}^{(3)}\Big\},$$

with φ being defined in Section 3.5. The constant $\beta_3^{(0)}$ will be found during the calculations of $\mathbf{u}^{(0)}(\mathbf{x},\varepsilon)$.

4.5.2 The term $\mathbf{u}^{(0)}$

According to the general representation (4.3.6), we have

$$\mathbf{u}^{(0)}(\mathbf{x},\varepsilon) = \varepsilon^{-4}\Big\{\sum_{i=1}^{2}\alpha_i^{(0)}\mathbf{e}^{(i)} + \varepsilon^2\alpha_3^{(0)}\mathbf{e}^{(3)} \qquad (4.5.2)$$

$$+ \Big(\varepsilon^2\sum_{i=1}^{2}\beta_i^{(0)}\mathbf{e}^{(i)} + \beta_3^{(0)}\mathbf{e}^{(3)}\Big)\times\mathbf{x}\Big\}$$

$$+ \mathfrak{X}(\mathbf{x},\varepsilon)\mathbf{u}_\Omega^{(0)}(\mathbf{x}) + \sum_{j=1}^{K}\Big\{\eta\Big(\frac{z}{\varepsilon}\Big)\mathfrak{U}^{(0,j)}(\zeta,z,\varepsilon)$$

$$+ \varepsilon^{-2}\mathcal{W}^{(0,j)}(\mathbf{X}^{(j)})\Xi(\mathbf{x}-\mathbf{a}^{(j)}) + \varepsilon^{-2}\mathcal{V}^{(0,j)}(\mathbf{Y}^{(j)})\Big\}.$$

In accordance with the recurrence procedure, described in Section 4.3.9, we find the functions $w_i^{(0,j)}$, $i=1,2$, which satisfy the boundary value problems (4.3.50)–(4.3.54) with homogeneous boundary conditions and

$$Q_i^{(0,j)} = -D_i^{-1}\Big\{\int_g \mathcal{F}_i^{(0,j)}(\zeta,z)d\zeta + \int_{\partial g}\mathcal{P}_i^{(0,j)}(\zeta,z)ds\Big\},\ i=1,2.$$

The equation (4.3.66) takes the form

$$\mathcal{A}\begin{pmatrix}\alpha_1^{(0)}\\ \alpha_2^{(0)}\\ \beta_3^{(0)}\end{pmatrix} = \begin{pmatrix}h_1^{(0)}\\ h_2^{(0)}\\ h_3^{(0)}\end{pmatrix}, \qquad (4.5.3)$$

where
$$h_i^{(0)} = \mathbb{F}_i(\mathbf{f}^{(0)}, \mathbf{p}^{(0)}; \Omega) + \sum_{j=1}^{K} \Big\{ \mathbb{F}_i(\mathbf{F}^{(1,j)}, \mathbf{P}^{(1,j)}; G)$$
$$- D_i \mathbb{Q}^{(III)}(Q_i^{(0,j)}, 0, 0, 0, 0) \Big\}, \quad i = 1, 2.$$

For $h_3^{(0)}$ we have
$$h_3^{(0)} = \mathbb{M}_3(\mathbf{f}^{(0)}, \mathbf{p}^{(0)}, \Omega) + \sum_{j=1}^{K} \Big\{ \mathbb{M}_3(\mathbf{F}^{(0,j)}, \mathbf{P}^{(0,j)}, G)$$
$$- a_1^{(j)} D_2 \mathbb{Q}^{(III)}(Q_2^{(0,j)}, 0, 0, 0, 0) + a_2^{(j)} D_1 \mathbb{Q}^{(III)}(Q_1^{(0,j)}, 0, 0, 0, 0)$$
$$+ \int_G (\mathbf{a}^{(j)} \times \mathbf{F}^{(1,j)})_3 d\mathbf{X} + \int_{\partial G} (\mathbf{a}^{(j)} \times \mathbf{P}^{(1,j)})_3 ds \Big\}.$$

Formulae (4.3.48) and (4.3.49) give expressions for $v_1^{(0,j)}$, $v_2^{(0,j)}$, $j = 1, \ldots, K$. Now, one can calculate the transverse components of forces and components of moments at junction points. Formulae (4.3.60)–(4.3.63), (4.3.65) and (4.3.66) give

$$\mathfrak{F}_1^{(0,j)} = -D_1 \Big\{ \frac{12}{l_j^3}(\alpha_1^{(0)} - a_2^{(j)} \beta_3^{(0)}) + \mathbb{Q}^{(III)}(Q_1^{(0,j)}, 0, 0, 0, 0) \Big\}$$
$$+ \mathbb{F}_1(\mathbf{F}^{(1,j)}, \mathbf{P}^{(1,j)}, G), \tag{4.5.4}$$

$$\mathfrak{F}_2^{(0,j)} = -D_2 \Big\{ \frac{12}{l_j^3}(\alpha_2^{(0)} + a_1^{(j)} \beta_3^{(0)}) + \mathbb{Q}^{(III)}(Q_2^{(0,j)}, 0, 0, 0, 0) \Big\}$$
$$+ \mathbb{F}_2(\mathbf{F}^{(1,j)}, \mathbf{P}^{(1,j)}, G), \tag{4.5.5}$$

$$\mathfrak{M}_1^{(0,j)} = -D_2 \Big\{ -\frac{6}{l_j^2}(\alpha_2^{(0)} + a_1^{(j)} \beta_3^{(0)}) + \mathbb{Q}^{(II)}(Q_2^{(0,j)}, 0, 0, 0, 0) \Big\}$$
$$+ \mathbb{M}_1(\mathbf{F}^{(0,j)}, \mathbf{P}^{(0,j)}, G), \tag{4.5.6}$$

$$\mathfrak{M}_2^{(0,j)} = D_1 \Big\{ -\frac{6}{l_j^2}(\alpha_1^{(0)} - a_2^{(j)} \beta_3^{(0)}) + \mathbb{Q}^{(II)}(Q_1^{(0,j)}, 0, 0, 0, 0) \Big\}$$
$$+ \mathbb{M}_2(\mathbf{F}^{(0,j)}, \mathbf{P}^{(0,j)}, G), \tag{4.5.7}$$

$$\mathfrak{M}_3^{(0,j)} = -D_4 \frac{\beta_3^{(0)}}{l_j} + \mathbb{M}_3(\mathbf{F}^{(0,j)}, \mathbf{P}^{(0,j)}, G). \tag{4.5.8}$$

Next, we determine the solution

$$\left\{ \boldsymbol{\mathcal{W}}^{(0,j)}; d_1^{(I,0,j)}, \ldots, d_6^{(I,0,j)}; d_1^{(II,0,j)}, \ldots, d_6^{(II,0,j)} \right\} \qquad (4.5.9)$$

of the junction layer problem (4.3.27) with $\tilde{\mathbf{F}}^{(m,j)}$, $\tilde{\mathbf{P}}^{(m,j)}$, $\mathfrak{M}^{(m,j)}$ replaced by $\mathbf{F}^{(0,j)}$, $\mathbf{P}^{(0,j)}$, $\mathfrak{M}^{(0,j)}$ and $\mathfrak{F}^{(-1,j)}$ being zero. Note that, from (4.3.4),
$$d_1^{(I,0,j)} = d_2^{(I,0,j)} = d_3^{(I,0,j)} = 0.$$

By Theorem 4.3, the vector-function $\boldsymbol{\mathcal{W}}^{(0,j)}$ decays exponentially in Π_+ and has the following asymptotics in the half-space:

$$\boldsymbol{\mathcal{W}}^{(0,j)}(\mathbf{X}) \sim \sum_{i=1}^{2} b_i^{(0,j)} \partial_{X_i} \boldsymbol{\mathcal{B}}^{(i)}(\mathbf{X}) + \frac{1}{2} b_3^{(0,j)} \left(\partial_{X_1} \boldsymbol{\mathcal{B}}^{(2)}(\mathbf{X}) + \partial_{X_2} \boldsymbol{\mathcal{B}}^{(1)}(\mathbf{X}) \right)$$
$$+ O(\|\mathbf{X}\|^{-3}) \text{ as } \|\mathbf{X}\| \to \infty, \mathbf{X} \in \mathbb{R}_-^3. \qquad (4.5.10)$$

Consider the base regions of the thin cylinders. The set

$$\left\{ \boldsymbol{\mathcal{V}}^{(0,j)}, c_1^{(0,j)}, \ldots, c_6^{(0,j)} \right\} \qquad (4.5.11)$$

solves the problem (4.3.36), (4.3.37), (4.3.47) with $m = 0$ and

$$\tilde{\boldsymbol{\Phi}}^{(0,j)}(\mathbf{Y}') = \boldsymbol{\Phi}^{(0,j)}(\mathbf{Y}') - \frac{dv_4^{(-1,j)}}{dz}(l_j)\boldsymbol{\xi}^{(4)}(\mathbf{Y}',0)$$
$$- \sum_{k=1}^{2} \frac{d^2 v_k^{(0,j)}}{dz^2}(l_j)\boldsymbol{\xi}^{(4+k)}(\mathbf{Y}',0).$$

Now, we determine the functions $w_i^{(0,j)}$, $i = 3, 4$, which satisfy the boundary value problems (4.3.57)–(4.3.59) with

$$Q_3^{(0,j)}(z) = D_3^{-1} \left\{ \int_g \mathcal{F}_3^{(0,j)}(\zeta, z) d\zeta + \int_{\partial g} \mathcal{P}_3^{(0,j)}(\zeta, z) ds \right\}, \ z \in (0, l_j),$$

$$Q_4^{(0,j)}(z) = D_4^{-1} \left\{ \int_g \left(\zeta_1 \mathcal{F}_2^{(0,j)}(\zeta, z) - \zeta_2 \mathcal{F}_1^{(0,j)}(\zeta, z) \right) d\zeta \right.$$
$$+ \int_{\partial g} \left(\zeta_1 \mathcal{P}_2^{(0,j)}(\zeta, z) - \zeta_2 \mathcal{P}_1^{(0,j)}(\zeta, z) \right) ds \Big\}.$$

To complete the calculation of the rigid-body displacements of the three-dimensional element Ω, we have to obtain the coefficients $\alpha_3^{(0)}$, $\beta_1^{(0)}$, $\beta_2^{(0)}$ in the ansatz (4.5.2). From (4.3.68), one has

THE LEADING ORDER APPROXIMATION

$$\begin{pmatrix} \alpha_3^{(0)} \\ \beta_1^{(0)} \\ \beta_2^{(0)} \end{pmatrix} = \mathcal{B}^{-1} \left[\begin{pmatrix} h_4^{(0)} \\ h_5^{(0)} \\ h_6^{(0)} \end{pmatrix} - \mathcal{D} \begin{pmatrix} \alpha_1^{(0)} \\ \alpha_2^{(0)} \\ \beta_3^{(0)} \end{pmatrix} \right], \qquad (4.5.12)$$

where

$$h_4^{(0)} = \mathbb{F}_3(\mathbf{f}^{(0)}, \mathbf{p}^{(0)}, \Omega) + \sum_{j=1}^{K} \left\{ D_3 \mathbb{Q}^{(I)}(Q_3^{(0,j)}, d_3^{(II,0,j)}, c_3^{(0,j)}) \right.$$
$$\left. + \mathbb{F}_3(\mathbf{F}^{(1,j)}, \mathbf{P}^{(1,j)}, G) \right\},$$

$$h_5^{(0)} = \mathbb{M}_1(\mathbf{f}^{(0)}, \mathbf{p}^{(0)}, \Omega)$$
$$+ \sum_{j=1}^{K} \left\{ - D_2 \mathbb{Q}^{(II)}(Q_2^{(0,j)}, 0, 0, 0, 0) + \mathbb{M}_1(\mathbf{F}^{(0,j)}, \mathbf{P}^{(0,j)}, G) \right.$$
$$\left. + a_2^{(j)} \left[D_3 \mathbb{Q}^{(I)}(Q_3^{(0,j)}, d_3^{(II,0,j)}, c_3^{(0,j)}) + \mathbb{F}_3(\mathbf{F}^{(1,j)}, \mathbf{P}^{(1,j)}, G) \right] \right\}$$

$$h_6^{(0)} = \mathbb{M}_2(\mathbf{f}^{(0)}, \mathbf{p}^{(0)}, \Omega)$$
$$+ \sum_{j=1}^{K} \left\{ D_1 \mathbb{Q}^{(II)}(Q_1^{(0,j)}, 0, 0, 0, 0) + \mathbb{M}_2(\mathbf{F}^{(0,j)}, \mathbf{P}^{(0,j)}, G) \right.$$
$$\left. - a_1^{(j)} \left[D_3 \mathbb{Q}^{(I)}(Q_3^{(0,j)}, d_3^{(II,0,j)}, c_3^{(0,j)}) + \mathbb{F}_3(\mathbf{F}^{(1,j)}, \mathbf{P}^{(1,j)}, G) \right] \right\}.$$

From (4.3.64), the longitudinal force at junction points $\mathbf{a}^{(j)}$, $j = 1, \ldots, K$, is given by

$$\mathfrak{F}_3^{(0,j)} = D_3 \left\{ -(\alpha_3^{(0)} + \beta_1^{(0)} a_2^{(j)} - \beta_2^{(0)} a_1^{(j)}) \frac{1}{l_j} \right.$$
$$\left. + \mathbb{Q}^{(I)}(Q_3^{(0,j)}, d_3^{(II,0,j)}, c_3^{(0,j)}) \right\}$$
$$+ \mathbb{F}_3(\mathbf{F}^{(1,j)}, \mathbf{P}^{(1,j)}, G). \qquad (4.5.13)$$

Following (4.3.8) with $k = 0$, we can obtain the singular part $\mathbf{u}_s^{(0)}$ of the displacement $\mathbf{u}_\Omega^{(0)}$ in Ω.

At this stage, we can find the regular part $\mathbf{u}_r^{(0)}$ of the term $\mathbf{u}_\Omega^{(0)}$ from (4.5.2). According to Sections 4.3.6, 4.3.5 and the boundary value problem (4.3.35), the vector $\mathbf{u}_r^{(0)}$ satisfies (4.2.10), (4.2.11) with $\mathbf{f}$, $\mathbf{p}$ being replaced by $\mathbf{f}^{(0)}$, $\mathbf{p}^{(0)}$ plus discrepancy terms brought in by the commutators of the cut-off function Θ and the differential operators of the Lamé system and the traction boundary conditions applied to the leading part of the asymptotic expansion (4.5.10). It should be noted that the principal force and moment of these discrepancy terms are equal to zero.

Using the formulae (4.3.55), (4.3.56) we determine the functions $v_3^{(0,j)}$ completely, and $v_4^{(0,j)}$ up to a linear term $\beta_3^{(1)} z/l_j$. The boundary value problem (4.3.18), (4.3.19) with $\mathbf{U}$, v_k, $\mathcal{F}$, $\mathcal{P}$ replaced by $\mathbf{U}^{(0,j)}$, $v_k^{(0,j)}$, $\mathcal{F}^{(0,j)}$, $\mathcal{P}^{(0,j)}$, respectively, provides the vector function $\mathbf{U}^{(0,j)}$. The constant $\beta_3^{(1)}$ is to be found on the next iteration.

4.6 Physical interpretation of the results

The characteristics of the physical model of the pile structure are the components of the rigid-body displacements of the pile cap and the reaction forces and moments at junction points. In the present section we analyse these quantities on the basis of the asymptotic algorithm developed for the region Ω_ε, which enables us to compare the pile structure model with the model of elastic multi-structure.

Let us assume that the load is applied to the region Ω only. Consider two cases.

4.6.1 The case $|\mathbb{M}_3| + |\mathbb{F}_1| + |\mathbb{F}_2| \neq 0$

In this case the leading order terms in the representation of rigid-body displacements of the pile cap Ω are given by

$$\varepsilon^{-4}(\alpha_1^{(0)} \mathbf{e}^{(1)} + \alpha_2^{(0)} \mathbf{e}^{(2)} + \beta_3^{(0)} \mathbf{e}^{(3)} \times \mathbf{x}), \qquad (4.6.1)$$

where, in accordance with (4.5.3),

$$\begin{pmatrix} \alpha_1^{(0)} \\ \alpha_2^{(0)} \\ \beta_3^{(0)} \end{pmatrix} = \mathcal{A}^{-1} \begin{pmatrix} \mathbb{F}_1(\mathbf{f}^{(0)}, \mathbf{p}^{(0)}, \Omega) \\ \mathbb{F}_2(\mathbf{f}^{(0)}, \mathbf{p}^{(0)}, \Omega) \\ \mathbb{M}_3(\mathbf{f}^{(0)}, \mathbf{p}^{(0)}, \Omega) \end{pmatrix}. \qquad (4.6.2)$$

The leading order terms of the components of the reaction forces and moments at junction points of the elastic multi–structure Ω_ε are specified by (see formulae (4.5.4)–(4.5.8) and (4.5.13))

$$\mathfrak{F}_1^{(0,j)} = -D_1 \frac{12}{l_j^3}(\alpha_1^{(0)} - a_2^{(j)}\beta_3^{(0)}),$$

$$\mathfrak{F}_2^{(0,j)} = -D_2 \frac{12}{l_j^3}(\alpha_2^{(0)} + a_1^{(j)}\beta_3^{(0)}),$$

$$\mathfrak{M}_3^{(0,j)} = -D_4 \frac{\beta_3^{(0)}}{l_j},$$

$$\mathfrak{M}_1^{(0,j)} = \frac{6D_2}{l_j^2}(\alpha_2^{(0)} + a_1^{(j)}\beta_3^{(0)}),$$

$$\mathfrak{M}_2^{(0,j)} = -\frac{6D_1}{l_j^2}(\alpha_1^{(0)} - a_2^{(j)}\beta_3^{(0)}),$$

$$\mathfrak{F}_3^{(0,j)} = -\frac{D_3}{l_j}(\alpha_3^{(0)} + a_2^{(j)}\beta_1^{(0)} - a_1^{(j)}\beta_2^{(0)} - d_3^{(II,0,j)} + c_3^{(0,j)}),$$

where, from (4.5.12), the coefficients $\alpha_3^{(0)}$, $\beta_1^{(0)}$, $\beta_2^{(0)}$ admit the representation

$$(\alpha_3^{(0)}, \beta_1^{(0)}, \beta_2^{(0)})^t = \boldsymbol{B}^{-1}\left\{(h_4^{(0)}, h_5^{(0)}, h_6^{(0)})^t - \boldsymbol{D}(\alpha_1^{(0)}, \alpha_2^{(0)}, \beta_3^{(0)})^t\right\}.$$

Here

$$h_4^{(0)} = \mathbb{F}_3(\mathbf{f}^{(0)}, \mathbf{p}^{(0)}, \Omega) + D_3 \sum_{j=1}^{K} \frac{d_3^{(II,0,j)} - c_3^{(0,j)}}{l_j},$$

$$h_5^{(0)} = \mathbb{M}_1(\mathbf{f}^{(0)}, \mathbf{p}^{(0)}, \Omega) + D_3 \sum_{j=1}^{K} a_2^{(j)} \frac{d_3^{(II,0,j)} - c_3^{(0,j)}}{l_j},$$

$$h_6^{(0)} = \mathbb{M}_2(\mathbf{f}^{(0)}, \mathbf{p}^{(0)}, \Omega) - D_3 \sum_{j=1}^{K} a_1^{(j)} \frac{d_3^{(II,0,j)} - c_3^{(0,j)}}{l_j}.$$

The constants $d_3^{(II,0,j)}$, $c_3^{(0,j)}$ come from the solutions of the junction layer and of the bottom layer problems (see (4.5.9) and (4.5.11)).

Comparing the asymptotic approach with the engineering model (see Section 4.2.7), we make the following comments:

(1) the coefficients $\alpha_1^{(0)}$, $\alpha_2^{(0)}$, $\beta_3^{(0)}$ in the representation of the leading order rigid-body displacements of Ω are the same for the elastic multi-structure Ω_ε and for the pile structure described in Section 4.2.7;

(2) the components $\mathfrak{F}_1^{(0,k)}$, $\mathfrak{F}_2^{(0,k)}$, $\mathfrak{M}_1^{(0,k)}$, $\mathfrak{M}_2^{(0,k)}$, $\mathfrak{M}_3^{(0,k)}$ of the transverse force and of the moment at junction points $\mathbf{a}^{(k)}$, $k = 1,\ldots, K$, of the multi-structure Ω_ε are determined by the same formulae (see (4.2.76), (4.2.78), (4.2.79) and (4.2.80)–(4.2.83)) as used for calculation of the leading part of lock forces and moments in the pile structure;

(3) the longitudinal components $\mathfrak{F}_3^{(0,k)}$ of the forces at junction points will differ from those calculated on the basis of the model of the pile structure (see Section 4.2.7); the group of coefficients $\alpha_3^{(0)}$, $\beta_1^{(0)}$, $\beta_2^{(0)}$ for the rigid-body displacements can be calculated by means of the same formulae as (4.2.90). However, for the case of the elastic multi-structure Ω_ε correction terms, involving $d_3^{(II,0,j)}$, $c_3^{(0,j)}$, occur in the right-hand side of the system of equations (see (4.5.12)).

Formula (4.6.2) shows that the leading order terms of the displacement field (see coefficients $\alpha_i^{(0)}$, $i = 1,2$, and $\beta_3^{(0)}$) in the pile cap outside neighbourhoods of the junction points follow the classical theory of pile structures where the pile cap is treated as a rigid body. However, high-order corrections are needed to evaluate the components $\alpha_3^{(0)}$, $\beta_i^{(0)}$, $i = 1,2$, and the third component $\mathfrak{F}_3^{(j)}$ of the lock forces. To get these values, we ought to perform an asymptotic study of the elastic field in the vicinity of the junction points and the thin rods. This procedure requires solutions of junction and bottom layer types.

4.6.2 The case $\mathbb{F}_1 = \mathbb{F}_2 = \mathbb{M}_3 = 0$

In this case one can deduce that the coefficients $\alpha_1^{(j)}$, $\alpha_2^{(j)}$, $\beta_3^{(j)}$, $j = 0, 1$, are equal to zero. Therefore, the leading order term of the displacement in Ω is specified by the rigid-body translations and rotations of the form

$$\varepsilon^{-2}\left\{\sum_{i=1}^{2}\alpha_i^{(2)}\mathbf{e}^{(i)} + \beta_3^{(2)}\mathbf{e}^{(3)} \times \mathbf{x} + \alpha_3^{(0)}\mathbf{e}^{(3)} + \sum_{i=1}^{2}\beta_i^{(0)}\mathbf{e}^{(i)} \times \mathbf{x}\right\}, \quad (4.6.3)$$

where the coefficients are given by

$$\begin{pmatrix}\alpha_3^{(0)} \\ \beta_1^{(0)} \\ \beta_2^{(0)}\end{pmatrix} = \mathcal{B}^{-1}\begin{pmatrix}\mathbb{F}_3\left(\mathbf{f}^{(0)},\mathbf{p}^{(0)},\Omega\right) \\ \mathbb{M}_1\left(\mathbf{f}^{(0)},\mathbf{p}^{(0)},\Omega\right) \\ \mathbb{M}_2\left(\mathbf{f}^{(0)},\mathbf{p}^{(0)},\Omega\right)\end{pmatrix}$$

PHYSICAL INTERPRETATION

$$\begin{pmatrix} \alpha_1^{(2)} \\ \alpha_2^{(2)} \\ \beta_3^{(2)} \end{pmatrix} = \mathcal{A}^{-1} \begin{pmatrix} h_1^{(2)} \\ h_2^{(2)} \\ h_3^{(2)} \end{pmatrix} - \mathcal{A}^{-1} \mathcal{D}^t \begin{pmatrix} \alpha_3^{(0)} \\ \beta_1^{(0)} \\ \beta_2^{(0)} \end{pmatrix},$$

where the matrices $\mathcal{A}, \mathcal{B}, \mathcal{D}$ are the same as in (4.3.67), (4.3.68), and the components $h_i^{(2)}$, $i = 1, 2, 3$, have the form

$$h_q^{(2)} = -\sum_j D_q \mathbb{Q}^{(III)}(0, 0, 0, d_{q+4}^{(II,1,j)}, c_{q+4}^{(1,j)}), \ q = 1, 2,$$

$$h_3^{(2)} = \sum_j \left\{ D_4 \mathbb{Q}^{(I)}(0, d_4^{(II,1,j)}, c_4^{(1,j)}) \right.$$
$$- a_1^{(j)} D_2 \mathbb{Q}^{(III)}(0, 0, 0, d_6^{(II,1,j)}, c_6^{(1,j)})$$
$$\left. + a_2^{(j)} D_1 \mathbb{Q}^{(III)}(0, 0, 0, d_5^{(II,1,j)}, c_5^{(1,j)}) \right\},$$

with the functionals $\mathbb{Q}^{(I)}$, $\mathbb{Q}^{(III)}$ being the same as in Section 4.2.6, and the constants $d_i^{(II,1,j)}$, $c_i^{(1,j)}$, $i = 5, 6$, brought in by the solutions of the junction layer and the bottom layer problems. The set (4.3.28) with $m = 1$ solves the boundary value problem (4.3.27) with

$$\tilde{\mathbf{F}}^{(1,j)} = 0, \ \tilde{\mathbf{P}}^{(1,j)} = 0, \ \mathfrak{M}^{(1,j)} = 0, \ \mathfrak{F}_1^{(0,j)} = \mathfrak{F}_2^{(0,j)} = 0,$$

and $\mathfrak{F}_3^{(0,j)}$ being defined by

$$\mathfrak{F}_3^{(0,j)} = -\frac{D_3}{l_j}(\alpha_3^{(0)} + \beta_1^{(0)} a_2^{(j)} - \beta_2^{(0)} a_1^{(j)}). \qquad (4.6.4)$$

The set (4.3.46) solves the boundary value problem (4.3.36), (4.3.37), (4.3.47) where

$$\mathcal{G}^{(1,j)} = 0, \ \mathbf{H}^{(1,j)} = 0,$$

and

$$\tilde{\mathbf{\Phi}}^{(1,j)} = \frac{1}{l_j} \left(\alpha_3^{(0)} + \beta_1^{(0)} a_2^{(j)} - \beta_2^{(0)} a_1^{(j)} \right) \boldsymbol{\xi}^{(3)}(\mathbf{Y}', 0).$$

In this case all components of the lock forces and moments for $m = 0$, except the longitudinal force, are equal to zero. The longitudinal force at a junction point $\mathbf{a}^{(j)}$ is given by (4.6.4).

Comparing the two models, we should mention that the quantities $\alpha_3^{(0)}$, $\beta_1^{(0)}$, $\beta_2^{(0)}$ and $\mathfrak{F}_3^{(0,j)}$ are the same both for the elastic multistructure Ω_ε and for the pile structure (see (4.2.90) and (4.2.77), (4.2.82)). The formulae for $\alpha_1^{(2)}$, $\alpha_2^{(2)}$, $\beta_3^{(2)}$ are similar for both models (compare with (4.2.92)). However, in the case of the elastic multi-structure Ω_ε the right-hand side of the system includes correction terms $h_q^{(2)}$, $q = 1, 2, 3$, that come from solutions of the bottom layer and of the junction layer problems.

For the general case of geometry of the cross-section g the quantities $d_{q+4}^{(II,1,j)}$, $c_{q+4}^{(1,j)}$, $q = 1, 2$, may differ from zero. However, if g is centrally symmetric, then the above coefficients are identically zero, and, therefore,

$$h_q^{(2)} = 0, \ q = 1, 2; \ h_3^{(2)} = \sum_j D_4 \mathbb{Q}^{(I)}(0, d_4^{(II,1,j)}, c_4^{(1,j)}).$$

Let us consider an example where the engineering approach works perfectly well. We assume that the cross-section g is centrally symmetric, and in addition suppose that the tractions and body forces are applied in such a way that the rigid-body rotation is absent, i.e. $\beta_1^{(0)} = \beta_2^{(0)} = \beta_3^{(2)} = 0$, and that the first two components of the principal force vector are zero. Then,

$$\alpha_1^{(2)} = \alpha_2^{(2)} = 0, \ \alpha_3^{(0)} = \frac{(\lambda + \mu) \mathbb{F}_3(\mathbf{f}^{(0)}, \mathbf{p}^{(0)}, \Omega)}{\mu(3\lambda + 2\mu) \mathrm{mes}_2 g \sum_{i=1}^K 1/l_i},$$

and

$$\mathfrak{F}_3^{(0,j)} = -\frac{\mathbb{F}_3(\mathbf{f}^{(0)}, \mathbf{p}^{(0)}, \Omega)}{l_j \sum_{k=1}^K 1/l_k},$$

which describes the leading part of the *longitudinal components of the lock forces*.

5
NON-DEGENERATE ELASTIC MULTI-STRUCTURES

The analysis of the elasticity problems in Chapter 4 shows that for multi-structures consisting of a set of parallel thin rods and a three-dimensional finite body the resistance to loads applied along thin rods is much larger than the resistance to forces associated with bending of thin rods. That is, with applied loads of the same order, the transverse displacements of the multi-structure are much larger compared with the order displacements along the rods. This phenomenon is connected with the special geometry of the multi-structure we dealt with.

In the present chapter we analyse elastic fields in a 1D–3D multi-structure for the case when the thin rods are not necessarily parallel to each other. Similar to Chapter 4, the skeleton of this structure is defined as the union of the body Ω and K one-dimensional rods attached to Ω at some points $\mathbf{a}^{(1)}, \ldots, \mathbf{a}^{(K)}$ and directed along the unit vectors $\boldsymbol{\mu}^{(1)}, \ldots, \boldsymbol{\mu}^{(K)}$. We specify the number of rods and their orientation in such a way that, with equal order values of the vertical and transverse external loads, the displacements of the multi-structure in all directions have the same order of magnitude. Further in the text, such multi-structures will be called *non-degenerate*; otherwise we use the word *degenerate* in relation to structures similar that studied in Chapter 4.

It will be shown that a non-degenerate multi-structure can be defined formally in such a way that the equalities

$$\boldsymbol{\mu}^{(j)} \cdot (\boldsymbol{\alpha} - \mathbf{a}^{(j)} \times \boldsymbol{\beta}) = 0, \ j = 1, \ldots, K,$$

imply that the vectors $\boldsymbol{\alpha}$ and $\boldsymbol{\beta}$ are zero. The multi-structure analysed in Chapter 4 does not satisfy this condition. (Arbitrary vectors $\boldsymbol{\alpha}$ and $\boldsymbol{\beta}$ with $\alpha_3 = 0$ and $\beta_1 = \beta_2 = 0$ solve the above equations.) Other examples of degenerate multi-structures will be considered in Section 5.1.7.

We construct a complete asymptotic expansion of the solution to a mixed boundary value problem of elasticity posed in the non-degenerate multi-structure. The asymptotic algorithm differs from that presented in Chapter 4. Some model problems used in the present chapter are similar to problems studied earlier, and whenever necessary we give references to appropriate sections of Chapter 4. We shall show that the representation of the principal term of the displacement vector in Ω agrees entirely with the engineering pile structure model described by a 6×6 system of linear algebraic equations with respect to rigid-body translations and rotations of the pile cap.

In Section 5.1 we present the pile structure model. A system of linear algebraic equations characterising a non-degenerate pile structure is studied. In the same section we present examples of degenerate and non-degenerate structures.

Formulation of the boundary value problem is given in Section 5.2.

Model problems and the properties of their solutions used in the asymptotic algorithm are discussed in Section 5.3.

Section 5.4 includes the total asymptotic expansion of the solution of the mixed boundary value problem posed in the multi-structure. The remainder estimate is obtained in Section 5.5.

Finally in Sections 5.6 and 5.7 we analyse the leading term of the asymptotic approximation and give its physical interpretation.

5.1 Pile structure model

In this section we describe a classical engineering model which regards Ω as a rigid (undeformable) body and the thin cylinders attached to Ω as elastic one-dimensional rods. Here, the multi-structure is replaced by its skeleton.

5.1.1 Skeleton of the multi-structure

Let $\mathbf{x} = (x_1, x_2, x_3)$ be Cartesian coordinates in $\mathbb{R}^3$ and $\mathbf{e}^{(1)}, \mathbf{e}^{(2)}$ and $\mathbf{e}^{(3)}$ the corresponding unit basis vectors.

As before, let Ω be a bounded domain in $\mathbb{R}^3$ with Lipschitz boundary $\partial\Omega$, and $\mathbf{a}^{(j)}, j = 1, \ldots, K$, be the points on $\partial\Omega$. We suppose that $\partial\Omega$ is flat in the vicinity of each point $\mathbf{a}^{(j)}$, but in contrast with Chapter 4 we do not suppose that each point $\mathbf{a}^{(j)}$ belongs to the same plane. We introduce one-dimensional segments r_j by

$$r_j = \left\{ \mathbf{x} \in \mathbb{R}^3 : \mathbf{x} = \mathbf{a}^{(j)} + t\boldsymbol{\mu}^{(j)}, \ 0 \leq t \leq l^{(j)} \right\}, \qquad (5.1.1)$$

with some unit vectors $\boldsymbol{\mu}^{(j)}$; it is assumed that $r_j \cap \bar{\Omega} = \mathbf{a}^{(j)}$. The second end point of the segment is denoted by $\mathbf{b}^{(j)}$:

$$\mathbf{b}^{(j)} = \mathbf{a}^{(j)} + l^{(j)} \boldsymbol{\mu}^{(j)}.$$

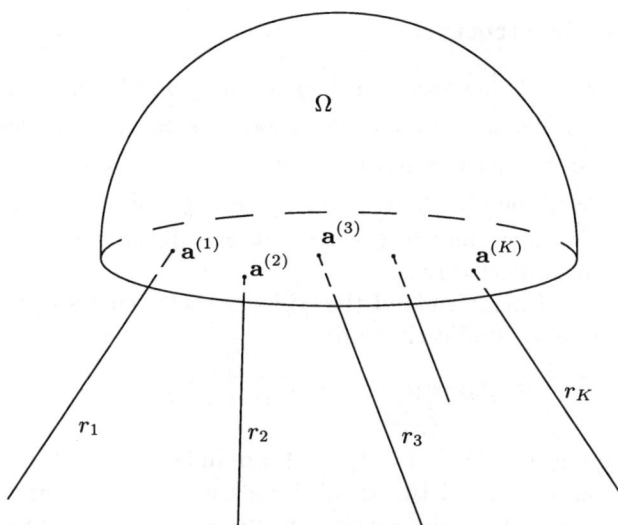

FIG. 5.1. Skeleton of a multi-structure.

Further, the points $\mathbf{a}^{(j)}$ and $\mathbf{b}^{(j)}$ will be called the *junction point* and the *base point*, respectively. *The skeleton of the multi-structure* is the union of Ω and $r_j, j = 1, \ldots, K$ (Fig. 5.1).

With every one-dimensional element r_j we associate an orthogonal system of coordinates $\mathbf{y} = (y_1^{(j)}, y_2^{(j)}, y_3^{(j)})$, with the origin at $\mathbf{a}^{(j)}$ and with the $y_3^{(j)}$ axis directed along $\boldsymbol{\mu}^{(j)}$. The new and old coordinates are related by

$$\mathbf{x} = \mathbf{a}^{(j)} + \boldsymbol{\Lambda}^{(j)} \mathbf{y}^{(j)}, \qquad (5.1.2)$$

where $\boldsymbol{\Lambda}^{(j)}$ is an orthogonal matrix, and

$$\boldsymbol{\mu}^{(j)} = \left(\boldsymbol{\Lambda}^{(j)}\right)^* \mathbf{e}^{(3)}, \qquad (5.1.3)$$

where $\left(\boldsymbol{\Lambda}^{(j)}\right)^*$ denotes the adjoint matrix of $\boldsymbol{\Lambda}^{(j)}$.

In what follows the segment r_j serves as the axis of a thin cylinder

$$\Pi_\varepsilon^{(j)} = \{\mathbf{y}^{(j)} : 0 < y_3^{(j)} < l^{(j)},\ \varepsilon^{-1}(y_1^{(j)}, y_2^{(j)}) \in g^{(j)}\},$$

where $g^{(j)}$ is a bounded two-dimensional domain with Lipschitz boundary. We direct the axes $Oy_1^{(j)}$ and $Oy_2^{(j)}$ along the inertia axes of $g^{(j)}$.

5.1.2 The pile structure

As in Section 4.2.7 the elastic properties of r_j will be characterised by the set of positive stiffness coefficients $C_k^{(j)}, k = 1, \ldots, 4$ (see Section 3.3). The first two coefficients $C_1^{(j)}$ and $C_2^{(j)}$ are related to transverse displacements in the directions of the $y_1^{(j)}$ and $y_2^{(j)}$ axes. The coefficients $C_3^{(j)}$ and $C_4^{(j)}$ correspond to the longitudinal displacement and to the axial rotation, respectively.

The orders of magnitude of the stiffness coefficients are specified by the formulae (also, see Section 4.6)

$$C_i^{(j)} = \varepsilon^4 D_i^{(j)}, \quad i = 1, 2, 4; \quad C_3^{(j)} = \varepsilon^2 D_3^{(j)}, \tag{5.1.4}$$

where the quantities $D_i^{(j)}$, $i = 1, \ldots, 4$, are independent of ε.

The displacements of the ends of rods at junction points and the rotations are equal to the displacements and the rotations of Ω at the same points. We assume that the base points of elastic rods are clamped. The same notations $\mathfrak{F}^{(j)}$ and $\mathfrak{M}^{(j)}$ are used for lock forces and lock moments applied at the junction points. The equilibrium equations have the form (4.2.67), (4.2.68), where $\mathfrak{F}$ and $\mathfrak{M}$ are the external force and moment applied to Ω. Further in the text we derive a system of linear algebraic equations with respect to components of rigid-body displacements of Ω and a representation for components of lock forces and moments at the junction points.

5.1.3 Mathematical model of the pile structure

The displacements of points of one-dimensional elastic rods r_j represented in the system of coordinates $(y_1^{(j)}, y_2^{(j)}, y_3^{(j)})$ are characterised by functions $v_1^{(j)}, v_2^{(j)}, v_3^{(j)}, v_4^{(j)}$ which correspond to transverse displacements, a longitudinal displacement and an axial rotation, respectively.

The rigid-body displacements of Ω are specified by

$$\boldsymbol{\alpha} + \boldsymbol{\beta} \times \mathbf{x}, \tag{5.1.5}$$

where $\boldsymbol{\alpha}$ and $\boldsymbol{\beta}$ are constant vectors.

The functions $v_i^{(j)}, i = 1, 2, 3, 4$, satisfy the ordinary differential equations

$$\frac{d^4 v_i^{(j)}}{dz^4}(z) = 0, \quad z \in (0, l^{(j)}), \quad i = 1, 2, \tag{5.1.6}$$

PILE STRUCTURE MODEL 217

$$\frac{d^2 v_i^{(j)}}{dz^2}(z) = 0, \quad z \in (0, l^{(j)}), \quad i = 3, 4, \qquad (5.1.7)$$

where $z = y_3^{(j)}$ within the thin cylinder $\Pi_\varepsilon^{(j)}$. The clamping conditions at the base points of the rods r_j are

$$v_i^{(j)}(l^{(j)}) = 0, \quad i = 1, 2, 3, 4, \qquad (5.1.8)$$

$$\frac{d v_i^{(j)}}{dz}(l^{(j)}) = 0, \quad i = 1, 2. \qquad (5.1.9)$$

The following notations are adopted:

$$\hat{\boldsymbol{\alpha}}^{(j)} = \Lambda^{(j)} \boldsymbol{\alpha}, \quad \hat{\boldsymbol{\beta}}^{(j)} = \Lambda^{(j)} \boldsymbol{\beta},$$
$$\hat{\mathfrak{F}}^{(j)} = \Lambda^{(j)} \mathfrak{F}^{(j)}, \quad \hat{\mathfrak{M}}^{(j)} = \Lambda^{(j)} \mathfrak{M}^{(j)}. \qquad (5.1.10)$$

The continuity conditions of rotations and translations at the junction points $\mathbf{a}^{(j)}, j = 1, \ldots, K$, can be written as

$$v_i^{(j)}(0) = \hat{\alpha}_i^{(j)} + (\hat{\boldsymbol{\beta}}^{(j)} \times \hat{\mathbf{a}}^{(j)})_i, \quad i = 1, 2, 3, \qquad (5.1.11)$$

$$v_4^{(j)}(0) = \hat{\beta}_3^{(j)}, \quad \frac{dv_1^{(j)}}{dz}(0) = \hat{\beta}_2^{(j)}, \quad \frac{dv_2^{(j)}}{dz}(0) = -\hat{\beta}_1^{(j)}. \qquad (5.1.12)$$

In the local coordinate system the components of forces and moments at junction points, exerted by elastic rods on the body Ω, are specified by

$$\hat{\mathfrak{F}}_i^{(j)} = -C_i^{(j)} \frac{d^3 v_i^{(j)}}{dz^3}(0), \quad i = 1, 2, \qquad (5.1.13)$$

$$\hat{\mathfrak{F}}_3^{(j)} = C_3^{(j)} \frac{dv_3^{(j)}}{dz}(0), \qquad (5.1.14)$$

$$\hat{\mathfrak{M}}_{3-i}^{(j)} = (-1)^{i+1} C_i^{(j)} \frac{d^2 v_i^{(j)}}{dz^2}(0), \quad i = 1, 2, \qquad (5.1.15)$$

$$\hat{\mathfrak{M}}_3^{(j)} = C_4^{(j)} \frac{dv_4^{(j)}}{dz}(0). \qquad (5.1.16)$$

5.1.4 Solution of the pile structure equations

According to Section 4.2.6 the solutions of (5.1.6)–(5.1.12) have the form

$$v_1^{(j)}(z) = \phi_I\left(\frac{z}{l^{(j)}}\right)(\hat{\alpha}_1 + (\hat{\boldsymbol{\beta}} \times \hat{\mathbf{a}}^{(j)})_1) + l^{(j)} \phi_{II}\left(\frac{z}{l^{(j)}}\right) \hat{\beta}_2, \qquad (5.1.17)$$

$$v_2^{(j)}(z) = \phi_I\left(\frac{z}{l^{(j)}}\right)(\hat{\alpha}_2 + (\hat{\boldsymbol{\beta}} \times \hat{\mathbf{a}}^{(j)})_2) - l^{(j)}\phi_{II}\left(\frac{z}{l^{(j)}}\right)\hat{\beta}_1, \quad (5.1.18)$$

$$v_3^{(j)}(z) = \left(1 - \frac{z}{l^{(j)}}\right)(\hat{\alpha}_3 + (\hat{\boldsymbol{\beta}} \times \hat{\mathbf{a}}^{(j)})_3), \quad (5.1.19)$$

$$v_4^{(j)}(z) = \left(1 - \frac{z}{l^{(j)}}\right)\hat{\beta}_3, \quad (5.1.20)$$

where
$$\phi_I(t) = (t-1)^2(2t+1), \quad \phi_{II}(t) = (t-1)^2 t.$$

Evaluating the right-hand sides of (5.1.13)–(5.1.16), we obtain

$$\hat{\mathfrak{F}}_1^{(j)} = -C_1^{(j)}\left\{\frac{12}{l^{(j)3}}(\hat{\alpha}_1 + (\hat{\boldsymbol{\beta}} \times \hat{\mathbf{a}}^{(j)})_1) + \frac{6}{l^{(j)2}}\hat{\beta}_2\right\}, \quad (5.1.21)$$

$$\hat{\mathfrak{F}}_2^{(j)} = -C_2^{(j)}\left\{\frac{12}{l^{(j)3}}(\hat{\alpha}_2 + (\hat{\boldsymbol{\beta}} \times \hat{\mathbf{a}}^{(j)})_2) - \frac{6}{l^{(j)2}}\hat{\beta}_1\right\}, \quad (5.1.22)$$

$$\hat{\mathfrak{F}}_3^{(j)} = -C_3^{(j)}(\hat{\alpha}_3 + (\hat{\boldsymbol{\beta}} \times \hat{\mathbf{a}}^{(j)})_3), \quad (5.1.23)$$

and

$$\hat{\mathfrak{M}}_1^{(j)} = C_2^{(j)}\left\{\frac{6}{l^{(j)2}}(\hat{\alpha}_2 + (\hat{\boldsymbol{\beta}} \times \hat{\mathbf{a}}^{(j)})_2) - \frac{4}{l^{(j)}}\hat{\beta}_1^{(j)}\right\}, \quad (5.1.24)$$

$$\hat{\mathfrak{M}}_2^{(j)} = C_1^{(j)}\left\{-\frac{6}{l^{(j)2}}(\hat{\alpha}_1 + (\hat{\boldsymbol{\beta}} \times \hat{\mathbf{a}}^{(j)})_1) - \frac{4}{l^{(j)}}\hat{\beta}_2^{(j)}\right\}, \quad (5.1.25)$$

$$\hat{\mathfrak{M}}_3^{(j)} = -C_4^{(j)}\frac{\hat{\beta}_3}{l^{(j)}}. \quad (5.1.26)$$

The above relations for the components of forces and moments together with (5.1.10) yield

$$\mathfrak{F}^{(j)} = -(\boldsymbol{\Lambda}^{(j)})^*\mathbf{R}^{(j)}\boldsymbol{\Lambda}^{(j)}(\boldsymbol{\alpha} + \boldsymbol{\beta} \times \mathbf{a}^{(j)}) - (\boldsymbol{\Lambda}^{(j)})^*\mathbf{Q}^{(j)}\boldsymbol{\Lambda}^{(j)}\boldsymbol{\beta}, \quad (5.1.27)$$

$$\mathfrak{M}^{(j)} = -(\boldsymbol{\Lambda}^{(j)})^*\mathbf{Q}^*\boldsymbol{\Lambda}^{(j)}(\boldsymbol{\alpha} + \boldsymbol{\beta} \times \mathbf{a}^{(j)}) - (\boldsymbol{\Lambda}^{(j)})^*\mathbf{T}^{(j)}\boldsymbol{\Lambda}^{(j)}\boldsymbol{\beta}, \quad (5.1.28)$$

where

$$\mathbf{R}^{(j)} = \text{diag}\left\{\frac{12C_1^{(j)}}{l^{(j)3}}, \frac{12C_2^{(j)}}{l^{(j)3}}, C_3^{(j)}\right\},$$

$$\mathbf{T}^{(j)} = \text{diag}\left\{\frac{4C_2^{(j)}}{l^{(j)}}, \frac{4C_1^{(j)}}{l^{(j)}}, \frac{C_4^{(j)}}{l^{(j)}}\right\},$$

PILE STRUCTURE MODEL

$$\mathbf{Q}^{(j)} = \frac{6}{l^{(j)2}} \begin{pmatrix} 0 & C_1^{(j)} & 0 \\ -C_2^{(j)} & 0 & 0 \\ 0 & 0 & 0 \end{pmatrix}.$$

5.1.5 Algebraic system for the pile structure model

We introduce the matrix

$$\mathbf{A}^{(j)} = \begin{pmatrix} 0 & -a_3^{(j)} & a_2^{(j)} \\ a_3^{(j)} & 0 & -a_1^{(j)} \\ -a_2^{(j)} & a_1^{(j)} & 0 \end{pmatrix}. \tag{5.1.29}$$

Then, for an arbitrary vector $\mathbf{b}$,

$$\mathbf{a}^{(j)} \times \mathbf{b} = \mathbf{A}^{(j)} \mathbf{b}. \tag{5.1.30}$$

Substituting the formulae (5.1.27), (5.1.28) into the balance relations (4.2.68) and using (5.1.31) we arrive at the following system of linear algebraic equations with respect to $\boldsymbol{\alpha}$ and $\boldsymbol{\beta}$:

$$\sum_{j=1}^{K} (\mathbf{\Lambda}^{(j)})^* \mathbf{R}^{(j)} \mathbf{\Lambda}^{(j)} \boldsymbol{\alpha} + \sum_{j=1}^{K} \{ (\mathbf{\Lambda}^{(j)})^* \mathbf{Q}^{(j)} \mathbf{\Lambda}^{(j)}$$
$$- (\mathbf{\Lambda}^{(j)})^* \mathbf{R}^{(j)} \mathbf{\Lambda}^{(j)} \mathbf{A}^{(j)} \} \boldsymbol{\beta} = \mathfrak{F}, \tag{5.1.31}$$
$$\sum_{j=1}^{K} \{ (\mathbf{\Lambda}^{(j)})^* \mathbf{Q}^{(j)} \mathbf{\Lambda}^{(j)} + \mathbf{A}^{(j)} (\mathbf{\Lambda}^{(j)})^* \mathbf{R}^{(j)} \mathbf{\Lambda}^{(j)} \} \boldsymbol{\alpha}$$
$$+ \sum_{j=1}^{K} \{ (\mathbf{\Lambda}^{(j)})^* \mathbf{T}^{(j)} \mathbf{\Lambda}^{(j)} - (\mathbf{\Lambda}^{(j)})^* (\mathbf{Q}^{(j)})^* \mathbf{\Lambda}^{(j)} \mathbf{A}^{(j)}$$
$$+ \mathbf{A}^{(j)} (\mathbf{\Lambda}^{(j)})^* \mathbf{Q}^{(j)} \mathbf{\Lambda}^{(j)}$$
$$- \mathbf{A}^{(j)} (\mathbf{\Lambda}^{(j)})^* \mathbf{R}^{(j)} \mathbf{\Lambda}^{(j)} \mathbf{A}^{(j)} \} \boldsymbol{\beta} = \mathfrak{M}. \tag{5.1.32}$$

Since $\mathbf{R}^{(j)}$ and $\mathbf{T}^{(j)}$ are diagonal and $(\mathbf{A}^{(j)})^* = -\mathbf{A}^{(j)}$ the matrix of the system (5.1.32), (5.1.33) is symmetric.

LEMMA 5.1

The matrix of the system (5.1.32), (5.1.33) is positive definite, and the following relation holds:

$$\mathfrak{F} \cdot \boldsymbol{\alpha} + \mathfrak{M} \cdot \boldsymbol{\beta} = \sum_{j=1}^{K} \int_0^{l^{(j)}} \sum_{i=1}^{2} \left\{ c_i^{(j)} \left(\frac{d^2 v_i^{(j)}}{dz^2} \right)^2 \right.$$

$$+ C_{i+2}^{(j)}\left(\frac{dv_{i+2}^{(j)}}{dz}\right)^2\Bigg\}dz. \tag{5.1.33}$$

Proof

By (4.2.67) and (4.2.68), the left-hand side of (5.1.34) can be written as

$$-\sum_{j=1}^{K}\{\mathfrak{F}^{(j)}\cdot\boldsymbol{\alpha} + (\mathfrak{M}^{(j)} + \mathbf{a}^{(j)}\times\mathfrak{F}^{(j)})\cdot\boldsymbol{\beta}\}$$

$$= -\sum_{j=1}^{K}\{\hat{\mathfrak{F}}^{(j)}\cdot\hat{\boldsymbol{\alpha}}^{(j)} + (\hat{\mathfrak{M}}^{(j)} + \hat{\mathbf{a}}^{(j)}\times\hat{\mathfrak{F}}^{(j)})\cdot\hat{\boldsymbol{\beta}}^{(j)}\}$$

$$= -\sum_{j=1}^{K}\{\hat{\mathfrak{F}}^{(j)}\cdot(\hat{\boldsymbol{\alpha}}^{(j)} + \hat{\mathbf{a}}^{(j)}\times\hat{\boldsymbol{\beta}}^{(j)}) + \hat{\mathfrak{M}}^{(j)}\times\hat{\boldsymbol{\beta}}^{(j)}\}.$$

Using (5.1.11)–(5.1.16) we can write this expression in the form

$$\sum_{j=1}^{K}\sum_{i=1}^{2}\Bigg\{C_i^{(j)}\left(\frac{d^3v_i^{(j)}}{dz^3}(0)v_i^{(j)}(0) - \frac{d^2v_i^{(j)}}{dz^2}(0)\frac{dv_i^{(j)}}{dz}(0)\right)$$

$$- C_{i+2}^{(j)}\frac{dv_{i+2}^{(j)}}{dz}(0)v_{i+2}^{(j)}(0)\Bigg\}.$$

Integrating the right-hand side (5.1.34) by parts and using the homogeneous equations (5.1.6), (5.1.7) and the homogeneous boundary conditions (5.1.8), (5.1.9) we arrive at (5.1.34).

The right-hand side of (5.1.34) is non-negative, and vanishes if and only if the vectors $\boldsymbol{\alpha}$ and $\boldsymbol{\beta}$ are zero. Therefore, the matrix of the system (5.1.32)–(5.1.33) is positive definite. □

It should be noticed that the quantity on the left-hand side of (5.1.34) has the physical meaning of strain energy of the pile structure.

By Lemma 5.1, system (5.1.32), (5.1.33) yields a unique set of the coefficients α_k, β_k, $k = 1,\ldots,6$, for the rigid-body displacements. Then, the components of lock forces and moments are evaluated by (5.1.27), (5.1.28). System (5.1.32), (5.1.33) is the basic relation in the pile structure model, and its matrix is called the *stiffness matrix of the pile structure*.

5.1.6 Non-degenerate and degenerate pile structures

By (5.1.4), up to leading order the system (5.1.32), (5.1.33) can be replaced by

$$\sum_{j=1}^{K} C_3^{(j)} \mathfrak{R}^{(j)} \alpha - \sum_{j=1}^{K} C_3^{(j)} \mathfrak{R}^{(j)} \mathbf{A}^{(j)} \beta = \mathfrak{F}, \qquad (5.1.34)$$

$$\sum_{j=1}^{K} C_3^{(j)} \mathbf{A}^{(j)} \mathfrak{R}^{(j)} \alpha - \sum_{j=1}^{K} C_3^{(j)} \mathbf{A}^{(j)} \mathfrak{R}^{(j)} \mathbf{A}^{(j)} \beta = \mathfrak{M}, \qquad (5.1.35)$$

where

$$\mathfrak{R}^{(j)} = (\mathbf{\Lambda}^{(j)})^* \operatorname{diag}\{0,0,1\} \mathbf{\Lambda}^{(j)}. \qquad (5.1.36)$$

LEMMA 5.2

The matrix of system (5.1.35), (5.1.36) is positive definite if and only if

$$\boldsymbol{\mu}^{(j)} \cdot (\boldsymbol{\alpha} - \mathbf{a}^{(j)} \times \boldsymbol{\beta}) = 0, \quad j = 1, \ldots, K, \qquad (5.1.37)$$

implies that $\alpha = \beta = 0$. *Here the unit vector* $\boldsymbol{\mu}^{(j)}$ *shows the direction of the rod* $r^{(j)}$.

Proof

The right-hand side of the identity

$$\sum_{j=1}^{K} C_3^{(j)} \{\boldsymbol{\alpha} \cdot \mathfrak{R}^{(j)} \boldsymbol{\alpha} - \boldsymbol{\alpha} \cdot \mathfrak{R}^{(j)} \mathbf{A}^{(j)} \boldsymbol{\beta} + \boldsymbol{\beta} \cdot \mathbf{A}^{(j)} \mathfrak{R}^{(j)} \boldsymbol{\alpha}$$

$$- \boldsymbol{\beta} \cdot \mathbf{A}^{(j)} \mathfrak{R}^{(j)} \mathbf{A}^{(j)} \boldsymbol{\beta}\} = \sum_{j=1}^{K} C_3^{(j)} (\boldsymbol{\alpha} - \mathbf{A}^{(j)} \boldsymbol{\beta}) \cdot \mathfrak{R}^{(j)} (\boldsymbol{\alpha} - \mathbf{A}^{(j)} \boldsymbol{\beta})$$

is non-negative and vanishes if and only if

$$\mathbf{e}^{(3)} \cdot \mathbf{\Lambda}^{(j)} (\boldsymbol{\alpha} - \mathbf{a}^{(j)} \times \boldsymbol{\beta}) = 0, \quad j = 1, \ldots, K.$$

The result follows. $\square$

The above lemma provides a motivation for the following definition.

DEFINITION The pile structure is said to be *non-degenerate* if the equalities (5.1.38) imply that $\boldsymbol{\alpha} = \boldsymbol{\beta} = \mathbf{0}$, or, equivalently, the matrix of the system (5.1.35), (5.1.36) is positive definite. Otherwise, the pile structure will be called *degenerate*.

REMARK 5.1

In order to describe the principal part of the displacement of the region Ω for the non-degenerate pile structure, one can use the simplified model where the hinge joints are imposed at junction points. Consequently, the rods r_j are subject to tension–compression load only, and the lock moments equal zero.

5.1.7 Examples

Non-degenerate pile structure

Assume that the number K of junction points equals six, and the set $\{r_j, j = 1, \ldots, 6\}$ involves three pairs of parallel rods (see Fig. 5.2). Namely, $\boldsymbol{\mu}^{(1)} = \boldsymbol{\mu}^{(2)}$, $\boldsymbol{\mu}^{(3)} = \boldsymbol{\mu}^{(4)}$, $\boldsymbol{\mu}^{(5)} = \boldsymbol{\mu}^{(6)}$. Then this pile structure satisfies the condition (5.1.38) if and only if $\boldsymbol{\mu}^{(1)}, \boldsymbol{\mu}^{(3)}, \boldsymbol{\mu}^{(5)}$ are linearly independent, and the vectors $\boldsymbol{\mu}^{(1)} \times (\mathbf{a}^{(1)} - \mathbf{a}^{(2)}), \boldsymbol{\mu}^{(3)} \times (\mathbf{a}^{(3)} - \mathbf{a}^{(4)}), \boldsymbol{\mu}^{(5)} \times (\mathbf{a}^{(5)} - \mathbf{a}^{(6)})$ span $\mathbb{R}^3$.

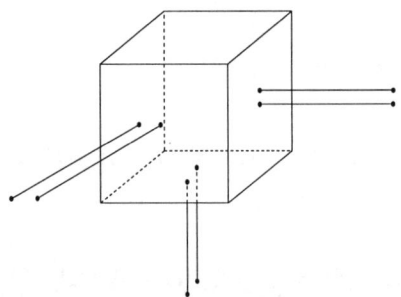

FIG. 5.2. Non-degenerate pile structure with six rods.

Degenerate pile structures

(1) Assume that $\boldsymbol{\mu}^{(1)}, \ldots, \boldsymbol{\mu}^{(K)}$ do not span $\mathbb{R}^3$, as shown in Fig. 5.3(a). Let $\boldsymbol{\alpha}$ be the vector orthogonal to the linear hull of $\boldsymbol{\mu}^{(1)}, \ldots, \boldsymbol{\mu}^{(K)}$,

and $\beta = 0$. Then the pair α, β gives a non-trivial solution of the homogeneous system (5.1.38). In particular, the pile structure with parallel elastic rods, considered in Chapter 4, belongs to this class, and, hence, it is degenerate.

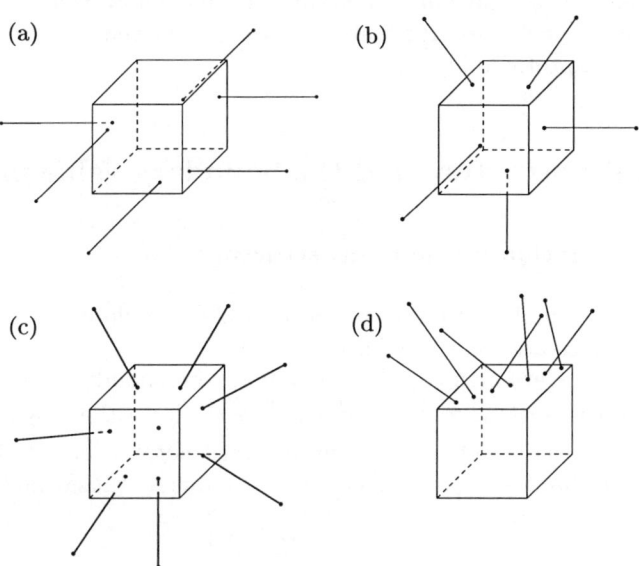

FIG. 5.3. Degenerate pile structures: (a) six rods orthogonal to the x_3 axis, (b) five arbitrarily directed rods, (c) radially directed rods, (d) junction points on the same straight line.

(2) Let $K < 6$ (see Fig. 5.3(b)). Then the number of equations (5.1.38) is less than the number of unknowns, and, consequently, there exists a non-trivial solution.

(3) Consider the case when the straight lines directed along $\boldsymbol{\mu}^{(j)}$, $j = 1, \ldots, K$, intersect in one point, as in Fig. 5.3(c). Then one can choose this point as the origin of the coordinate system, and consequently obtain

$$\boldsymbol{\mu}^{(j)} \times \mathbf{a}^{(j)} = \mathbf{0}, \quad j = 1, \ldots, K.$$

Hence, any pair of the form $\boldsymbol{\alpha} = \mathbf{0}, \boldsymbol{\beta} \neq \mathbf{0}$ gives a non-trivial solution of the system (5.1.38).

(4) Finally, suppose that all junction points are located on the same

straight line (see Fig. 5.3(d)). One can use the coordinate system with the origin which belongs to this line. Taking $\alpha = 0$, and $\boldsymbol{\beta} = \mathbf{a}^{(1)} - \mathbf{a}^{(2)}$, we obtain the non-trivial solution of (5.1.38).

Next, we shall formulate the boundary value problem for the Lamé operator in the multi-structure Ω_ε and show that the simple engineering model, described above, can be derived on the basis of rigorous asymptotic analysis.

5.2 Multi-structure and the boundary value problem

5.2.1 Description of the multi-structure

We deal with a multi-structure Ω_ε with the skeleton described in Section 5.1.1 (examples are shown in Figs. 5.2 and 5.1).

We need a Cartesian coordinate system corresponding to a junction point. Let the system $\mathbf{x}^{(j)} = (x_1^{(j)}, x_2^{(j)}, x_3^{(j)})$ have the origin at $\mathbf{a}^{(j)}$, and the $x_3^{(j)}$ axis directed along the outward normal on $\partial\Omega$. Then the coordinates $\mathbf{x}$ and $\mathbf{x}^{(j)}$ are connected by the linear transformation

$$\mathbf{x} = \mathbf{a}^{(j)} + \mathcal{R}^{(j)}\mathbf{x}^{(j)}$$

with an orthogonal matrix $\mathcal{R}^{(j)}$. If the rods r_j are perpendicular to $\partial\Omega$ at the junction points then one can suppose that $\mathbf{x}^{(j)} = \mathbf{y}^{(j)}$ and, hence, $\mathcal{R}^{(j)} = \mathbf{\Lambda}^{(j)}$.

Also, we shall use the scaled coordinates

$$\mathbf{X}^{(j)} = (X_1^{(j)}, X_2^{(j)}, X_3^{(j)}) = \varepsilon^{-1}\mathbf{x}^{(j)},$$

$$\boldsymbol{\mathcal{Y}}^{(j)} = (\mathcal{Y}_1^{(j)}, \mathcal{Y}_2^{(j)}, \mathcal{Y}_3^{(j)}) = \varepsilon^{-1}\mathbf{y}^{(j)},$$

to describe Ω_ε in the vicinity of junction points, and

$$\mathbf{Y}^{(j)} = (Y_1^{(j)}, Y_2^{(j)}, Y_3^{(j)}) = \varepsilon^{-1}(y_1^{(j)}, y_2^{(j)}, y_3^{(j)} - l^{(j)})$$

to describe Ω_ε in neighbourhoods of the base points $\mathbf{b}^{(j)}$. Clearly, $Y_1^{(j)} = \mathcal{Y}_1^{(j)}$, $Y_2^{(j)} = \mathcal{Y}_2^{(j)}$ and

$$\mathbf{X}^{(j)} = \mathcal{L}^{(j)}\boldsymbol{\mathcal{Y}}^{(j)},$$

where $\mathcal{L}^{(j)} = (\mathcal{R}^{(j)})^*\mathbf{\Lambda}^{(j)}$ with $\mathbf{\Lambda}^{(j)}$ as in Section 5.1.1.

The following set of model domains, independent of ε, will be used for a description of the multi-structure.

(1) The domain Ω corresponds to 'the pile cap' of the multi-structure.

(2) The segments r_j and two-dimensional bounded domains $g^{(j)}$ with Lipschitz boundaries describe the scaled cross-sections of thin rods.

(3) To each junction point $\mathbf{a}^{(j)}$ we assign a domain $G^{(j)}$ with Lipschitz boundary such that in the exterior of the unit ball it coincides with a disjoint union of a half-space $\mathbb{R}^3_-$ and a semi-cylinder with a cross-section $g^{(j)}$. It is supposed that the cylinder axis is transverse to $\partial \mathbb{R}^3_-$ (see Fig. 5.4).

The coordinates $\mathbf{X}^{(j)}$ and $\mathbf{\mathcal{Y}}^{(j)}$ are used to describe $G^{(j)}$. In addition, we suppose that the $O\mathcal{Y}_3^{(j)}$ axis is directed along the cylinder, and the axes $O\mathcal{Y}_1^{(j)}$, $O\mathcal{Y}_2^{(j)}$ coincide with the inertia axes of $g^{(j)}$, and

$$\mathbb{R}^3_- = \left\{ \mathbf{X}^{(j)} : X_3^{(j)} < 0 \right\}.$$

For sufficiently large positive ρ the following relation holds:

$$\left\{ \mathbf{\mathcal{Y}}^{(j)} : \mathcal{Y}_3^{(j)} > \rho \right\} \cap G^{(j)} = \left\{ \mathbf{\mathcal{Y}}^{(j)} : \mathcal{Y}_3^{(j)} > \rho, \left(\mathcal{Y}_1^{(j)}, \mathcal{Y}_2^{(j)} \right) \in g^{(j)} \right\}.$$

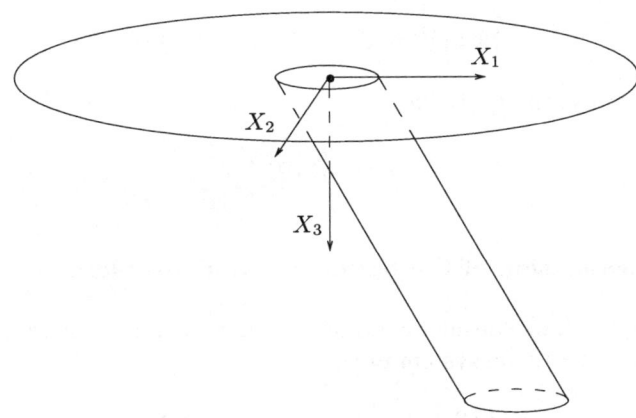

FIG. 5.4. A junction layer region.

(4) Also, we use the notation $\Pi_-^{(j)}$ for a semi-infinite cylinder

$$\left\{ \mathbf{Y}^{(j)} : Y_3^{(j)} < 0, (Y_1^{(j)}, Y_2^{(j)}) \in g^{(j)} \right\}.$$

Now we are in a position to describe the multi-structure Ω_ε. Let $B(a, d)$ denote a ball of radius d with centre at $\mathbf{a}$. There exists a positive quantity ρ independent of ε such that

$$\Omega_\varepsilon = \Omega \cup \left(\cup_j \left(B(\mathbf{a}^{(j)}, \rho) \cap \Omega_\varepsilon \right) \right) \cup \left(\cup_j \Pi_\varepsilon^{(j)} \right),$$

where

$$\Pi_\varepsilon^{(j)} = \left\{ \mathbf{y}^{(j)} : 0 < y_3^{(j)} < l^{(j)},\ \varepsilon^{-1}(y_1^{(j)}, y_2^{(j)}) = (Y_1^{(j)}, Y_2^{(j)}) \in g^{(j)} \right\}$$

and the following relation holds:

$$B(\mathbf{a}^{(j)}, \rho) \cap \Omega_\varepsilon = \left\{ \mathbf{x} : \varepsilon^{-1} \mathbf{x}^{(j)} \in G^{(j)} \cap B(0, \varepsilon^{-1}\rho) \right\}.$$

Moreover,

$$B(\mathbf{b}^{(j)}, \rho) \cap \Omega_\varepsilon = \left\{ \mathbf{y}^{(j)} : \varepsilon^{-1}(\mathbf{y}^{(j)} - (0, 0, l^{(j)})) \in \Pi^{(j)} \cap B(0, \varepsilon^{-1}\rho) \right\}.$$

We introduce the base regions

$$S_\varepsilon^{(j)} = \{ \mathbf{x} : y_3^{(j)} = l^{(j)},\ \varepsilon^{-1}(y_1^{(j)}, y_2^{(j)}) \in g^{(j)} \},$$

and we also use the notation

$$S_\varepsilon = \cup_{j=1}^K S_\varepsilon^{(j)}.$$

5.2.2 Formulation of the boundary value problem

As in Chapter 4, we consider a mixed Dirichlet–Neumann boundary value problem for the Lamé system in Ω_ε:

$$-\mathbf{L}(\partial_x)\mathbf{u}(\mathbf{x}, \varepsilon) = \mathbf{F}(\mathbf{x}, \varepsilon),\ \mathbf{x} \in \Omega_\varepsilon, \qquad (5.2.1)$$

where $\mathbf{L}$ is the Lamé differential operator, and $\mathbf{u}$ is the displacement vector.

The traction boundary condition is

MODEL PROBLEMS

$$\sigma^{(n)}(\mathbf{u};\mathbf{x}) := \sigma(\mathbf{u};\mathbf{x})\mathbf{n} = \mathbf{P}(\mathbf{x},\varepsilon), \; \mathbf{x} \in \partial\Omega_\varepsilon \setminus \overline{S}_\varepsilon, \qquad (5.2.2)$$

and the displacement boundary condition is given by

$$\mathbf{u}(\mathbf{x},\varepsilon) = \mathbf{\Phi}^{(j)}(\mathbf{x},\varepsilon), \; \mathbf{x} \in S_\varepsilon^{(j)}, \; j = 1,\ldots,K, \qquad (5.2.3)$$

where $\mathbf{n}$ is the outward unit normal on $\partial\Omega_\varepsilon$.

We seek a displacement field with finite elastic energy. Theorem 4.8 also holds for problem (5.2.1)–(5.2.3) and therefore this problem has a unique solution which satisfies estimate (4.4.5).

This problem is uniquely solvable in $(H^1(\Omega_\varepsilon))^3$ if $\mathbf{F} \in (L_2(\Omega_\varepsilon))^3, \mathbf{P} \in (L_2(\partial\Omega \setminus S_\varepsilon))^3$ and $\mathbf{\Phi}^{(j)} \in (H^{1/2}(S_\varepsilon^{(j)}))^3$. The solution satisfies estimate (4.4.5). The proof of these facts is the same as that of Theorem 4.8.

5.3 Model problems

The asymptotic algorithm involves solutions of certain model boundary value problems independent of ε, which are stated in the present section.

5.3.1 Junction layer

Consider a domain G with Lipschitz boundary specified as a union of the half-space $\mathbb{R}_-^3 = \{\mathbf{X} : X_3 < 0\}$ and an infinite region, located in $\mathbb{R}_+^3$ and coinciding (outside a bounded neighbourhood of the origin) with a semi-infinite cylinder Π_+. We assume that the axis of this cylinder is transverse to the boundary of the half-space. Introduce an additional system of coordinates $\mathcal{Y} = (\mathcal{Y}_1, \mathcal{Y}_2, \mathcal{Y}_3)$ centred at $\mathbf{X} = 0$ and with the axis $O\mathcal{Y}_3$ coinciding with the axis of the cylinder, and being directed outwards from the half-space. The axes $O\mathcal{Y}_1$, $O\mathcal{Y}_2$ have the same orientation as the inertia axes of the cross-section g of the cylinder Π_+. The coordinates $\mathbf{X}$ and $\mathcal{Y}$ are related by

$$\mathbf{X} = \mathcal{L}\mathcal{Y}, \qquad (5.3.1)$$

where $\mathcal{L}$ is a rotation matrix. In the text below the region G will coincide with one of $G^{(j)}, j = 1,\ldots,k$, describing the junction regions. In this case the matrix $\mathcal{L} = \mathcal{L}^{(j)}$ is connected with the matrices $\mathcal{R}^{(j)}, \Lambda^{(j)}$ by

$$\mathcal{L}^{(j)} = (\mathcal{R}^{(j)})^* \Lambda^{(j)}.$$

A vector function $\mathbf{W}$ is said to belong to the space $\mathbf{E}(G)$ if $\mathbf{W} \in (\mathbf{H}^1(G \cap B_r))^3$ for all positive r and

$$\mathcal{E}(\mathbf{W};G) + \int_{\mathbb{R}^3_-} \|\mathbf{W}\|^2 \frac{d\mathbf{X}}{\|\mathbf{X}\|^2} < \infty,$$

where the functional $\mathcal{E}$ is defined in Section 3.8 by (3.8.1). Similar to Lemma 4.1 one can deduce

LEMMA 5.3

The functional $\mathbf{W} \to (\mathcal{E}(\mathbf{W};G))^{1/2}$ defines a norm in the space $\mathbf{E}(G)$, and the estimate (4.2.12) holds with $\mathbf{X}$ being replaced by $\mathcal{Y}$ in the integrals over Π_+.

Now, we consider the boundary value problem in G with prescribed tractions on ∂G:

$$-\mathbf{L}(\partial_\mathbf{X})\mathbf{W}(\mathbf{X}) = \mathbf{F}(\mathbf{X}), \quad \mathbf{X} \in G, \tag{5.3.2}$$
$$\sigma^{(n)}(\mathbf{W};\mathbf{X}) = \mathbf{P}(\mathbf{X}), \quad \mathbf{X} \in \partial G. \tag{5.3.3}$$

By a solution of this problem we mean a vector function $\mathbf{W} \in \mathbf{E}(G)$ such that

$$\sum_{i,j=1}^{3} \int_G \sigma_{ij}(\mathbf{W};\mathbf{X})\varepsilon_{ij}(\mathcal{V};\mathbf{X})d\mathbf{X} = h(\mathcal{V}), \tag{5.3.4}$$

for any $\mathcal{V} \in \mathbf{E}(G)$, where

$$h(\mathcal{V}) = \int_G \mathfrak{F}(\mathbf{X}) \cdot \mathcal{V}(\mathbf{X})d\mathbf{X} + \int_{\partial G} \mathbf{P}(\mathbf{X}) \cdot \mathcal{V}(\mathbf{X})dS_X. \tag{5.3.5}$$

The following statement deals with the solvability of problem (5.3.2), (5.3.3) and the asymptotic representation of its solution.

THEOREM 5.1

Let $\mathbf{F}$ and $\mathbf{P}$ be vector-valued functions from $(\mathbf{L}_2(G))^3$ and $(\mathbf{L}_2(\partial G))^3$ with compact supports.

(i) The problem (5.3.2), (5.3.3) is uniquely solvable in $\mathbf{E}(G)$.

(ii) The solution $\mathbf{W}$ admits the asymptotic representation (4.2.17) in the half-space $\mathbb{R}^3_-$.

(iii) In the cylindrical extension Π_+

$$\mathbf{W}(\mathbf{X}) = \mathcal{L}(\mathbf{a} + \mathbf{b} \times \mathcal{Y} + \mathbf{R}(\mathcal{Y})) \tag{5.3.6}$$

MODEL PROBLEMS

where **a**, **b** *are constant vectors, and*

$$\|\partial_{\mathcal{Y}_3}^j \mathbf{R}(\cdot, \mathcal{Y}_3)\|_{(H^1(g))^3} \leq C_j e^{-\gamma \mathcal{Y}_3}, \quad \gamma > 0, \tag{5.3.7}$$

for large positive $\mathcal{Y}_3$ *and some constant* C_j, $j = 0, 1, \ldots$.

The proof is identical to that of Theorem 4.2.

Theorem 5.1 gives a solution $\mathcal{W}$ which does not decay in the cylindrical extension Π_+ and vanishes like $O(\|\mathbf{X}\|^{-1})$ at infinity in the half-space $\mathbb{R}^3_-$. Similar to Section 4.2.3, we change the boundary value problem by introducing 12 constants in its formulation in such a way that the solution decays in both directions.

A smooth function η is introduced in such a way that $\eta(z) = 0$ for $z < C$ and $\eta(z) = 1$ for $z > 2C$, for some positive sufficiently large C. Consider the boundary value problem

$$-\mathbf{L}(\partial_{\mathbf{X}})\mathcal{W}(\mathbf{X}) = \mathbf{F}(\mathbf{X}) + \mathcal{L} \sum_{j=1}^{6} \left\{ d_j^{(I)} \mathbf{L}(\partial_{\mathbf{y}}) \eta(\mathcal{Y}_3) \boldsymbol{\xi}^{(j)}(\mathcal{Y}) \right.$$

$$\left. + d_j^{(II)} \mathbf{L}(\partial_{\mathbf{y}}) \eta(\mathcal{Y}_3) \boldsymbol{\psi}^{(j)}(\mathcal{Y}) \right\} \text{ in } G, \tag{5.3.8}$$

$$\sigma^{(n)}(\mathcal{W}; \mathbf{X}) = \mathbf{P}(\mathbf{X}) - \mathcal{L} \sum_{j=1}^{6} \left\{ d_j^{(I)} \sigma^{(n)}(\eta \boldsymbol{\xi}^{(j)}; \mathcal{Y}) \right.$$

$$\left. + d_j^{(II)} \sigma^{(n)}(\eta \boldsymbol{\psi}^{(j)}; \mathcal{Y}) \right\} \text{ on } \partial G, \tag{5.3.9}$$

where $d_j^{(I)}$, $d_j^{(II)}$, $j = 1, 2, \ldots, 6$, are constants.

Then the statement analogous to Theorem 4.3 from Section 4.2.3 is given by

THEOREM 5.2

Let $\mathbf{F} \in (\mathbf{L}_2(G))^3$, $\mathbf{P} \in (\mathbf{L}_2(\partial G))^3$ *have compact supports. There exists a unique set*

$$\{\mathcal{W}; d_1^{(I)}, \ldots, d_6^{(I)}; d_1^{(II)}, \ldots, d_6^{(II)}\} \tag{5.3.10}$$

which satisfies the boundary value problem (5.3.8), (5.3.9), *where* $\mathcal{W} \in \mathbf{E}(G)$ *admits the asymptotic expansion* (4.2.17) *in* $\mathbb{R}^3_-$ *with* $\mathfrak{F} = 0$, $\mathfrak{M} = 0$, *and decays exponentially in* Π_+. *The constants* $d_j^{(I)}$ *are given by*

$$d_q^{(I)} = D_q^{-1} (\mathcal{L}\mathbb{F})_q (\mathbf{F}, \mathbf{P}, G), \quad q = 1, 2, \tag{5.3.11}$$

$$d_3^{(I)} = -D_3^{-1}(\mathcal{L}\mathbb{F})_3(\mathbf{F},\mathbf{P},G), \qquad (5.3.12)$$

$$d_4^{(I)} = -D_4^{-1}(\mathcal{L}\mathbb{M})_3(\mathbf{F},\mathbf{P},G), \qquad (5.3.13)$$

$$d_5^{(I)} = -D_1^{-1}(\mathcal{L}\mathbb{M})_2(\mathbf{F},\mathbf{P},G), \qquad (5.3.14)$$

$$d_6^{(I)} = D_2^{-1}(\mathcal{L}\mathbb{M})_1(\mathbf{F},\mathbf{P},G). \qquad (5.3.15)$$

Note that the components of the vectors $\mathbb{F}$ and $\mathbb{M}$ are evaluated in coordinates $\mathbf{X}$. The proof follows the same pattern as in Theorem 4.3.

REMARK 5.2

If we take a different function $\eta_1 = \eta_1(Y_3)$ in (5.3.8), (5.3.9) (with the same properties as η) then we can check directly that the collection

$$\{\mathcal{W}^{(1)}; d_1^{(I)}, \ldots, d_6^{(I)}; d_1^{(II)}, \ldots, d_6^{(II)}\},$$

where

$$\mathcal{W}^{(1)} = \mathcal{W} + \sum_{j=1}^{6} \left(d_j^{(I)}(\eta - \eta_1)\boldsymbol{\xi}^{(j)} + d_j^{(II)}(\eta - \eta_1)\boldsymbol{\psi}^{(j)} \right),$$

will be a solution of (5.3.8), (5.3.9) (with η replaced by η_1). Thus, the constants $d_j^{(I)}, d_j^{(II)}, j = 1, \ldots, 6$, do not depend on the choice of η.

5.3.2 Remaining model problems

Here we list a number of model problems independent of ε with appropriate references to Section 4.2.

Model problem for Ω. We refer to the boundary value problem presented in Section 4.2.2 for the Lamé system in Ω with traction boundary conditions and admit that the tractions may be singular at the junction points $\mathbf{a}^{(j)}$, $j = 1, \ldots, K$. The description of the model problem follows the same pattern as in Section 4.2.2.

Model problem for the bottom layer. Consider a semi-infinite cylinder Π_- with a cross-section g and a base region S (it represents one of the cylinders $\Pi_-^{(j)}$ introduced above). Then, a mixed boundary value problem is posed for the Lamé operator, where the Dirichlet data are specified on the base region, and tractions are given on the lateral surface of the cylinder (see Section 4.2.4).

Model problem for a bounded two-dimensional domain.
Formulation of the model boundary value problem in a two-dimensional domain is given in Section 2.2.

Model problems on the axis of a rod.
These are considered in Section 4.2.6.

Stiffness matrix for the skeleton.
When constructing the asymptotics of $\mathbf{u}(\cdot,\varepsilon)$ we shall make use of the following. We introduce a matrix which depends on the geometry of the multi-structure and on the normalised longitudinal stiffness coefficients $D_3^{(j)}$, $j=1,\ldots,K$, of elastic rods:

$$\mathcal{A} = \sum_{j=1}^{K} D_3^{(j)} \begin{pmatrix} \mathfrak{R}^{(j)} & -\mathfrak{R}^{(j)}\mathbf{A}^{(j)} \\ \mathbf{A}^{(j)}\mathfrak{R}^{(j)} & -\mathbf{A}^{(j)}\mathfrak{R}^{(j)}\mathbf{A}^{(j)} \end{pmatrix}, \qquad (5.3.16)$$

where $\mathfrak{R}^{(j)}$ and $\mathbf{A}^{(j)}$ are given by (5.1.37), (5.1.30). After multiplication by ε^2 this matrix coincides with the matrix of system (5.1.35), (5.1.36) which appeared in the previous analysis of the pile structure model. Since we are dealing with the non-degenerate multi-structure, it follows from Section 5.1.6 that the matrix $\mathcal{A}$ is positive definite.

5.4 Asymptotic expansion of the solution

In this section an asymptotic series is constructed for the solution of a mixed boundary value problem in the non-degenerate multi-structure which consists of the union of a three-dimensional cap Ω and a set of cylindrical extensions $\Pi_\varepsilon^{(j)}$, $j=1,\ldots,K$.

Local coordinates are defined in Section 5.4.1. A multi-scaled asymptotic representation of the right-hand sides is specified in Section 5.4.2. In Section 5.4.3 we describe the asymptotic expansion of the displacement field in Ω_ε. The general term of this asymptotic series is constructed in Sections 5.4.4–5.4.7. In Section 5.4.9 we present the sequence of recurrent relations.

5.4.1 Cut-off functions

We collect the coordinate systems used for the description of Ω_ε. The global system of coordinates $\mathbf{x} = (x_1, x_2, x_3)$ will be used for the description of elastic fields in the entire multi-structure. However, in the vicinity of junction points, the end points $b^{(j)}$ and the middle regions of thin rods we use other local coordinates.

For the junction regions we use the coordinates

$$\mathbf{X}^{(j)} = \varepsilon^{-1}\mathbf{x}^{(j)} = \varepsilon^{-1}\left(\mathcal{R}^{(j)}\right)^*(\mathbf{x} - \mathbf{a}^{(j)})$$

and

$$\mathcal{Y}^{(j)} = \varepsilon^{-1}\mathbf{y}^{(j)} = \varepsilon^{-1}\left(\Lambda^{(j)}\right)^*(\mathbf{x} - \mathbf{a}^{(j)}),$$

introduced in Section 5.2.1. Note that the $\mathcal{Y}_3^{(j)}$ axis is directed along the cylindrical extension of $G^{(j)}$, and the axes $\mathcal{Y}_1^{(j)}$, $\mathcal{Y}_2^{(j)}$ coincide with the inertia axes of the cross-section $g^{(j)}$.

The coordinate system

$$(\zeta_1^{(j)}, \zeta_2^{(j)}, z^{(j)}) = (\mathcal{Y}_1^{(j)}, \mathcal{Y}_2^{(j)}, \mathcal{Y}_3^{(j)})$$

is introduced in $\Pi_\varepsilon^{(j)}$.

The scaled variables

$$\mathbf{Y}^{(j)} = (\mathcal{Y}_1^{(j)}, \mathcal{Y}_2^{(j)}, \varepsilon^{-1}(y_3^{(j)} - l^{(j)}))$$

are used in the vicinity of the end $\mathbf{b}^{(j)}$ of the thin cylinder

$$\Pi_\varepsilon^{(j)} = \{\mathbf{y}^{(j)} : 0 \leq y_3^{(j)} \leq l^{(j)}, \varepsilon^{-1}(y_1^{(j)}, y_2^{(j)}) \in g^{(j)}\}.$$

We define the following cut-off function with support in Ω:

$$\mathfrak{X}(\mathbf{x}, \varepsilon) = \chi_\Omega(\mathbf{x}) \prod_{j=1}^K \chi(\mathbf{X}^{(j)}),$$

where χ_Ω is the characteristic function of Ω, χ is smooth, $\chi(\mathbf{X}) = 0$ when $\|\mathbf{X}\| < C$ and $\chi(\mathbf{X}) = 1$ when $\|\mathbf{X}\| > 2C$ for some constant C independent of ε; the constant C is chosen in such a way that for any $j = 1, \ldots, K$, the region $G^{(j)} \setminus B(0, C)$ is represented by a disjoint union of $\mathbb{R}^3_- \setminus B(0, C)$ and a semi-infinite cylindrical extension. Note that $\mathfrak{X}$ is smooth in Ω_ε, and vanishes outside Ω.

Also, in the thin cylinders we use the smooth cut-off function $\eta(Z)$ which equals 0 for $Z < 1$ and $\eta(Z) = 1$ for $Z > 2$.

The third cut-off function $\Xi^{(j)} = \Xi^{(j)}(\mathbf{x})$ for the junction layer is defined in $\mathbb{R}^2$ in such a way that $\Xi^{(j)}$ is equal to 1 in $\Omega \cap B(\mathbf{a}^{(j)}, d/4)$. It vanishes in $\Omega \setminus B(\mathbf{a}^{(j)}, d/2)$ and is extended by 1 in $\Pi_\varepsilon^{(j)}$ and by 0 in all other rods $\Pi_\varepsilon^{(i)}$, $i \neq j$. Here $d = \min_{i,j} \text{dist}(\mathbf{a}^{(j)}, \mathbf{a}^{(i)})$.

5.4.2 Asymptotic representation of the right-hand sides for the case of a non-degenerate multi-structure

The right-hand sides of the boundary value problem (5.2.1)–(5.2.3) are represented as asymptotic series involving four types of terms which correspond to the region Ω, the junction regions, thin rods and the bottom region. That is, $\mathbf{F}, \mathbf{P}$ and $\Phi^{(j)}$, $j = 1, \ldots, K$, in (5.2.1)–(5.2.3) admit the asymptotic expansions

$$\mathbf{F}(\mathbf{x},\varepsilon) \sim \sum_{k=0}^{\infty} \varepsilon^k \bigg\{ \mathfrak{X}(\mathbf{x},\varepsilon)\mathbf{f}^{(k)}(\mathbf{x}) + \sum_{j=1}^{K} \bigg[\mathcal{R}^{(j)} \varepsilon^{-4} \mathbf{F}^{(k,j)}(\mathbf{X}^{(j)})$$
$$+ \Lambda^{(j)} \varepsilon^{-2} \mathcal{F}^{(k,j)}(\zeta^{(j)}, z^{(j)}) \eta\Big(\frac{z^{(j)}}{\varepsilon}\Big) + \Lambda^{(j)} \varepsilon^{-4} \mathcal{G}^{(k,j)}(\mathbf{Y}^{(j)}) \bigg] \bigg\} \quad (5.4.1)$$

for $\mathbf{x} \in \Omega_\varepsilon$,

$$\mathbf{P}(\mathbf{x},\varepsilon) \sim \sum_{k=0}^{\infty} \varepsilon^k \bigg\{ \mathfrak{X}(\mathbf{x},\varepsilon)\mathbf{p}^{(k)}(\mathbf{x}) + \sum_{j=1}^{K} \bigg[\mathcal{R}^{(j)} \varepsilon^{-3} \mathbf{P}^{(k,j)}(\mathbf{X}^{(j)})$$
$$+ \Lambda^{(j)} \Big(\varepsilon^{-1} \mathcal{P}^{(k,j)}(\zeta^{(j)}, z^{(j)}) \eta\Big(\frac{z^{(j)}}{\varepsilon}\Big) + \varepsilon^{-3} \mathbf{H}^{(k,j)}(\mathbf{Y}^{(j)})\Big) \bigg] \bigg\} (5.4.2)$$

for $\mathbf{x} \in \partial\Omega_\varepsilon \setminus \bar{S}_\varepsilon$,

$$\Phi^{(j)}(\mathbf{x},\varepsilon) \sim \Lambda^{(j)} \sum_{k=0}^{\infty} \varepsilon^{k-2} \Phi^{(k,j)}(Y_1^{(j)}, Y_2^{(j)}), \ \mathbf{x} \in S_\varepsilon. \quad (5.4.3)$$

It is assumed that $\mathbf{f}^{(k)} \in (L_2(\Omega))^3$ and $\mathbf{p}^{(k)} \in (L_2(\partial\Omega))^3$ are smooth in fixed neighbourhoods of the junction points. The vector-valued functions $\mathfrak{F}^{(k,j)}$, $\mathbf{P}^{(k,j)}$ belong to $(L_2(G^{(j)}))^3$, $(L_2(\partial G^{(j)}))^3$ and are supported in a ball centred at the origin; $\mathcal{F}^{(k,j)} \in C^\infty([0,l^{(j)}];(L_2(g^{(j)}))^3)$, $\mathcal{P}^{(k,j)} \in C^\infty([0,l^{(j)}];(L_2(\partial g^{(j)}))^3)$.

The vector functions $\mathcal{G}^{(k,j)} \in L_2(\Pi_-^{(j)})$, $\mathbf{H}^{(k,j)} \in L_2(\partial\Pi_-^{(j)} \setminus \bar{S}^{(j)})$ have bounded supports located in a fixed ball centred at the origin, and $\Phi^{(k,j)} \in H^{1/2}(S^{(j)})$.

5.4.3 Structure of the asymptotic series for the displacement field in Ω_ε

The displacement field $\mathbf{u}(\mathbf{x},\varepsilon)$ will be sought in the form of the asymptotic series

$$\mathbf{u}(\mathbf{x},\varepsilon) \sim \sum_{k=0}^{\infty} \varepsilon^k \mathbf{u}^{(k)}(\mathbf{x},\varepsilon), \qquad (5.4.4)$$

with

$$\mathbf{u}^{(k)}(\mathbf{x},\varepsilon) = \varepsilon^{-2}(\boldsymbol{\alpha}^{(k)} + \boldsymbol{\beta}^{(k)} \times \mathbf{x}) + \mathfrak{X}(\mathbf{x},\varepsilon)\mathbf{u}_\Omega^{(k)}(\mathbf{x}) \qquad (5.4.5)$$
$$+ \sum_{j=1}^{K} \Big\{ \Lambda^{(j)} \eta(z^{(j)}/\varepsilon) \mathfrak{U}^{(k,j)}(\zeta^{(j)}, z^{(j)}, \varepsilon)$$
$$+ \mathcal{R}^{(j)} \varepsilon^{-2} \Xi^{(j)}(\mathbf{x}) \mathcal{W}^{(k,j)}(\mathbf{X}^{(j)}) + \Lambda^{(j)} \varepsilon^{-2} \eta(z^{(j)}/\varepsilon) \mathcal{V}^{(k,j)}(\mathbf{Y}^{(j)}) \Big\}.$$

Here $\alpha_i^{(k)}$, $\beta_i^{(k)}$, $i = 1, 2, 3$, are constant quantities, and the terms in the first brackets in (5.4.5) represent rigid-body displacements; these coefficients are defined by the system of linear equations of the form (5.1.35), (5.1.36). The vector-valued function $\mathbf{u}_\Omega^{(k)}$ solves the traction boundary value problem in Ω, and can be represented as

$$\mathbf{u}_\Omega^{(k)}(\mathbf{x}) = \mathbf{u}_r^{(k)}(\mathbf{x}) + \mathbf{u}_s^{(k)}(\mathbf{x}). \qquad (5.4.6)$$

The field $\mathbf{u}_r^{(k)} \in (H^1(\Omega))^3$ is a solution of the traction boundary value problem with self-balanced right-hand sides from $(L_2(\Omega))^3$ which are smooth in the vicinity of the junction points. The vector function $\mathbf{u}_s^{(k)}$ is defined by

$$\mathbf{u}_s^{(k)}(\mathbf{x})$$
$$= \sum_{j=1}^{K} \Big\{ \mathcal{N}(\mathbf{x}, \mathbf{a}^{(j)}) \mathfrak{F}^{(k,j)} + \frac{1}{2} (\nabla_{\mathbf{x}^*} \times \mathcal{N}^t(\mathbf{x}, \mathbf{x}^*)|_{\mathbf{x}^* = \mathbf{a}^{(i)}})^t \mathfrak{M}^{(k,j)} \Big\}, \quad (5.4.7)$$

where $\mathcal{N}$ is Green's matrix for the Lamé system; $\mathfrak{F}^{(k,j)}$, $\mathfrak{M}^{(k,i)}$ are constant vectors which satisfy the balance relations

$$\mathbb{F}(\mathbf{f}^{(k)}, \mathbf{p}^{(k)}, \Omega) + \sum_{j=1}^{K} \mathfrak{F}^{(k,j)} = 0, \qquad (5.4.8)$$

$$\mathbb{M}(\mathbf{f}^{(k)}, \mathbf{p}^{(k)}, \Omega) + \sum_{j=1}^{K} \Big\{ \mathfrak{M}^{(k,j)} + \mathbf{a}^{(j)} \times \mathfrak{F}^{(k,j)} \Big\} = 0, \quad (5.4.9)$$

with $\mathbb{F}$ and $\mathbb{M}$ introduced in (4.2.5), (4.2.6).

The product
$$\mathfrak{X}(\mathbf{x},\varepsilon)\mathbf{u}_\Omega^{(k)}$$
is extended as zero into thin cylinders.

The vector function $\mathfrak{U}^{(k,j)}$ corresponds to the displacement in a thin cylinder $\Pi_\varepsilon^{(j)}$, and is given by

$$\mathfrak{U}^{(k,j)}(\zeta^{(j)}, z^{(j)}, \varepsilon) = \varepsilon^{-4} \sum_{i=1}^{2} \sum_{q=0}^{3} \varepsilon^q \frac{d^q}{dz^{(j)q}} v_i^{(k,j)}(z^{(j)}) \phi^{(i,q)}(\zeta^{(j)})$$
$$+ \varepsilon^{-2} \sum_{i=3}^{4} \sum_{q=0}^{1} \varepsilon^q \frac{d^q}{dz^{(j)q}} v_i^{(k,j)}(z^{(j)}) \phi^{(i,q)}(\zeta^{(j)})$$
$$+ \mathbf{U}^{(k,j)}(\zeta^{(j)}, z^{(j)}) \qquad (5.4.10)$$

(see Section 4.3.3). In particular, the functions $v_i^{(k,j)}$ depend only on the longitudinal variable and solve ordinary differential equations (4.3.15)–(4.3.17): $v_1^{(k,j)}$, $v_2^{(k,j)}$ satisfy the differential equation of fourth order and describe bending displacements of the rod, whereas $v_3^{(k,j)}$, $v_4^{(k,j)}$ satisfy the differential equations of second order and correspond to the longitudinal displacement and the axial rotation; the vector function $\mathbf{U}^{(k,j)} \in C^\infty([0, l^{(j)}]; (H^1(g))^3)$ is specified in the same way as in Theorem 4.6; and the vector functions $\phi^{(i,q)}$ represent the eigenvectors and generalised eigenvectors of the spectral problem (3.6.11), (3.6.12) corresponding to the zero eigenvalue.

The junction layer $\mathcal{W}^{(k,j)}$ and the constants $d_i^{(I,k,j)}$, $d_i^{(II,k,j)}$, $i = 1, 2, \ldots, 6$, give a solution of the boundary value problem of the type (5.3.8), (5.3.9) in $G^{(j)}$, and $\mathcal{W}^{(k,j)}$ decays at infinity (see Theorem 5.2).

The bottom layer $\mathcal{V}^{(k,j)}$ which decays at infinity and the constants $c_q^{(k,j)}$, $q = 1, 2, \ldots, 6$, are specified in a way identical to that given in Section 4.3.

The constants $d_i^{(I,k,j)}$, $d_i^{(II,k,j)}$, $c_i^{(k,j)}$ contribute to the boundary conditions for the functions $v_i^{(k,j)}$.

In the sequel we describe an algorithm for evaluation of the term $\mathbf{u}^{(m)}$ in expansion (5.4.4). Treatment of the junction points and of fields in the thin rods and base regions is very similar to that presented in Section 4.3. However, the order of magnitude of the rigid-body translations and rotations in the 'pile cap' Ω differs from the case of degenerate multi-structure. In order to simplify the presentation, we shall adopt the notations $\mathbf{X}, \mathcal{Y}, \zeta, z$ and $\mathbf{Y}$ without upper index.

5.4.4 The junction layer

The junction layer $\mathcal{W}^{(m,j)}$ should compensate for the error associated with the right-hand sides $\mathbf{F}^{(m,j)}$, $\mathbf{P}^{(m,j)}$ and the discrepancy terms caused by the fields in Ω and in thin cylinders. This vector function solves the boundary value problem

$$\begin{pmatrix} -\mathbf{L}(\partial \mathbf{x}) \\ \mathbf{T}(\mathbf{X}, \partial \mathbf{x}) \end{pmatrix} \boldsymbol{\mathcal{W}}^{(m,j)}(\mathbf{X}) = \begin{pmatrix} \tilde{\mathbf{F}}^{(m,j)}(\mathbf{X}) \\ \tilde{\mathbf{P}}^{(m,j)}(\mathbf{X}) \end{pmatrix}$$

$$+ \begin{pmatrix} [\mathbf{L}(\partial \mathbf{x}), \chi(\mathbf{X})] \\ -[\mathbf{T}(\mathbf{X}, \partial \mathbf{x}), \chi(\mathbf{X})] \end{pmatrix} \Big\{ \boldsymbol{\mathcal{M}}(\mathbf{X}, 0) \boldsymbol{\mathcal{R}}^{(j)} \mathfrak{F}^{(m-1,j)}$$

$$+ \frac{1}{2} (\nabla_{\mathbf{X}'} \times \boldsymbol{\mathcal{M}}^t(\mathbf{X}, \mathbf{X}')|_{\mathbf{X}'=0})^t \boldsymbol{\mathcal{R}}^{(j)} \mathfrak{M}^{(m,j)} \Big\}$$

$$+ \mathcal{L}^{(j)} \begin{pmatrix} [\mathbf{L}(\partial \boldsymbol{y}), \eta(\boldsymbol{y}_3)] \\ -[\mathbf{T}(\boldsymbol{y}, \partial \boldsymbol{y}), \eta(\boldsymbol{y}_3)] \end{pmatrix} \sum_{k=1}^{6} \Big\{ d_k^{(I,m,j)} \boldsymbol{\xi}^{(k)}(\boldsymbol{\mathcal{Y}})$$

$$+ d_k^{(II,m,j)} \boldsymbol{\psi}^{(k)}(\boldsymbol{\mathcal{Y}}) \Big\}, \quad \text{on } G \times \partial G, \qquad (5.4.11)$$

where $\mathbf{L}$ denotes the Lamé operator, $\mathbf{T}$ stands for the operator of tractions defined in Section 4.3.5, and $\boldsymbol{\mathcal{M}}(\mathbf{X}, \mathbf{X}')$ is the Mindlin solution (see Section 3.4.2). The vector-valued functions $\tilde{\mathbf{F}}^{(m,j)}$, $\tilde{\mathbf{P}}^{(m,j)}$ depend on $\mathbf{F}^{(m,j)}$, $\mathbf{P}^{(m,j)}$ and two kinds of discrepancy terms: first, the terms coming from the expansion of the displacement in a thin cylinder and containing $\mathfrak{U}^{(k,j)}$, $k \leq m - 2$, and their derivatives, and, consequently, $\mathcal{F}^{(k,j)}, \mathcal{P}^{(k,j)}$ for $k \leq m - 2$. Second, terms coming from the expansion of the bounded part of $\mathbf{u}_\Omega^{(k)}$, $k \leq m - 2$, at junction points.

The vector functions $\boldsymbol{\psi}^{(k)}$, $\boldsymbol{\xi}^{(k)}$, $k = 1, \ldots, 6$, represent polynomial (with respect to the longitudinal variable) solutions (the same as in Section 3.6) of the homogeneous traction boundary value problem for the Lamé system in an infinite cylinder.

Using Theorem 5.2 we obtain that there exists a unique set

$$\{ \boldsymbol{\mathcal{W}}^{(m,j)}; d_1^{(I,m,j)}, \ldots, d_6^{(I,m,j)}; d_1^{(II,m,j)}, \ldots, d_6^{(II,m,j)} \}, \quad (5.4.12)$$

such that the vector function $\boldsymbol{\mathcal{W}}^{(m,j)}$ decays at infinity.

We relate the values $v_k^{(s,j)}$ and their derivatives to the coefficients $d_k^{(i,m,j)}$, $k = 1, \ldots, 6$; $i = I, II$. Using (5.3.11)–(5.3.15) we deduce that

$$v_q^{(m+m_q-2,j)}(0) = d_q^{(II,m,j)}, \quad q = 1, 2, 3, 4; \qquad (5.4.13)$$

$$\frac{dv_q^{(m+1,j)}}{dz}(0) = d_{4+q}^{(II,m,j)}, \quad q = 1, 2; \qquad (5.4.14)$$

$$\frac{d^3 v_q^{(m-1,j)}}{dz^3}(0) = d_q^{(I,m,j)} = (D_q^{(j)})^{-1}\Big\{-(\Lambda^{(j)}\mathfrak{F}^{(m-1,j)})_q$$
$$+ (\mathcal{L}^{(j)*}\mathbb{F})_q(\tilde{\mathbf{F}}^{(m,j)}, \tilde{\mathbf{P}}^{(m,j)}, G)\Big\}, \quad q = 1, 2, \quad (5.4.15)$$

$$\frac{dv_3^{(m-1,j)}}{dz}(0) = d_3^{(I,m,j)} = -(D_3^{(j)})^{-1}\Big\{-(\Lambda^{(j)}\mathfrak{F}^{(m-1,j)})_3$$
$$+ (\mathcal{L}^{(j)*}\mathbb{F})_3(\tilde{\mathbf{F}}^{(m,j)}, \tilde{\mathbf{P}}^{(m,j)}, G)\Big\}, \quad (5.4.16)$$

$$\frac{dv_4^{(m-1,j)}}{dz}(0) = d_4^{(I,m,j)} = -(D_4^{(j)})^{-1}\Big\{-(\Lambda^{(j)}\mathfrak{M}^{(m,j)})_3$$
$$+ (\mathcal{L}^{(j)*}\mathbb{M})_3(\tilde{\mathbf{F}}^{(m,j)}, \tilde{\mathbf{P}}^{(m,j)}, G)\Big\}, \quad (5.4.17)$$

$$\frac{d^2 v_q^{(m,j)}}{dz^2}(0) = d_{4+q}^{(I,m,j)} = (-1)^q (D_q^{(j)})^{-1}\Big\{-(\Lambda^{(j)}\mathfrak{M}^{(m,j)})_{(3-q)}$$
$$+ (\mathcal{L}^{(j)*}\mathbb{M})_{3-q}(\tilde{\mathbf{F}}^{(m,j)}, \tilde{\mathbf{P}}^{(m,j)}, G)\Big\}, \quad q = 1, 2, \quad (5.4.18)$$

where the vectors $\mathbb{F}$, $\mathbb{M}$ are evaluated in the scaled coordinates $\mathbf{X}$.

5.4.5 Displacement in Ω

The boundary value problem for the vector function $\mathbf{u}_\Omega^{(m)}$ is formulated as

$$\begin{pmatrix} -\mathbf{L}(\partial_\mathbf{x}) \\ \mathbf{T}(\mathbf{x}, \partial_\mathbf{x}) \end{pmatrix} \mathbf{u}_\Omega^{(m)}(\mathbf{x}) = \begin{pmatrix} \tilde{\mathbf{f}}^{(m)}(\mathbf{x}) \\ \tilde{\mathbf{p}}^{(m)}(\mathbf{x}) \end{pmatrix} \quad (5.4.19)$$

$$+ \sum_{j=1}^{K} \begin{pmatrix} \big[\mathbf{L}(\partial_\mathbf{x}), \Xi(\mathbf{x}-\mathbf{a}^{(j)})\big] \\ -\big[\mathbf{T}(\mathbf{x}, \partial_\mathbf{x}), \Xi(\mathbf{x}-\mathbf{a}^{(j)})\big] \end{pmatrix} \mathcal{R}^{(j)*}\Big\{\sum_{q=1}^{2} b_q^{(m,j)} \partial_{x_q} \mathcal{B}^{(q)}(\mathbf{x}^{(j)})$$

$$+ \frac{1}{2} b_3^{(m,j)}(\partial_{x_1}\mathcal{B}^{(2)}(\mathbf{x}^{(j)}) + \partial_{x_2}\mathcal{B}^{(1)}(\mathbf{x}^{(j)}))\Big\} \text{ in } \begin{pmatrix} \Omega \\ \partial\Omega\setminus\cup_j \mathbf{a}^{(j)} \end{pmatrix}.$$

Here $b_q^{(m,j)}$ are the coefficients from expansion (4.2.17) for the function $\mathcal{W}^{(m,j)}$; the principal force and moment vectors for the last sum in (5.4.19) are equal to zero. The vectors $\tilde{\mathbf{f}}^{(m)}$, $\tilde{\mathbf{p}}^{(m)}$ are given by the sum of $\mathbf{f}^{(m)}$, $\mathbf{p}^{(m)}$ and the self-balanced junction layer discrepancies from the previous steps of the asymptotic algorithm. These error terms are brought by the commutators of the cut-off functions and differential operators of the Lamé system and traction boundary conditions applied to corresponding terms of order $O(\|\mathbf{X}\|^{-k})$, $k \geq 3$, of the asymptotic expansion of the junction layer $\mathcal{W}^{(k,j)}$, $k < m$, at infinity in the half-space. The principal force and moment vectors for the right-hand side of (5.4.19) are the same as for $\mathbf{f}^{(m)}$, $\mathbf{p}^{(m)}$.

5.4.6 The bottom layer

The special solution $\mathfrak{U}^{(m,j)}$ of the elasticity problem in a thin cylinder and the rigid-body translations and rotations from the expansion (5.4.5) give the error terms in the Dirichlet boundary condition on $S_\varepsilon^{(j)}$. The boundary value problem for the bottom layer has the form

$$-\mathbf{L}(\partial_\mathbf{Y})\mathcal{V}^{(m,j)}(\mathbf{Y}) = \mathcal{G}^{(m,j)}(\mathbf{Y}), \quad \mathbf{Y} \in \Pi_-, \qquad (5.4.20)$$

$$\sigma^{(n)}(\mathcal{V}^{(m,j)};\mathbf{Y}) = \mathbf{H}^{(m,j)}(\mathbf{Y}), \quad \mathbf{Y} \in \partial\Pi_- \setminus S, \qquad (5.4.21)$$

$$\mathcal{V}^{(m,j)}(\mathbf{Y}) = \tilde{\Phi}^{(m,j)}(\mathbf{Y}) - \sum_{k=1}^{2} v_k^{(m+2,j)}(l^{(j)})\psi^{(k)} - v_3^{(m,j)}(l^{(j)})\psi^{(3)}$$

$$- v_4^{(m,j)}(l^{(j)})\psi^{(4)}(\mathbf{Y}) - \sum_{k=1}^{2} \left\{ \frac{dv_k^{(m+1,j)}}{dz}(l^{(j)})\psi^{(4+k)}(\mathbf{Y}) + \hat{\alpha}_k^{(m,j)}\mathbf{e}^{(k)} \right\}$$

$$- \hat{\alpha}_3^{(m,j)}\mathbf{e}^{(3)} - \hat{\beta}^{(m,j)} \times \hat{\mathbf{b}}^{(j)} - \left(\sum_{k=1}^{2} \hat{\beta}_k^{(m-1,j)}\mathbf{e}^{(k)} + \hat{\beta}_3^{(m-1,j)}\mathbf{e}^{(3)} \right) \times \mathbf{Y},$$

$$\text{for } \mathbf{Y} \in S, \qquad (5.4.22)$$

where $\tilde{\Phi}^{(m,j)}$ involves $\Phi^{(m,j)}$ and traces on S_ε of the functions $v_1^{(k,j)}$, $v_2^{(k,j)}$, $k \leq m$, $v_3^{(k,j)}, v_4^{(k,j)}$, $k \leq m-1$, and the quantities $\hat{\alpha}_k^{(j)}$, $\hat{\beta}_k^{(j)}$ represent coefficients in the asymptotic expansions of components of the rigid-body translation and rotation evaluated in the local system of coordinates,

$$\hat{\alpha}^{(m,j)} = \Lambda^{(j)}\alpha^{(m)}, \quad \hat{\beta}^{(m,j)} = \Lambda^{(j)}\beta^{(m)};$$

$\Lambda^{(j)}$ is the rotation matrix corresponding to the orientation of the thin rod $\Pi_\varepsilon^{(j)}$ with respect to the rigid body Ω.

$$\frac{d^3 v_q^{(m-1,j)}}{dz^3}(0) = d_q^{(I,m,j)} = (D_q^{(j)})^{-1}\Big\{ -(\Lambda^{(j)}\mathfrak{F}^{(m-1,j)})_q$$
$$+ (\mathcal{L}^{(j)*}\mathbb{F})_q(\tilde{\mathbf{F}}^{(m,j)}, \tilde{\mathbf{P}}^{(m,j)}, G)\Big\}, \quad q = 1, 2, \quad (5.4.15)$$

$$\frac{dv_3^{(m-1,j)}}{dz}(0) = d_3^{(I,m,j)} = -(D_3^{(j)})^{-1}\Big\{ -(\Lambda^{(j)}\mathfrak{F}^{(m-1,j)})_3$$
$$+ (\mathcal{L}^{(j)*}\mathbb{F})_3(\tilde{\mathbf{F}}^{(m,j)}, \tilde{\mathbf{P}}^{(m,j)}, G)\Big\}, \quad (5.4.16)$$

$$\frac{dv_4^{(m-1,j)}}{dz}(0) = d_4^{(I,m,j)} = -(D_4^{(j)})^{-1}\Big\{ -(\Lambda^{(j)}\mathfrak{M}^{(m,j)})_3$$
$$+ (\mathcal{L}^{(j)*}\mathbb{M})_3(\tilde{\mathbf{F}}^{(m,j)}, \tilde{\mathbf{P}}^{(m,j)}, G)\Big\}, \quad (5.4.17)$$

$$\frac{d^2 v_q^{(m,j)}}{dz^2}(0) = d_{4+q}^{(I,m,j)} = (-1)^q (D_q^{(j)})^{-1}\Big\{ -(\Lambda^{(j)}\mathfrak{M}^{(m,j)})_{(3-q)}$$
$$+ (\mathcal{L}^{(j)*}\mathbb{M})_{3-q}(\tilde{\mathbf{F}}^{(m,j)}, \tilde{\mathbf{P}}^{(m,j)}, G)\Big\}, \quad q = 1, 2, \quad (5.4.18)$$

where the vectors $\mathbb{F}$, $\mathbb{M}$ are evaluated in the scaled coordinates $\mathbf{X}$.

5.4.5 Displacement in Ω

The boundary value problem for the vector function $\mathbf{u}_\Omega^{(m)}$ is formulated as

$$\begin{pmatrix} -\mathbf{L}(\partial_\mathbf{x}) \\ \mathbf{T}(\mathbf{x}, \partial_\mathbf{x}) \end{pmatrix} \mathbf{u}_\Omega^{(m)}(\mathbf{x}) = \begin{pmatrix} \tilde{\mathbf{f}}^{(m)}(\mathbf{x}) \\ \tilde{\mathbf{p}}^{(m)}(\mathbf{x}) \end{pmatrix} \quad (5.4.19)$$

$$+ \sum_{j=1}^{K} \begin{pmatrix} [\mathbf{L}(\partial_\mathbf{x}), \Xi(\mathbf{x}-\mathbf{a}^{(j)})] \\ -[\mathbf{T}(\mathbf{x}, \partial_\mathbf{x}), \Xi(\mathbf{x}-\mathbf{a}^{(j)})] \end{pmatrix} \mathcal{R}^{(j)*} \Big\{ \sum_{q=1}^{2} b_q^{(m,j)} \partial_{x_q} \mathcal{B}^{(q)}(\mathbf{x}^{(j)})$$

$$+ \frac{1}{2} b_3^{(m,j)} (\partial_{x_1} \mathcal{B}^{(2)}(\mathbf{x}^{(j)}) + \partial_{x_2} \mathcal{B}^{(1)}(\mathbf{x}^{(j)})) \Big\} \text{ in } \Big(\frac{\Omega}{\partial\Omega \setminus \cup_j \mathbf{a}^{(j)}} \Big).$$

Here $b_q^{(m,j)}$ are the coefficients from expansion (4.2.17) for the function $\mathcal{W}^{(m,j)}$; the principal force and moment vectors for the last sum in (5.4.19) are equal to zero. The vectors $\tilde{\mathbf{f}}^{(m)}$, $\tilde{\mathbf{p}}^{(m)}$ are given by the sum of $\mathbf{f}^{(m)}$, $\mathbf{p}^{(m)}$ and the self-balanced junction layer discrepancies from the previous steps of the asymptotic algorithm. These error terms are brought by the commutators of the cut-off functions and differential operators of the Lamé system and traction boundary conditions applied to corresponding terms of order $O(\|\mathbf{X}\|^{-k})$, $k \geq 3$, of the asymptotic expansion of the junction layer $\mathcal{W}^{(k,j)}$, $k < m$, at infinity in the half-space. The principal force and moment vectors for the right-hand side of (5.4.19) are the same as for $\mathbf{f}^{(m)}$, $\mathbf{p}^{(m)}$.

5.4.6 The bottom layer

The special solution $\mathfrak{U}^{(m,j)}$ of the elasticity problem in a thin cylinder and the rigid-body translations and rotations from the expansion (5.4.5) give the error terms in the Dirichlet boundary condition on $S_\varepsilon^{(j)}$. The boundary value problem for the bottom layer has the form

$$-\mathbf{L}(\partial_\mathbf{Y})\mathcal{V}^{(m,j)}(\mathbf{Y}) = \mathcal{G}^{(m,j)}(\mathbf{Y}), \quad \mathbf{Y} \in \Pi_-, \tag{5.4.20}$$

$$\sigma^{(n)}(\mathcal{V}^{(m,j)}; \mathbf{Y}) = \mathbf{H}^{(m,j)}(\mathbf{Y}), \quad \mathbf{Y} \in \partial\Pi_- \setminus S, \tag{5.4.21}$$

$$\mathcal{V}^{(m,j)}(\mathbf{Y}) = \tilde{\Phi}^{(m,j)}(\mathbf{Y}) - \sum_{k=1}^{2} v_k^{(m+2,j)}(l^{(j)})\psi^{(k)} - v_3^{(m,j)}(l^{(j)})\psi^{(3)}$$

$$- v_4^{(m,j)}(l^{(j)})\psi^{(4)}(\mathbf{Y}) - \sum_{k=1}^{2}\left\{\frac{dv_k^{(m+1,j)}}{dz}(l^{(j)})\psi^{(4+k)}(\mathbf{Y}) + \hat{\alpha}_k^{(m,j)}\mathbf{e}^{(k)}\right\}$$

$$- \hat{\alpha}_3^{(m,j)}\mathbf{e}^{(3)} - \hat{\boldsymbol{\beta}}^{(m,j)} \times \hat{\mathbf{b}}^{(j)} - \left(\sum_{k=1}^{2}\hat{\beta}_k^{(m-1,j)}\mathbf{e}^{(k)} + \hat{\beta}_3^{(m-1,j)}\mathbf{e}^{(3)}\right) \times \mathbf{Y},$$

$$\text{for } \mathbf{Y} \in S, \tag{5.4.22}$$

where $\tilde{\Phi}^{(m,j)}$ involves $\Phi^{(m,j)}$ and traces on S_ε of the functions $v_1^{(k,j)}$, $v_2^{(k,j)}$, $k \leq m$, $v_3^{(k,j)}, v_4^{(k,j)}$, $k \leq m-1$, and the quantities $\hat{\alpha}_k^{(j)}$, $\hat{\beta}_k^{(j)}$ represent coefficients in the asymptotic expansions of components of the rigid-body translation and rotation evaluated in the local system of coordinates,

$$\hat{\boldsymbol{\alpha}}^{(m,j)} = \Lambda^{(j)}\boldsymbol{\alpha}^{(m)}, \quad \hat{\boldsymbol{\beta}}^{(m,j)} = \Lambda^{(j)}\boldsymbol{\beta}^{(m)};$$

$\Lambda^{(j)}$ is the rotation matrix corresponding to the orientation of the thin rod $\Pi_\varepsilon^{(j)}$ with respect to the rigid body Ω.

We use Theorem 4.5 and deduce that there exists a unique set

$$\left\{\mathcal{V}^{(m,j)};\ v_k^{(m+2,j)}(l^{(j)}),\ \frac{dv_k^{(m+1,j)}}{dz}(l^{(j)}),\ v_{k+2}^{(m,j)}(l^{(j)}),\ k=1,2\right\}, \quad (5.4.23)$$

which satisfies the problem (5.4.20)–(5.4.22), and the field $\mathcal{V}^{(m,j)}$ decays exponentially at infinity.

The constant elements of the set (5.4.23) are evaluated as

$$v_i^{(m+2,j)}(l^{(j)}) = -(\hat{\alpha}_i^{(m,j)} + (\hat{\boldsymbol{\beta}}^{(m,j)} \times \hat{\mathbf{b}}^{(j)})_i) + c_i^{(m,j)}, \quad (5.4.24)$$
$$i = 1, 2,$$

$$v_3^{(m,j)}(l^{(j)}) = -(\hat{\alpha}_3^{(m,j)} + (\hat{\boldsymbol{\beta}}^{(m,j)} \times \hat{\mathbf{b}}^{(j)})_3) + c_3^{(m,j)}, \quad (5.4.25)$$

$$v_4^{(m,j)}(l^{(j)}) = -\hat{\beta}_3^{(m-1,j)} + c_4^{(m,j)}, \quad (5.4.26)$$

$$\frac{dv_1^{(m+1,j)}}{dz}(l^{(j)}) = -\hat{\beta}_2^{(m-1,j)} + c_5^{(m,j)}, \quad (5.4.27)$$

$$\frac{dv_2^{(m+1,j)}}{dz}(l^{(j)}) = \hat{\beta}_1^{(m-1,j)} + c_6^{(m,j)}. \quad (5.4.28)$$

Then the set

$$\{\mathcal{V}^{(m,j)}; c_q^{(m,j)},\ q = 1, 2, \ldots, 6\} \quad (5.4.29)$$

is the unique solution of the system (5.4.20) with the traction boundary condition (5.4.21) and the displacement boundary condition

$$\mathcal{V}^{(m,j)}(\mathbf{Y}) = \tilde{\boldsymbol{\Phi}}^{(m,j)}(Y_1, Y_2) - \sum_{q=1}^{6} c_q^{(m,j)} \psi^{(q)}(\mathbf{Y}), \quad \mathbf{Y} \in S. \quad (5.4.30)$$

We note that the set (5.4.29) depends on $\mathcal{G}^{(m,j)}$, $\mathbf{H}^{(m,j)}$, $\boldsymbol{\Phi}^{(m,j)}$ and $\mathcal{F}^{(k,j)}$, $\mathcal{P}^{(k,j)}$, $k \leq m - 2$, and the functions $v_j^{(k,j)}$, $i = 1, 2, 3, 4$, $k \leq m - 1$.

5.4.7 Functions $v_i^{(m,j)}$

Similar to Section 4.3.8, the functions $v_i^{(m,j)}$, $i = 1, 2$, can be written in the form

$$v_1^{(m,j)}(z) = -(\hat{\alpha}_1^{(m-2,j)} + (\hat{\boldsymbol{\beta}}^{(m-2,j)} \times \hat{\mathbf{b}}^{(j)})_1)\psi_I\left(\frac{z}{l^{(j)}}\right)$$

$$- \hat{\beta}_2^{(m-2,j)} l^{(j)} \psi_{II}\left(\frac{z}{l^{(j)}}\right) + w_1^{(m,j)}(z), \qquad (5.4.31)$$

$$v_2^{(m,j)}(z) = -(\hat{\alpha}_2^{(m-2,j)} + (\hat{\boldsymbol{\beta}}^{(m-2,j)} \times \hat{\mathbf{b}}^{(j)})_2) \psi_I\left(\frac{z}{l^{(j)}}\right)$$

$$+ \hat{\beta}_1^{(m-2,j)} l^{(j)} \psi_{II}\left(\frac{z}{l^{(j)}}\right) + w_2^{(m,j)}(z), \qquad (5.4.32)$$

where
$$\psi_I(z) = z^2(3 - 2z), \quad \psi_{II}(z) = z^2(z - 1),$$

and the functions $w_i^{(m,j)}, i = 1, 2$, satisfy the boundary value problems of the form (4.3.50)–(4.3.54). The quantities $v_3^{(m,j)}$ and $v_4^{(m,j)}$ are given by

$$v_3^{(m,j)}(z) = -(\hat{\alpha}_3^{(m,j)} + (\hat{\boldsymbol{\beta}}^{(m,j)} \times \hat{\mathbf{b}}^{(j)})_3)\frac{z}{l^{(j)}} + w_3^{(m,j)}(z) \quad (5.4.33)$$

$$v_4^{(m,j)}(z) = -\hat{\beta}_3^{(m-1,j)} \frac{z}{l^{(j)}} + w_4^{(m,j)}(z), \qquad (5.4.34)$$

where the functions $w_3^{(m,j)}$ and $w_4^{(m,j)}$ satisfy boundary value problems of the form (4.3.57)–(4.3.59).

The right-hand sides $Q_i^{(m,j)}$ in equations (4.3.50), (4.3.57) are specified by

$$Q_k^{(m,j)} = -D_k^{-1}\left\{\int_g \tilde{\mathcal{F}}_k^{(m,j)}(\zeta, z)d\zeta + \int_{\partial g} \tilde{\mathcal{P}}_k^{(m,j)}(\zeta, z)dl_\zeta\right\}, \quad (5.4.35)$$

for $k = 1, 2$, and

$$Q_3^{(m,j)} = D_3^{-1}\left\{\int_g \tilde{\mathcal{F}}_3^{(m,j)}(\zeta, z)d\zeta + \int_{\partial g} \tilde{\mathcal{P}}_3^{(m,j)}(\zeta, z)dl_\zeta\right\}, \quad (5.4.36)$$

$$Q_4^{(m,j)} = D_4^{-1}\left\{\int_g (\zeta_1 \tilde{\mathcal{F}}_2^{(m,j)}(\zeta, z) - \zeta_2 \tilde{\mathcal{F}}_1^{(m,j)}(\zeta, z))d\zeta \right.$$

$$\left. + \int_{\partial g} (\zeta_1 \tilde{\mathcal{P}}_2^{(m,j)}(\zeta, z) - \zeta_2 \tilde{\mathcal{P}}_1^{(m,j)}(\zeta, z))dl_\zeta\right\}. \quad (5.4.37)$$

Here, the vector functions $\tilde{\boldsymbol{\mathcal{F}}}^{(m,j)}$, $\tilde{\boldsymbol{\mathcal{P}}}^{(m,j)}$ involve the coefficients $\boldsymbol{\mathcal{F}}^{(m,j)}, \boldsymbol{\mathcal{P}}^{(m,j)}$ in the expansion of the load applied in thin cylinders plus discrepancy terms associated with the thin elastic rod problem from the previous steps of the asymptotic algorithm.

5.4.8 Evaluation of the lock forces and moments at junction points

It follows from (5.4.15)–(5.4.18) and (5.4.31)–(5.4.34) that the components of the lock forces and moments satisfy

$$\hat{\mathfrak{F}}_i^{(m,j)} = (\mathcal{L}^{(j)*}\mathbf{F}(\tilde{\mathbf{F}}^{(m+1,j)}, \tilde{\mathbf{P}}^{(m+1,j)}; G))_i \qquad (5.4.38)$$

$$- D_i^{(j)} \left\{ \frac{12}{(l^{(j)})^3} (\hat{\alpha}_i^{(m-2,j)} + (\hat{\boldsymbol{\beta}}^{(m-2,j)} \times \hat{\mathbf{b}}^{(j)})_i) - \frac{6}{(l^{(j)})^2} \hat{\beta}_{3-i}^{(m-2,j)} \right.$$

$$\left. + \mathbb{Q}^{(III)}(Q_i^{(m,j)}, d_i^{(II,m-2,j)}, c_i^{(m-2,j)}, d_{4+i}^{(II,m-1,j)}, c_{4+i}^{(m-1,j)}) \right\},$$

for $i = 1, 2$,

$$\hat{\mathfrak{F}}_3^{(m,j)} = (\mathcal{L}^{(j)*}\mathbf{F}(\tilde{\mathbf{F}}^{(m,j)}, \tilde{\mathbf{P}}^{(m,j)}; G))_3$$

$$+ D_3^{(j)} \left\{ \frac{-1}{l^{(j)}} (\hat{\alpha}_3^{(m,j)} + (\hat{\boldsymbol{\beta}}^{(m,j)} \times \hat{\mathbf{b}}^{(j)})_3) \right.$$

$$\left. + \mathbb{Q}^{(I)}(Q_3^{(m,j)}, d_3^{(II,m,j)}, c_3^{(m,j)}) \right\}, \qquad (5.4.39)$$

$$\hat{\mathfrak{M}}_3^{(m,j)} = (\mathcal{L}^{(j)*}\mathbf{M}(\tilde{\mathbf{F}}^{(m,j)}, \tilde{\mathbf{P}}^{(m,j)}; G))_3$$

$$+ D_4^{(j)} \left\{ \frac{-\hat{\beta}_3^{(m-2,j)}}{l^{(j)}} + \mathbb{Q}^{(I)}(Q_4^{(m-1,j)}, d_4^{(II,m-1,j)}, c_4^{(m-1,j)}) \right\}, \qquad (5.4.40)$$

$$\hat{\mathfrak{M}}_1^{(m,j)} = (\mathcal{L}^{(j)*}\mathbf{M}(\tilde{\mathbf{F}}^{(m,j)}, \tilde{\mathbf{P}}^{(m,j)}; G))_1 \qquad (5.4.41)$$

$$- D_2^{(j)} \left\{ \frac{-6}{(l^{(j)})^2} (\hat{\alpha}_2^{(m-2,j)} + (\hat{\boldsymbol{\beta}}^{(m-2,j)} \times \hat{\mathbf{b}}^{(j)})_2) + 4\frac{\hat{\beta}_1^{(m-2,j)}}{l^{(j)}} \right.$$

$$\left. + \mathbb{Q}^{(II)}(Q_2^{(m,j)}, d_2^{(II,m-2,j)}, c_2^{(m-2,j)}, d_6^{(II,m-1,j)}, c_6^{(m-1,j)}) \right\},$$

$$\hat{\mathfrak{M}}_2^{(m,j)} = (\mathcal{L}^{(j)*}\mathbf{M}(\tilde{\mathbf{F}}^{(m,j)}, \tilde{\mathbf{P}}^{(m,j)}; G))_2 \qquad (5.4.42)$$

$$+ D_1^{(j)} \left\{ \frac{-6}{(l^{(j)})^2} (\hat{\alpha}_1^{(m-2,j)} + (\hat{\boldsymbol{\beta}}^{(m-2,j)} \times \hat{\mathbf{b}}^{(j)})_1) - 4\frac{\hat{\beta}_2^{(m-2,j)}}{l^{(j)}} \right.$$

$$\left. + \mathbb{Q}^{(II)}(Q_1^{(m,j)}, d_1^{(II,m-2,j)}, c_1^{(m-2,j)}, d_5^{(II,m-1,j)}, c_5^{(m-1,j)}) \right\},$$

where the functionals $\mathbb{Q}^{(m)}, m = I, II, III$, are the same as in (4.2.56)–(4.2.58), where l should be replaced by $l^{(j)}$.

5.4.9 Algebraic system for $\alpha^{(m)}, \beta^{(m)}$

Using the balance equations (5.4.8), (5.4.9) for the principal force and moment vectors we obtain the system of recurrent relations with respect to the components of $\alpha^{(m)}$ and $\beta^{(m)}$

$$\mathcal{A}\begin{pmatrix}\alpha^{(m)}\\\beta^{(m)}\end{pmatrix}=\begin{pmatrix}\mathbb{F}(\mathbf{f}^{(m)},\mathbf{p}^{(m)};\Omega)\\\mathbb{M}(\mathbf{f}^{(m)},\mathbf{p}^{(m)};\Omega)\end{pmatrix}+\mathcal{H}^{(m)}$$
$$+\begin{pmatrix}\mathcal{S}_I^{(m-2)}\\\mathcal{S}_{II}^{(m-2)}(\alpha^{(m-2)},\beta^{(m-2)})\end{pmatrix}. \qquad (5.4.43)$$

ere $\mathbb{F}(\mathbf{f}^{(m)}, \mathbf{p}^{(m)}, \Omega)$, $\mathbb{M}(\mathbf{f}^{(m)}, \mathbf{p}^{(m)}, \Omega)$ are the principal force and moment vectors for the region Ω, the matrix $\mathcal{A}$ is given by (5.3.16), and the two last terms of (5.4.43) are specified as follows:

$$\mathcal{H}^{(m)}=\sum_{j=1}^{K}\left\{\begin{pmatrix}\boldsymbol{\Lambda}^{(j)*}(\mathbf{I}-\boldsymbol{\mathcal{J}})\boldsymbol{\mathcal{L}}^{(j)*}\mathbb{F}(\tilde{\mathbf{F}}^{(m+1,j)},\tilde{\mathbf{P}}^{(m+1,j)};G)\\ \boldsymbol{\Lambda}^{(j)*}\boldsymbol{\mathcal{L}}^{(j)*}\mathbb{M}(\tilde{\mathbf{F}}^{(m,j)},\tilde{\mathbf{P}}^{(m,j)};G)\end{pmatrix}\right.$$
$$\left.+\begin{pmatrix}\boldsymbol{\Lambda}^{(j)*}\boldsymbol{\mathcal{J}}\boldsymbol{\mathcal{L}}^{(j)*}\mathbb{F}(\tilde{\mathbf{F}}^{(m,j)},\tilde{\mathbf{P}}^{(m,j)};G)+\mathfrak{H}_I^{(m,j)}\\ \mathfrak{H}_{II}^{(m,j)}\end{pmatrix}\right\}, \qquad (5.4.44)$$

where $\boldsymbol{\mathcal{J}} = \text{diag}\{0,0,1\}$ and $\mathbf{I}$ is the identity matrix;

$$(\mathfrak{H}_I^{(m,j)})_p =$$

$$\sum_{s=1}^{3}\Lambda_{sp}^{(j)}\left\{\delta_{s1}\mathbb{Q}^{(III)}(Q_1^{(m,j)},d_1^{(II,m-2,j)},c_1^{(m-2,j)},d_5^{(II,m-1,j)},c_5^{(m-1,j)})\right.$$
$$+\delta_{s2}\mathbb{Q}^{(III)}(Q_2^{(m,j)},d_2^{(II,m-2,j)},c_2^{(m-2,j)},d_6^{(II,m-1,j)},c_6^{(m-1,j)})$$
$$\left.+\delta_{s3}\mathbb{Q}^{(I)}(Q_3^{(m,j)},d_3^{(II,m,j)},c_3^{(m,j)})\right\},$$

$$(\mathfrak{H}_{II}^{(m,j)})_p =$$

$$\sum_{s=1}^{3}\Lambda_{sp}^{(j)}\left\{\delta_{s1}\mathbb{Q}^{(II)}(Q_2^{(m,j)},d_2^{(II,m-2,j)},c_2^{(m-2,j)},d_6^{(II,m-1,j)},c_6^{(m-1,j)})\right.$$
$$+\delta_{s2}\mathbb{Q}^{(II)}(Q_1^{(m,j)},d_1^{(II,m-2,j)},c_1^{(m-2,j)},d_5^{(II,m-1,j)},c_5^{(m-1,j)})$$
$$\left.+\delta_{s3}\mathbb{Q}^{(I)}(Q_4^{(m-1,j)},d_4^{(II,m-1,j)},c_4^{(m-1,j)})\right\},$$

and

$$(\mathcal{S}_I)_p^{(m)} = -\sum_{j=1}^{K}\sum_{s=1}^{3}\Lambda_{sp}^{(j)}\bigg\{\delta_{s1}D_1^{(j)}\bigg[\frac{12}{(l^{(j)})^3}(\hat{\alpha}_1^{(m,j)})$$
$$+ (\hat{\boldsymbol{\beta}}^{(m,j)} \times \hat{\mathbf{b}}^{(j)})_1) - \frac{6}{(l^{(j)})^2}\hat{\beta}_2^{(m,j)}\bigg]$$
$$+ \delta_{s2}D_2^{(j)}\bigg[\frac{12}{(l^{(j)})^3}(\hat{\alpha}_2^{(m,j)} + (\hat{\boldsymbol{\beta}}^{(m,j)} \times \hat{\mathbf{b}}^{(j)})_2) + \frac{6}{(l^{(j)})^2}\hat{\beta}_1^{(m,j)}\bigg]\bigg\}, \quad (5.4.45)$$

$$(\mathcal{S}_{II})_p^{(m)} = \sum_{j=1}^{K}\sum_{s=1}^{3}\Lambda_{sp}^{(j)}\bigg\{-\delta_{s1}D_2^{(j)}\bigg[\frac{-6}{(l^{(j)})^2}(\hat{\alpha}_2^{(m,j)}$$
$$+ (\hat{\boldsymbol{\beta}}^{(m,j)} \times \hat{\mathbf{b}}^{(j)})_2) + 4\frac{\hat{\beta}_1^{(m,j)}}{l^{(j)}}\bigg]$$
$$+ \delta_{s2}D_1^{(j)}\bigg[\frac{-6}{(l^{(j)})^2}(\hat{\alpha}_1^{(m,j)} + (\hat{\boldsymbol{\beta}}^{(m,j)} \times \hat{\mathbf{b}}^{(j)})_1) - 4\frac{\hat{\beta}_2^{(m,j)}}{l^{(j)}}\bigg]$$
$$- \delta_{s1}D_4^{(j)}\frac{\hat{\beta}_3^{(m,j)}}{l^{(j)}}\bigg\}. \quad (5.4.46)$$

Note that the matrix $\mathcal{A}$ of the algebraic system (5.4.43) is the same as the stiffness matrix (see (5.4.43)) for the engineering pile structure model.

5.4.10 The recurrent procedure for the asymptotic expansion

In Sections 5.4.3–5.4.7 we derived recurrent relations for the terms of the asymptotic expansion (5.4.4). Here, we describe the sequence of steps of the asymptotic algorithm, which enables one to find these terms.

(1) Using (5.4.40)–(5.4.42) we find $\hat{\mathfrak{M}}_k^{(m,j)}$, $k = 1, 2, 3$.

(2) We solve the boundary value problem (5.4.11) and find $\mathcal{W}^{(m,j)}$ and the constants $d_k^{(I,m,j)}$ and $d_k^{(II,m,j)}$, $k = 1, 2, \ldots, 6$.

(3) Then we solve the problem (5.4.20), (5.4.21) and (5.4.30), and find the functions $\mathcal{V}^{(m,j)}$ and the constants $c_q^{(m,j)}$, $q = 1, 2, \ldots, 6$.

(4) Next, we solve the system (5.4.43) of algebraic equations with respect to $\boldsymbol{\alpha}^{(m)}, \boldsymbol{\beta}^{(m)}$.

(5) Using equations (5.4.38), (5.4.39) we specify $\hat{\mathfrak{F}}_k^{(m,j)}$, $k = 1, 2, 3$, and hence find the singular part $\mathbf{u}_s^{(m)}$ of $\mathbf{u}_\Omega^{(m)}$ (see (5.4.6)).

(6) By solving the problem (5.4.19) we find the regular part $\mathbf{u}_r^{(m)}$ of $\mathbf{u}_\Omega^{(m)}$.

(7) Following Section 5.4.7 we evaluate $v_i^{(m,j)}$, $i = 1, 2, 3, 4$.

(8) The boundary value problem (4.2.44), (4.2.45) with $\mathcal{F}, \mathcal{P}$, being replaced by $\tilde{\mathcal{F}}^{(m,j)}, \tilde{\mathcal{P}}^{(m,j)}$, and v_i replaced by $v_i^{(m,j)}$, gives the vector function $\mathbf{U}^{(m,j)}$.

Applying this procedure for $m = 1, 2, \ldots$ we find all the terms of the asymptotic expansion (5.4.4).

5.5 Estimate for the remainder of the asymptotic expansion

The next statement provides a justification of the asymptotic expansion (5.4.4).

THEOREM 5.3

The solution $\mathbf{u}(\mathbf{x}, \varepsilon)$ of the problem (5.2.1)–(5.2.3) with the right-hand sides of the form (5.4.1)–(5.4.3) is approximated by the asymptotic expansion (5.4.4) as

$$\left\| \mathbf{u} - \sum_{j=0}^{N} \varepsilon^j \mathbf{u}^{(j)}(\mathbf{x}, \varepsilon) \right\|_{(H^1(\Omega_\varepsilon))^3} \leq \text{Const } \varepsilon^{N-1}, \qquad (5.5.1)$$

where the constant does not depend on ε.

Note that the right-hand side of (5.5.1) includes ε^{N-1} instead of ε^{N-3} as in Theorem 4.9.

The proof follows the same pattern as in Section 4.4. The only difference is that the sum

$$s_N = \sum_{m=0}^{N} \varepsilon^m \mathbf{u}^{(m)}(\mathbf{x}, \varepsilon)$$

satisfies the estimate

$$\|s_n - s_N\|^2_{(H^1(\Omega_\varepsilon))^3} \leq \text{Const } \varepsilon^{N-1} \qquad (5.5.2)$$

for $n > N$.

5.6 Analysis of the leading term

In this section we construct the leading term $\mathbf{u}^{(0)}$ in the asymptotic expansion (5.4.4). For simplicity we assume that $\mathbf{F}^{(0,j)}, \mathbf{P}^{(0,j)}$, $\mathcal{F}^{(0,j)}, \mathcal{P}^{(0,j)}$ and $\mathbf{G}^{(0,j)}, \mathbf{H}^{(0,j)}$ in the asymptotic expansions (5.4.1)–(5.4.3) are equal to zero, but the terms $\mathbf{f}^{(0)}, \mathbf{p}^{(0)}$ can be non-zero. We follow the iteration procedure given in Section 5.4.9. It follows from (5.4.40)–(5.4.42) that $\hat{\mathfrak{M}}_k^{(0,j)} = 0$, $k = 1, 2, 3$. Then in the boundary value problem (5.4.11) all the right-hand sides are equal to zero, and hence $\mathcal{W}^{(0,j)}$ and the constants $d_k^{(I,0,j)}, d_k^{(II,0,j)}$, $k = 1, 2, \ldots, 6$, vanish. In a similar way we find that the functions $\mathcal{V}^{(0,j)}$ and the constants $c_q^{(0,j)}$, $q = 1, 2, \ldots, 6$, are equal to zero. System (5.4.43) becomes

$$\mathcal{A}\begin{pmatrix}\boldsymbol{\alpha}^{(0)} \\ \boldsymbol{\beta}^{(0)}\end{pmatrix} = \begin{pmatrix}\mathbb{F}(\mathbf{f}^{(0)}, \mathbf{p}^{(0)}; \Omega) \\ \mathbb{M}(\mathbf{f}^{(0)}, \mathbf{p}^{(0)}; \Omega)\end{pmatrix} + \mathcal{H}^{(0)}, \qquad (5.6.1)$$

where, according to (5.4.44),

$$\mathcal{H}^{(0)} = \sum_{j=1}^K \begin{pmatrix}\Lambda^{(j)*}(\mathbf{I}-\mathcal{J})\mathcal{L}^{(j)*}\mathbb{F}(\mathbf{F}^{(1,j)}, \mathbf{P}^{(1,j)}; G) \\ 0\end{pmatrix}.$$

Solving system (5.6.1) we find $\boldsymbol{\alpha}^{(0)}$ and $\boldsymbol{\beta}^{(0)}$. By (5.4.38), (5.4.39)

$$\hat{\mathcal{F}}_k^{(0,j)} = (\mathcal{L}^{(j)*}\mathbb{F}(\mathbf{F}^{(1,j)}, \mathbf{P}^{(1,j)}; G))_k, \ k = 1, 2, \qquad (5.6.2)$$

and

$$\hat{\mathcal{F}}_3^{(0,j)} = -\frac{D_3^{(j)}}{l^{(j)}}(\hat{\alpha}_3^{(0,j)} + (\hat{\boldsymbol{\beta}}^{(0,j)} \times \hat{\mathbf{b}}^{(j)})_3), \qquad (5.6.3)$$

and hence we find the singular part $\mathbf{u}_s^{(0)}$ of $\mathbf{u}_\Omega^{(0)}$ (see (5.4.6)). Problem (5.4.19) becomes

$$-\mathbf{L}\mathbf{u}_r^{(0)} = \mathbf{L}\mathbf{u}_s^{(0)} + \mathbf{f}^{(0)} \quad \text{in } \Omega,$$

and

$$\sigma^{(n)}(\mathbf{u}_r^{(0)}; \mathbf{x}) = -\sigma^{(n)}(\mathbf{u}_s^{(0)}; \mathbf{x}) + \mathbf{p}^{(0)} \quad \text{on } \partial\Omega.$$

The problem is solvable, and it yields $\mathbf{u}_r^{(0)}$. Since all the functions $w_k^{(0,j)}$, $k = 1, 2, 3, 4$, are zero, we deduce from (5.4.31), (5.4.32) and (5.4.34) that the functions $v_1^{(0,j)}, v_2^{(0,j)}$ and $v_4^{(0,j)}$ vanish. By (5.4.33)

$$v_3^{(0,j)}(z) = -\frac{z}{l(j)}(\alpha_3^{(0,j)} + (\hat{\beta}^{(0,j)} \times \hat{\mathbf{b}}^{(j)})_3).$$

The right-hand sides of problem (4.3.18), (4.3.19) for $\mathbf{U}^{(0,j)}$ are equal to zero. Hence $\mathbf{U}^{(0,j)} = \mathbf{0}$.

Finally, we obtain

$$\mathbf{u}^{(0)}(\mathbf{x}, \varepsilon) = \varepsilon^{-2}(\boldsymbol{\alpha}^{(0)} + \boldsymbol{\beta}^{(0)} \times \mathbf{x}) + \mathfrak{X}(\mathbf{x}, \varepsilon)\mathbf{u}_\Omega^{(0)}(\mathbf{x})$$

$$+ \varepsilon^{-2} \sum_{j=1}^{K} \Lambda^{(j)} \eta\left(\frac{z^{(j)}}{\varepsilon}\right) \sum_{q=0}^{1} \varepsilon^q \frac{d^q}{dz^{(j)q}} v_3^{(0,j)}(z^{(j)}) \phi^{(3,q)}(\zeta^{(j)}).$$

5.7 Physical interpretation

For the sake of simplicity assume that the body force and surface tractions are applied only to the region Ω. Then, $\mathcal{H}^{(0)} = \mathbf{0}$ and from (5.4.43) we obtain that the vectors $\boldsymbol{\alpha}^{(0)}$ and $\boldsymbol{\beta}^{(0)}$, representing rigid-body translations and rotations, solve the following system of linear algebraic equations

$$\mathcal{A}\begin{pmatrix} \boldsymbol{\alpha}^{(0)} \\ \boldsymbol{\beta}^{(0)} \end{pmatrix} = \begin{pmatrix} \mathbb{F}(\mathbf{f}^{(0)}, \mathbf{p}^{(0)}) \\ \mathbb{M}(\mathbf{f}^{(0)}, \mathbf{p}^{(0)}) \end{pmatrix}, \qquad (5.7.1)$$

with the positive definite matrix $\mathcal{A}$. This system is uniquely solvable and all components of rigid-body displacements have the same order of magnitude. Outside the vicinity of junction points, the displacement of the region Ω is specified by

$$\mathbf{u}(\mathbf{x}, \varepsilon) \sim \varepsilon^{-2}(\boldsymbol{\alpha}^{(0)} + \boldsymbol{\beta}^{(0)} \times \mathbf{x}). \qquad (5.7.2)$$

The situation is different from the case of the degenerate pile structure with parallel rods, where, as shown in Section 4.6, one has two groups of components of different order of magnitude.

It follows from the formulae (5.4.38)–(5.4.42) that only the longitudinal components of lock forces $\hat{\mathfrak{F}}_3^{(0,j)}$ (see (5.6.3)) matter in the leading order approximation. Remaining components are small. The leading order approximation of the solution corresponds to the case when thin rods are subject to a longitudinal tension–compression deformation.

Hence, one can see that the leading order approximation agrees entirely with the engineering pile structure model described in Section 5.1.

Note that the factor ε^{-2} in the expression (5.7.2) should be regarded as the normalisation parameter, and in terms of physical quantities it means that the applied load must be small in order to provide displacements appropriate for models of linear elasticity.

6

SPECTRAL ANALYSIS FOR 3D–1D MULTI-STRUCTURES

The asymptotic analysis developed in previous chapters can be applied to the study of spectral properties of boundary value problems in multi-structures. In this chapter we show this for the multi-structure depicted in Fig. 2.1, which was studied in Chapters 2 and 4. Here, we obtain asymptotic formulae for the first eigenvalues of the Laplace and Lamé operators with the same boundary conditions as before.

The construction of asymptotics is based upon an operator theoretical scheme described in Section 6.1. Some two-sided estimates are derived for the eigenvalues of a self-adjoint operator in the Hilbert space by eigenvalues of a certain matrix.

In Section 6.2, for the first eigenvalue of the Laplacian we obtain the asymptotic representation

$$\lambda_1 = \varepsilon^2 (\mathrm{mes}_3 \Omega)^{-1} \sum_{j=1}^{K} (l^{(j)})^{-1} \mathrm{mes}_2 g^{(j)} + O(\varepsilon^3),$$

where $g^{(j)}$ is the scaled cross-section and $l^{(j)}$ is the length of the cylinder $\Pi_\varepsilon^{(j)}$. We also obtain an asymptotic formula for the first eigenfunction.

Section 6.3 includes the analysis of a spectral problem of linear elasticity. We derive the formula for the first three eigenvalues:

$$\lambda_j = \varepsilon^4 \rho_j + O(\varepsilon^5), \ j = 1, 2, 3,$$

where ρ_j are eigenvalues of a certain matrix. It is shown that the remaining eigenvalues are greater than Const ε^2.

In Section 6.4 we consider the case where the material of the elastic multi-structure is inhomogeneous: the material of the rods is supposed to have a low mass density in comparison with the material of the region Ω. This additional assumption enables one to obtain explicit asymptotic formulae for the first six eigenvalues characterising the natural frequencies of oscillation in translational and rotational modes.

6.1 An abstract scheme for the asymptotics of eigenvalues

Let $\mathfrak{X}$ and $\mathcal{H}$ be Hilbert spaces with the scalar products $(\cdot,\cdot)_{\mathfrak{X}}$, $(\cdot,\cdot)_{\mathcal{H}}$ and with the norms

$$\|x\|_{\mathfrak{X}} = (x,x)_{\mathfrak{X}}^{1/2}, \quad \|h\|_{\mathcal{H}} = (h,h)_{\mathcal{H}}^{1/2}.$$

We suppose that the space $\mathcal{H}$ is densely and compactly embedded in the space $\mathfrak{X}$.

Let us consider a sesquilinear form $a(\cdot,\cdot)$ on $\mathcal{H}$ (see Birman and Solomjak (1987), Chapter 2, Section 4) such that for all $h \in \mathcal{H}$

$$a(h,h) \geq 0, \qquad (6.1.1)$$

and

$$a(h,h) \geq c_0 \|h\|_{\mathcal{H}}^2 - c_1 \|h\|_{\mathfrak{X}}^2, \qquad (6.1.2)$$

where c_0, c_1 are positive constants. We adopt the notation

$$\mathfrak{X}_a = \{h \in \mathcal{H} : a(h,h) = 0\}.$$

LEMMA 6.1

The set $\mathfrak{X}_a$ is linear and has a finite dimension.

Proof

Since the form a is non-negative, we have

$$a(h_1 + h_2, h_1 + h_2) \leq 2(a(h_1,h_1) + a(h_2,h_2))$$

which yields the linearity of $\mathfrak{X}_a$.

From the second property of the form a we deduce that

$$\|h\|_{\mathcal{H}}^2 \leq c_0^{-1} c_1 \|h\|_{\mathfrak{X}}^2, \quad \text{for all } h \in \mathfrak{X}_a. \qquad (6.1.3)$$

Since the embedding $\mathcal{H} \hookrightarrow \mathfrak{X}$ is compact, the last inequality may only hold on a finite-dimensional space. In fact, let $\{x_k\}$ be an infinite basis in $\mathfrak{X}_a$, orthogonal and normalised in $\mathcal{H}$. Denote by $\{y_k\}$ a subsequence of $\{x_k\}$ which converges in $\mathfrak{X}$. Then by (6.1.3)

$$2 = \|y_k - y_l\|_{\mathcal{H}}^2 \leq c_0^{-1} c_1 \|y_k - y_l\|_{\mathfrak{X}}^2 \to 0,$$

as $k, l \to \infty$. The result follows by contradiction. □

Let $\mathbf{H}$ be a closed subspace in $\mathcal{H}$, dense in $\mathfrak{X}$. The form a with the domain $\mathbf{H}$ corresponds to a self-adjoint operator $\mathcal{A}$ with the domain $\mathbf{D}_{\mathcal{A}}$, dense in $\mathbf{H}$, such that

$$(\mathcal{A}h, v)_{\mathfrak{X}} = a(h, v), \quad h \in \mathbf{D}_{\mathcal{A}}, \ v \in \mathbf{H}.$$

This operator is non-negative and its spectrum is discrete and consists of isolated eigenvalues of finite multiplicity (see, for example, Birman and Solomjak (1987), Chapter 10, Section 1). We enumerate the eigenvalues (taking into account multiplicities) in non-decreasing order

$$0 \leq \lambda_1 \leq \lambda_2 \leq \ldots.$$

These eigenvalues can be determined by the max–min principle

$$\lambda_m = \max_{\Phi \subset \mathbf{H}} \min_{h \in \Phi \setminus \{0\}} \frac{a(h, h)}{\|h\|_{\mathfrak{X}}^2}, \qquad (6.1.4)$$

where max is taken over all subspaces of $\mathbf{H}$ of codimension less than or equal to $m - 1$ (see Birman and Solomjak (1987), Chapter 10, Section 2).

Our objective is to obtain the estimates for λ_m, $m = 1, 2, \ldots$, in terms of eigenvalues of a certain finite-dimensional spectral problem.

Consider an n-dimensional subspace $\mathfrak{X}_0$ of $\mathfrak{X}_a$ and denote by P the orthogonal projector from $\mathfrak{X}$ onto $\mathfrak{X}_0$. We suppose that

$$\|h - Ph\|_{\mathfrak{X}}^2 \leq q a(h, h) \quad \text{for all } h \in \mathbf{H}, \qquad (6.1.5)$$

where q is a positive constant. We also need the subspace

$$\mathbf{H}_1 = \{h \in \mathbf{H} : Ph = 0\}$$

as well as the subspace

$$\mathbf{H}_0 = \{h \in \mathbf{H} : a(h, g) = 0, \text{ for all } g \in \mathbf{H}_1\}. \qquad (6.1.6)$$

LEMMA 6.2

(i) *The space* $\mathbf{H}$ *can be represented as the direct sum* (Rudin (1991), p.106) *of* $\mathbf{H}_0$ *and* $\mathbf{H}_1$.

(ii) $\dim \mathbf{H}_0 = n$.

Proof

(i) First, we note that $\mathbf{H}_0$ and $\mathbf{H}_1$ are closed in $\mathbf{H}$. Therefore, it is sufficient to show that $\mathbf{H}_0 \cap \mathbf{H}_1 = \{0\}$ and that every element $h \in \mathbf{H}$ can be represented as $h = h_0 + h_1$, where $h_0 \in \mathbf{H}_0$ and $h_1 \in \mathbf{H}_1$.

If $h \in \mathbf{H}_0 \cap \mathbf{H}_1$ then $a(h,h) = 0$ and $Ph = 0$. Hence $h = 0$ by (6.1.5). Let $h \in \mathbf{H}$, and let h_1 denote the solution of the following problem:

$$a(h_1, g) = a(h, g), \quad \text{for all } g \in \mathbf{H}_1.$$

Since by (6.1.5) the form $a(\cdot, \cdot)$ is non-degenerate on $\mathbf{H}_1$ this problem is uniquely solvable (by the Riesz theorem). It follows that $h_0 = h - h_1 \in \mathbf{H}_0$.

(ii) By definition of $\mathbf{H}_1$ the dimension of every direct complement of $\mathbf{H}_1$ to $\mathbf{H}$ is equal to n. Therefore the result follows from (i). □

Now, we introduce a finite-dimensional spectral problem induced by the ratio

$$\frac{a(h,h)}{\|h\|_\mathbf{X}^2}, \quad h \in \mathbf{H}_0, \qquad (6.1.7)$$

and denote by ν_m (here $0 \leq \nu_1 \leq \cdots \leq \nu_n$) its eigenvalues specified by the max–min principle, i.e.

$$\nu_m = \max_{\Psi \subset \mathbf{H}_0} \min_{h \in \Psi \setminus \{0\}} \frac{a(h,h)}{\|h\|_\mathbf{X}^2}, \qquad (6.1.8)$$

where max is taken over all subspaces of $\mathbf{H}_0$ of codimension less than or equal to $m - 1$.

THEOREM 6.1

The inequalities

$$\frac{\nu_m}{1 + q\nu_m} \leq \lambda_m \leq \nu_m \qquad (6.1.9)$$

hold for $m = 1, 2, \ldots, n$, *and*

$$\lambda_{n+1} \geq 1/q, \qquad (6.1.10)$$

where q is the constant in (6.1.5).

Proof

By (6.1.4) and by codim $\mathbf{H}_1 = n$ we have

$$\lambda_{n+1} \geq \min_{h \in \mathbf{H}_1 \setminus \{0\}} \frac{a(h,h)}{\|h\|_{\tilde{x}}^2}$$

and (6.1.10) follows from (6.1.5) and the definition of $\mathbf{H}_1$.

Next, we consider the inequality on the right in (6.1.9). Let Φ be an extremal subspace in (6.1.4) for λ_m. Then

$$\lambda_m = \min_{h \in \Phi \setminus \{0\}} \frac{a(h,h)}{\|h\|_{\tilde{x}}^2} \leq \min_{h \in (\Phi \cap \mathbf{H}_0) \setminus \{0\}} \frac{a(h,h)}{\|h\|_{\tilde{x}}^2}$$

and the result follows from (6.1.8).

It remains to prove the inequality on the left in (6.1.9) for the case when $\nu_m > 0$. Let Ψ be an extremal subspace in (6.1.8) for ν_m. Then

$$\|h\|_{\tilde{x}}^2 \leq \frac{1}{\nu_m} a(h,h) \quad \text{for all } h \in \Psi. \qquad (6.1.11)$$

We take $\Phi = \Psi + \mathbf{H}_1$. Since $\text{codim}(\Phi) \leq m-1$, then by (6.1.4)

$$\lambda_m \geq \min_{h \in \Phi \setminus \{0\}} \frac{a(h,h,)}{\|h\|_{\tilde{x}}^2}. \qquad (6.1.12)$$

Let $h = h_0 + h_1$ with $h_0 \in \Psi$ and $h_1 \in \mathbf{H}_1$. We have

$$\|h\|_{\tilde{x}}^2 = \|h - Ph\|_{\tilde{x}}^2 + \|Ph_0\|_{\tilde{x}}^2$$
$$\leq \|h - Ph\|_{\tilde{x}}^2 + \|h_0\|_{\tilde{x}}^2.$$

Hence by (6.1.4) and (6.1.11)

$$\|h\|_{\tilde{x}}^2 \leq q a(h,h) + \frac{1}{\nu_m} a(h_0, h_0).$$

By definition of $\mathbf{H}_0$,
$$a(h_0, h_0) \leq a(h,h)$$

which implies that

$$\frac{a(h,h)}{\|h\|_{\mathfrak{X}}^2} \geq \left(q + \frac{1}{\nu_m}\right)^{-1}$$

for all $h \in \Phi \setminus \{0\}$. The result follows from (6.1.12).

□

6.2 Spectral problem for the Laplacian

As in Chapter 2, we consider the multi-structure Ω_ε with the thin rods parallel to each other. Let λ_1 and u_1 be the smallest positive eigenvalue and the corresponding real eigenfunction for the spectral problem

$$-\Delta u(x,\varepsilon) = \lambda u(x,\varepsilon), \quad x \in \Omega_\varepsilon, \qquad (6.2.1)$$

$$\frac{\partial}{\partial n} u(x,\varepsilon) = 0, \quad x \in \partial\Omega_\varepsilon \setminus \cup_i \bar{S}_\varepsilon^{(i)}, \qquad (6.2.2)$$

$$u(x,\varepsilon) = 0, \quad x \in S_\varepsilon^{(i)}, \; i = 1, \ldots, K. \qquad (6.2.3)$$

The eigenfunction will be normalised by

$$\|u_1\|_{L_2(\Omega_\varepsilon)} = 1.$$

6.2.1 The first eigenvalue

The following theorem gives the asymptotic approximation for the first eigenvalue λ_1 of (6.2.1)–(6.2.3).

THEOREM 6.2

The eigenvalue λ_1 satisfies

$$\lambda_1 = \varepsilon^2 B + O(\varepsilon^3), \qquad (6.2.4)$$

where

$$B = (\mathrm{mes}_3 \Omega_0)^{-1} \sum_{j=1}^{K} (l^{(j)})^{-1} \mathrm{mes}_2 g^{(j)}. \qquad (6.2.5)$$

All remaining eigenvalues are bounded from below by some positive constant independent of ε.

Proof

We shall use Theorem 6.1. In our case

$$\mathfrak{X} = L_2(\Omega_\varepsilon), \quad \mathcal{H} = \mathbf{H}^1(\Omega_\varepsilon),$$

$$\mathbf{H} = \{u \in \mathbf{H}^1(\Omega_\varepsilon) : u = 0 \text{ on } S_\varepsilon\},$$

$$a(u,v) = \int_{\Omega_\varepsilon} \nabla u \cdot \nabla \bar{v} dx.$$

Inequalities (6.1.1) and (6.1.2) are obviously valid. Furthermore, $\mathfrak{X}_0 = \mathfrak{X}_a$ is the space of complex constants. By definition of $\mathbf{H}_0$ the function $u \in \mathbf{H}_0$ if and only if $u \in \mathbf{H}$ and

$$\int_{\Omega_\varepsilon} \nabla u \cdot \nabla v d\mathbf{x} = 0$$

for all $v \in \mathbf{H}$ orthogonal to 1 in $L_2(\Omega_\varepsilon)$. Hence, $\mathbf{H}_0$ coincides with the space of functions Const A, where A is the solution of the problem (2.7.1)–(2.7.3).

The estimate (6.1.5) takes the form of the Poincaré inequality for the domain Ω_ε

$$\int_{\Omega_\varepsilon} \left| u(\mathbf{x}) - \frac{1}{\operatorname{mes}_3 \Omega_\varepsilon} \int_{\Omega_\varepsilon} u(\mathbf{x}') d\mathbf{x}' \right|^2 d\mathbf{x} \leq q \int_{\Omega_\varepsilon} |\nabla u|^2 d\mathbf{x}. \qquad (6.2.6)$$

We shall verify that this estimate holds with a positive constant q independent of ε. By Lemma 2.7, for all $u \in \mathbf{H}$

$$\int_{\Omega_\varepsilon} |u|^2 d\mathbf{x} \leq C\varepsilon^{-2} \int_{\Omega_\varepsilon} |\nabla u|^2 d\mathbf{x}, \qquad (6.2.7)$$

where the constant C is independent of ε. Applying inequality (2.6.3) we obtain

$$\int_{\Pi_\varepsilon} |u|^2 d\mathbf{x} \leq C_1 \int_{\Pi_\varepsilon} |\nabla u|^2 d\mathbf{x}, \qquad (6.2.8)$$

with C_1 independent of ε. We shall also use the Poincaré inequality for Ω:

$$\int_{\Omega} \left| u(\mathbf{x}) - \frac{1}{\operatorname{mes}_3 \Omega} \int_{\Omega} u(\mathbf{x}') d\mathbf{x}' \right|^2 d\mathbf{x} \leq C_2 \int_{\Omega} |\nabla u|^2 d\mathbf{x}, \qquad (6.2.9)$$

where C_2 is a constant (independent of ε). For the domain Ω_ε we have

SPECTRAL PROBLEM FOR THE LAPLACIAN

$$\int_{\Omega_\varepsilon} \left| u(\mathbf{x}) - \frac{1}{\mathrm{mes}_3 \Omega_\varepsilon} \int_{\Omega_\varepsilon} u(\mathbf{x}')d\mathbf{x}' \right|^2 d\mathbf{x}$$

$$\leq 4 \int_{\Omega} \left| u(\mathbf{x}) - \frac{1}{\mathrm{mes}_3 \Omega} \int_{\Omega} u(\mathbf{x}')d\mathbf{x}' \right|^2 d\mathbf{x}$$

$$+ 2 \int_{\Pi_\varepsilon} \left| u(\mathbf{x}) - \frac{1}{\mathrm{mes}_3 \Omega_\varepsilon} \int_{\Pi_\varepsilon} u(\mathbf{x}')d\mathbf{x}' \right|^2 d\mathbf{x} + C_3 \varepsilon^2 \left(\int_{\Omega} u d\mathbf{x} \right)^2.$$

Then applying (6.2.7), (6.2.8), (6.2.9) we arrive at (6.2.6).

By Theorem 6.1 we derive

$$\frac{\nu}{1+q\nu} \leq \lambda_1 \leq \nu, \qquad (6.2.10)$$

where

$$\nu = \frac{\|\nabla A\|^2_{L_2(\Omega_\varepsilon)}}{\|A\|^2_{L_2(\Omega_\varepsilon)}}. \qquad (6.2.11)$$

By the same theorem, all remaining eigenvalues $\lambda_2, \lambda_3, \ldots$ are greater than $1/q$.

By (2.7.1), the relation (6.2.11) can be reduced to the form

$$\nu = \frac{(\mathrm{mes}_3 \Omega)^{-1} \int_{\Omega_\varepsilon} A(x,\varepsilon)d\mathbf{x}}{\int_{\Omega_\varepsilon} A(x,\varepsilon)^2 d\mathbf{x}}. \qquad (6.2.12)$$

According to Theorem 2.2

$$\int_{\Omega_\varepsilon} |\nabla (A - \mathcal{U})|^2 d\mathbf{x} + \varepsilon \int_{\Omega_\varepsilon} |A - \mathcal{U}|^2 d\mathbf{x} \leq \mathrm{Const}\, \varepsilon^{-1}, \qquad (6.2.13)$$

where $\mathcal{U}$ is given by (2.7.8). Hence, it follows from (6.2.11) that

$$\nu = \varepsilon^2 (\mathrm{mes}_3 \Omega)^{-1} \sum_{j=1}^{K} (l^{(j)})^{-1} \mathrm{mes}_2 g^{(j)} + O(\varepsilon^3), \qquad (6.2.14)$$

which together with (6.2.10) gives (6.2.4). □

6.2.2 The first eigenfunction

Let us denote by u_1 the normalised eigenfunction of problem (6.2.1)–(6.2.3).

LEMMA 6.3

The eigenfunction u_1 can be represented in the form

$$u_1 = \pm \|\mathcal{U}\|_{L_2(\Omega_\varepsilon)}^{-1} \mathcal{U} + w, \qquad (6.2.15)$$

where $\mathcal{U}$ is given by (2.7.8). The remainder w satisfies

$$\|w\|_{\mathbf{H}^1(\Omega_\varepsilon)} \leq \text{Const } \varepsilon, \qquad (6.2.16)$$

with a constant independent of ε.

Proof

First, we prove that

$$u_1 = \pm \|A\|_{L_2(\Omega_\varepsilon)}^{-1} A + w_1, \qquad (6.2.17)$$

where

$$\|w_1\|_{\mathbf{H}^1(\Omega_\varepsilon)} \leq \text{Const } \varepsilon^{3/2}. \qquad (6.2.18)$$

We denote by $\{\lambda_j\}_{j\geq 1}$ a non-decreasing sequence of eigenvalues of (6.2.1)–(6.2.3) and by $\{u_j\}$ the corresponding sequence of the orthogonal normed, real eigenfunctions.

We represent the function A in the form of the Fourier series

$$A = b \sum_{j=1}^{\infty} a_j u_j,$$

with $b = \|A\|_{L_2(\Omega_\varepsilon)}$. Thus,

$$\sum_{j=1}^{\infty} |a_j|^2 = 1. \qquad (6.2.19)$$

Using (6.2.11) we have

$$\nu = \sum_{j=1}^{\infty} \lambda_j |a_j|^2,$$

which along with (6.2.4) and (6.2.14) gives

$$(\varepsilon^2 B + O(\varepsilon^3))|a_1|^2 + \sum_{j=2}^{\infty} \lambda_j |a_j|^2 = \varepsilon^2 B + O(\varepsilon^3). \qquad (6.2.20)$$

Relations (6.2.19) and (6.2.20) imply that

$$\sum_{j=2}^{\infty} |a_j|^2 = O(\varepsilon^2), \quad |a_1|^2 = 1 + O(\varepsilon^2),$$

and from (6.2.20) we derive the asymptotic equality

$$\sum_{j=2}^{\infty} \lambda_j |a_j|^2 = O(\varepsilon^3). \qquad (6.2.21)$$

Using (6.2.19) and the inequality $\lambda_2 \geq c_0$ we obtain

$$\sum_{j=2}^{\infty} |a_j|^2 = O(\varepsilon^3), \quad |a_1|^2 = 1 + O(\varepsilon^3). \qquad (6.2.22)$$

Hence, we get

$$\frac{1}{b} A = a_1 u_1 - w_2,$$

and

$$\|w_2\|_{L_2(\Omega_\varepsilon)}^2 + \|\nabla w_2\|_{L_2(\Omega_\varepsilon)}^2 \leq C\varepsilon^3,$$

since

$$\|w_2\|_{L_2(\Omega_\varepsilon)}^2 = \sum_{j=2}^{\infty} |a_j|^2, \quad \|\nabla w_2\|_{L_2(\Omega_\varepsilon)}^2 = \sum_{j=2}^{\infty} \lambda_j |a_j|^2.$$

Now (6.2.17) follows from (6.2.22). The final representation (6.2.15) is a consequence of (6.2.17) combined with (6.2.13) and the relation $\|A\|_{L_2(\Omega_\varepsilon)}^{-1} = O(\varepsilon^2)$. $\square$

6.3 Asymptotics of first eigenvalues of the Lamé operator

In this section we consider a spectral problem for the equations of linear elasticity in the same multi-structure Ω_ε. We obtain explicit asymptotic formulae for the first three eigenvalues by using the abstract Theorem 6.1.

6.3.1 Spectral problem

Consider the spectral problem

$$-\mathbf{L}(\partial_\mathbf{x})\mathbf{u} = \lambda \mathbf{u}, \text{ in } \Omega_\varepsilon, \qquad (6.3.1)$$
$$\sigma^{(n)}(\mathbf{u};\mathbf{x}) = 0, \text{ on } \partial\Omega \setminus S_\varepsilon, \qquad (6.3.2)$$
$$\mathbf{u}(\mathbf{x}) = 0, \text{ on } S_\varepsilon. \qquad (6.3.3)$$

In our case the notations introduced in Section 6.1 are interpreted as

$$\mathfrak{X} = (L_2(\Omega_\varepsilon))^3, \ \mathcal{H} = (H^1(\Omega_\varepsilon))^3.$$

The sesquilinear form $a(\cdot,\cdot)$ is given by

$$a(\mathbf{u},\mathbf{v}) = \sum_{i,j=1}^3 \int_{\Omega_\varepsilon} \sigma_{ij}(\mathbf{u};\mathbf{x})\overline{\varepsilon_{ij}(\mathbf{v};\mathbf{x})}d\mathbf{x}. \qquad (6.3.4)$$

Note that

$$a(\mathbf{u},\mathbf{u}) = 2\mathcal{E}(\mathbf{u};\Omega_\varepsilon),$$

where the functional $\mathcal{E}$ was defined by (3.8.1). The inequality (6.1.1) is obvious and (6.1.2) follows from Korn's inequality (3.8.2).

The notation $\mathbf{H}$ is used for a space of functions from $(H^1(\Omega_\varepsilon))^2$ vanishing on S_ε. The space $\mathfrak{X}_a$ consists of vector functions $\mathbf{b} + \mathbf{c} \times \mathbf{x}$, where $\mathbf{b}, \mathbf{c} \in \mathbb{C}^3$.

The spaces $\mathfrak{X}_0$ and $\mathbf{H}_0$ will be specified in Section 6.3.4.

6.3.2 Korn-type inequalities

We assume that all points $\mathbf{a}^{(1)}, \ldots, \mathbf{a}^{(K)}$ do not lie on the same straight line and that the centre of mass of the body Ω is located on the Ox_3 axis, i.e.

$$\int_\Omega x_i d\mathbf{x} = 0, \ i = 1, 2.$$

We need linearly independent vector functions $\mathbf{\Psi}^{(1)}, \mathbf{\Psi}^{(2)}, \mathbf{\Psi}^{(3)}$ subject to

$$\int_\Omega \mathbf{e}^{(3)} \cdot \mathbf{\Psi}^{(j)} d\mathbf{x} = \int_\Omega (\mathbf{e}^{(1)} \times \mathbf{x}) \cdot \mathbf{\Psi}^{(j)} d\mathbf{x}$$
$$= \int_\Omega (\mathbf{e}^{(2)} \times \mathbf{x}) \cdot \mathbf{\Psi}^{(j)} d\mathbf{x} = 0, \qquad (6.3.5)$$

where $\mathbf{e}^{(1)}, \mathbf{e}^{(2)}$ and $\mathbf{e}^{(3)}$ are the directions of the coordinate axes. One can verify by direct calculations that these orthogonality conditions are satisfied by the vector functions

$$\boldsymbol{\Psi}^{(1)}(\mathbf{x}) = \mathbf{e}^{(1)} - (I_2 I_1 - Q_3^2)^{-1} \kappa (I_1 \mathbf{e}^{(1)} + Q_3 \mathbf{e}^{(2)}) \times \mathbf{x}$$

$$\boldsymbol{\Psi}^{(2)}(\mathbf{x}) = \mathbf{e}^{(2)} + (I_2 I_1 - Q_3^2)^{-1} \kappa (I_2 \mathbf{e}^{(1)} + Q_3 \mathbf{e}^{(2)}) \times \mathbf{x}$$

$$\boldsymbol{\Psi}^{(3)}(\mathbf{x}) = \mathbf{e}^{(3)} \times \mathbf{x} + (I_2 I_1 - Q_3^2)^{-1}[(I_1 Q_1 + Q_2 Q_3)\mathbf{e}^{(1)}$$
$$+ (I_2 Q_2 + Q_1 Q_3)\mathbf{e}^{(2)}] \times \mathbf{x},$$

where

$$I_i = \int_\Omega (\|\mathbf{x}\|^2 - x_i^2) d\mathbf{x}, \ i = 1, 2, 3, \tag{6.3.6}$$

are the moments of inertia, and

$$Q_j = \int_\Omega x_j^{-1} x_1 x_2 x_3 d\mathbf{x}, \ \kappa = \int_\Omega x_3 d\mathbf{x}. \tag{6.3.7}$$

LEMMA 6.4

If $\mathbf{u} \in \mathbf{H}$ satisfies

$$\int_\Omega \mathbf{u} \cdot \boldsymbol{\Psi}^{(j)} d\mathbf{x} = 0, \ j = 1, 2, 3, \tag{6.3.8}$$

then

$$\int_{\Omega_\varepsilon} \|\mathbf{u}\|^2 d\mathbf{x} \leq C \varepsilon^{-2} a(\mathbf{u}, \mathbf{u}), \tag{6.3.9}$$

with a positive constant C independent of ε.

Proof

Denote by Z the set of functions $\mathbf{v} \in (H^1(\Omega))^3$ subject to

$$\int_\Omega \mathbf{v}(\mathbf{x}) \cdot (\mathbf{a} + \mathbf{b} \times \mathbf{x}) d\mathbf{x} = 0,$$

for all $\mathbf{a}, \mathbf{b} \in \mathbb{C}^3$. By C we shall denote a positive constant independent of ε, which may be different in different estimates. By Lemma 3.5(ii), for $\mathbf{v} \in Z$

$$\int_\Omega \|\mathbf{v}(\mathbf{x})\|^2 d\mathbf{x} + \int_{\partial\Omega} \|\mathbf{v}(\mathbf{x})\|^2 ds \leq C\mathcal{E}(\mathbf{v};\Omega). \qquad (6.3.10)$$

From (6.3.5), in the region Ω the vector function $\mathbf{u}$ admits the representation

$$\mathbf{u} = \mathbf{v} + \mathbf{w}, \qquad (6.3.11)$$

where $\mathbf{v} \in Z$,

$$\mathbf{w}(\mathbf{x}) = \alpha \mathbf{e}^{(3)} + \beta \mathbf{e}^{(1)} \times \mathbf{x} + \gamma \mathbf{e}^{(2)} \times \mathbf{x}, \qquad (6.3.12)$$

with constant α, β and γ, and $\mathbf{v}$, $\mathbf{w}$ satisfy

$$\int_\Omega \mathbf{v} \cdot \mathbf{w} d\mathbf{x} = 0. \qquad (6.3.13)$$

Using (6.3.10) and (6.3.13) we deduce that

$$\int_\Omega \|\mathbf{u}(\mathbf{x})\|^2 d\mathbf{x} \leq C\mathcal{E}(\mathbf{v};\Omega) + \int_\Omega \|\mathbf{w}\|^2 d\mathbf{x}. \qquad (6.3.14)$$

Let Γ_ε denote the junction region of the boundary $\partial\Omega$:

$$\Gamma_\varepsilon = \{\mathbf{x} \in \partial\Omega : x_3 = 0, \varepsilon^{-1}(x_1 - a_1^{(j)}, x_2 - a_2^{(j)}) \in g,\ j = 1,\ldots,K\}.$$

Let

$$d(\varepsilon) = \sup \int_\Omega \|\mathbf{w}\|^2 d\mathbf{x} \left(\int_{\Gamma_\varepsilon} |w_3|^2 dx_1 dx_2\right)^{-1}$$

where the supremum is taken over all $\mathbf{w}$ of the form (6.3.12) with $\alpha^2 + \beta^2 + \gamma^2 = 1$. By (6.3.12), we have $w_3 = \alpha + \beta y - \gamma x$. Since the points $\mathbf{a}^{(j)}, j = 1,\ldots, K$, do not lie on the same straight line it follows that $\|w_3\| \geq \text{Const}$ at one of the junction points, and therefore

$$d(\varepsilon) \leq C\varepsilon^{-2}.$$

Using the definition of $d(\varepsilon)$, (6.3.14), and the equality $\varepsilon_{ij}(\mathbf{w}) = 0$ we derive

$$\int_\Omega \|\mathbf{u}\|^2 d\mathbf{x} \leq C\left(\mathcal{E}(\mathbf{u};\Omega) + \varepsilon^{-2} \int_{\Gamma_\varepsilon} |w_3|^2 dx_1 dx_2\right). \qquad (6.3.15)$$

By (6.3.10) and (6.3.11),

$$\int_{\Gamma_\varepsilon} |w_3|^2 d\mathbf{x} \leq 2 \int_{\Gamma_\varepsilon} (|u_3|^2 + |v_3|^2) d\mathbf{x} \leq 2 \int_{\Gamma_\varepsilon} |u_3|^2 d\mathbf{x} + c\mathcal{E}(\mathbf{u};\Omega). \qquad (6.3.16)$$

Combining (6.3.15) and (6.3.16) we arrive at

$$\int_\Omega \|\mathbf{u}\|^2 d\mathbf{x} \leq C\varepsilon^{-2}\left(\mathcal{E}(\mathbf{u};\Omega) + \int_{\Gamma_\varepsilon} |u_3|^2 dx_1 dx_2\right).$$

Since $u_3 = 0$ on S_ε, we have the inequality

$$\int_{\Gamma_\varepsilon} |u_3|^2 dx_1 dx_2 \leq C \sum_{j=1}^N \int_{\Pi_\varepsilon^{(j)}} |\partial_{x_3} u_3|^2 d\mathbf{x}.$$

Hence

$$\int_\Omega \|\mathbf{u}\|^2 d\mathbf{x} \leq C\varepsilon^{-2} \mathcal{E}(\mathbf{u};\Omega_\varepsilon). \qquad (6.3.17)$$

Inequality (3.8.18) and the boundary condition $\mathbf{u} = 0$ on S_ε provide the estimate

$$\sum_{j=1}^K \int_{\Pi_\varepsilon^{(j)}} \|\mathbf{u}\|^2 d\mathbf{x} \leq C\varepsilon^{-2} \mathcal{E}(\mathbf{u}; \cup_j \Pi_\varepsilon^{(j)}). \qquad (6.3.18)$$

Using (6.3.17), (6.3.18) we arrive at (6.3.9). □

We show that inequality (6.3.9) is still valid if the domain Ω is replaced by Ω_ε in the orthogonality condition (6.3.8).

LEMMA 6.5

If $\mathbf{u} \in \mathbf{H}$ *satisfies*

$$\int_{\Omega_\varepsilon} \mathbf{u} \cdot \mathbf{\Psi}^{(j)} d\mathbf{x} = 0, \ j = 1, 2, 3, \qquad (6.3.19)$$

then the inequality (6.3.9) holds.

Proof

By the linear independence of $\mathbf{\Psi}^{(1)}, \mathbf{\Psi}^{(2)}$ and $\mathbf{\Psi}^{(3)}$ there exists a unique constant vector (c_1, c_2, c_3) satisfying

$$\int_\Omega \left(\mathbf{u} - \sum_{k=1}^3 c_k \mathbf{\Psi}^{(k)}\right) \cdot \mathbf{\Psi}^{(j)} d\mathbf{x} = 0, \ j = 1, 2, 3,$$

and such that

$$\sum_{k=1}^{3}|c_k| \leq C\sum_{k=1}^{3}\left|\int_{\Omega}\mathbf{u}\cdot\mathbf{\Psi}^{(j)}d\mathbf{x}\right| = C\sum_{k=1}^{3}\left|\int_{\Pi_{\varepsilon}^{(j)}}\mathbf{u}\cdot\mathbf{\Psi}^{(j)}d\mathbf{x}\right|.$$

This implies that

$$\sum_{k=1}^{3}|c_k| \leq C\varepsilon\|\mathbf{u}\|_{(L_2(\Omega_{\varepsilon}))^3}, \qquad (6.3.20)$$

where C does not depend on ε and $\mathbf{u}$.

Let $\eta = \eta(z)$ be a smooth function defined for $z \geq 0$ in such a way that $\eta(0) = 1$ and $\eta(z) = 0$ for $z > \delta$, where δ is a positive number which does not exceed $\min_j(l^{(j)}/2)$. We represent $\mathbf{u}$ as the sum

$$\mathbf{u} = \mathbf{v} + \mathbf{w},$$

where

$$\mathbf{v}(\mathbf{x}) = \mathbf{u}(\mathbf{x}) - \sum_{k=1}^{3}c_k\mathbf{\Psi}^{(k)}(\mathbf{x}) \quad \text{in } \Omega,$$

and

$$\mathbf{v}(\mathbf{x}) = \mathbf{u}(\mathbf{x}) - \sum_{k=1}^{3}c_k\left(\mathbf{\Psi}^{(k)}(\mathbf{x}) - \eta(l^{(j)} - x_3)\mathbf{\Psi}^{(k)}(\mathbf{x}',l^{(j)})\right) \quad \text{in } \Pi_{\varepsilon}^{(j)}.$$

Here, $\mathbf{v} \in \mathbf{H}$ and

$$\int_{\Omega}\mathbf{v}\cdot\mathbf{\Psi}^{(j)}d\mathbf{x} = 0 \quad \text{for } j = 1,2,3.$$

Therefore, by Lemma 6.4,

$$\|\mathbf{u}\|_{(L_2(\Omega_{\varepsilon}))^3}^2 \leq C\left(\varepsilon^{-2}a(\mathbf{v},\mathbf{v}) + \|\mathbf{w}\|_{(L_2(\Omega_{\varepsilon}))^3}^2\right). \qquad (6.3.21)$$

Using (6.3.20) one has

$$\|\mathbf{w}\|_{(L_2(\Omega_{\varepsilon}))^3}^2 \leq C\varepsilon^2\|\mathbf{u}\|_{(L_2(\Omega_{\varepsilon}))^3}^2. \qquad (6.3.22)$$

The definition of $\mathbf{v}$ provides the estimate

$$a(\mathbf{v},\mathbf{v}) \leq 2\left(a(\mathbf{u},\mathbf{u}) + 2\sum_{j=1}^{K}\sum_{k=1}^{3}|c_k|^2\mathcal{E}(\mathbf{w}^{(j,k)};\Pi_{\varepsilon}^{(j)})\right),$$

where

$$\mathbf{w}^{(j,k)}(\mathbf{x}) = \eta(l_j - x_3)\mathbf{\Psi}^{(k)}(\mathbf{x}', l_j).$$

Owing to the obvious inequality

$$\mathcal{E}(\mathbf{w}^{(j,k)}; \Pi_\varepsilon^{(j)}) \leq C\varepsilon^2$$

and (6.3.20), we arrive at

$$a(\mathbf{v},\mathbf{v}) \leq 2a(\mathbf{u},\mathbf{u}) + C\varepsilon^4\|\mathbf{u}\|^2_{(L_2(\Omega_\varepsilon))^3}. \tag{6.3.23}$$

The estimate (6.3.21) together with (6.3.22) and (6.3.23) lead to (6.3.9). □

6.3.3 Spaces $\mathfrak{X}_0$ and $\mathbf{H}_0$

The space $\mathfrak{X}_0$ will be chosen as a linear hull of the vector functions $\mathbf{\Psi}^{(1)}$, $\mathbf{\Psi}^{(2)}$, $\mathbf{\Psi}^{(3)}$, and the inequality (6.3.9), proved in Lemma 6.5, yields (6.1.5) with $q = \text{Const } \varepsilon^{-2}$.

According to the definition (6.1.6), the space $\mathbf{H}_0$ consists of the vector functions $\mathbf{u} \in \mathbf{H}$ such that

$$\int_{\Omega_\varepsilon} \sum_{i,j=1}^{3} \sigma_{ij}(\mathbf{u})\varepsilon_{ij}(\mathbf{v})d\mathbf{x} = 0$$

holds for all $\mathbf{v} \in \mathbf{H}$ subject to

$$\int_{\Omega_\varepsilon} \mathbf{v} \cdot \mathbf{\Psi}^{(j)} d\mathbf{x} = 0, \quad j = 1,2,3.$$

Let $\mathfrak{N}(\mathbf{\Psi})$ be the solution of the boundary value problem (4.1.1)–(4.1.3) with $\mathbf{F} = \mathbf{\Psi}$, $\mathbf{P} = 0$, $\mathbf{\Phi}^{(k)} = 0$, $k = 1,\ldots,K$. Then the space $\mathbf{H}_0$ is the linear hull of vector functions $\mathfrak{N}(\mathbf{\Psi}^{(k)})$, $k = 1,2,3$.

6.3.4 Asymptotic formula for the eigenvalues

Let $\rho_1 \leq \rho_2 \leq \rho_3$ be eigenvalues of the problem

$$\mathcal{A}\mathbf{q} = \rho \mathbf{J}\mathbf{q}, \tag{6.3.24}$$

where $\mathbf{q} \in \mathbb{C}^3$, and

$$\mathbf{J} = \text{diag}\{\text{mes}_3\Omega, \text{mes}_3\Omega, I_3\}. \tag{6.3.25}$$

The matrix $\mathcal{A}$ is given by

$$\mathcal{A} = \sum_{k=1}^{K} \frac{12}{(l^{(k)})^3} \begin{pmatrix} D_1 & 0 & -D_1 a_2^{(k)} \\ 0 & D_2 & D_2 a_1^{(k)} \\ -D_1 a_2^{(k)} & D_2 a_1^{(k)} & (a_1^{(k)})^2 D_2 + (a_2^{(k)})^2 D_1 \\ & & + (l^{(k)})^2 (D_4/12) \end{pmatrix} \qquad (6.3.26)$$

(see (4.2.59)), where the coordinates $a_1^{(k)}$ and $a_2^{(k)}$ of junction points are regarded as the Cartesian system of coordinates such that Ox_1, Ox_2 are parallel to the principal inertia axes of the region g, all junction points are located in the plane $x_3 = 0$, and the centre of inertia of the body Ω is placed on the Ox_3 axis. The constants D_1, D_2 and D_4 are defined by the equalities

$$D_k = \mu \frac{2\mu + 3\lambda}{\lambda + \mu} \int_g X_k^2 d\mathbf{X}', \quad k = 1, 2$$

and

$$D_4 = \mu \int_g \|\nabla \varphi - X_2 \mathbf{e}^{(1)} + X_1 \mathbf{e}^{(2)}\|^2 d\mathbf{X}',$$

where φ is the torsion potential (see Section 3.5.1).

Finally, I_3 is the moment of inertia of Ω with respect to the Ox_3 axis specified by (6.3.6). Since $\mathcal{A}$ is positive definite (see (4.2.63)), the eigenvalues ρ_i are positive.

The following statement gives the asymptotics for the first three eigenvalues of the spectral problem (6.3.1)–(6.3.3).

THEOREM 6.3

The first three eigenvalues $\lambda_1 \leq \lambda_2 \leq \lambda_3$ of the spectral problem (6.3.1)–(6.3.3) admit the asymptotic representation

$$\lambda_j = \varepsilon^4 \rho_j + O(\varepsilon^5). \qquad (6.3.27)$$

Other eigenvalues are greater than $C\varepsilon^2$ with some positive constant C independent of ε.

Proof

Here, we use the notations $\mathfrak{X}$, $\mathfrak{X}_0$, $\mathcal{H}$, $\mathbf{H}$ and $\mathbf{H}_0$ introduced in the previous sections.

The ratio (6.1.7) of quadratic forms can be written as

$$\frac{2\mathcal{E}(\mathfrak{N}(\Psi);\Omega_\varepsilon)}{\|\mathfrak{N}(\Psi)\|^2_{(L_2(\Omega_\varepsilon))^3}}, \quad \Psi \in \mathfrak{X}_0,$$

or, equivalently,

$$\frac{\int_{\Omega_\varepsilon} \mathfrak{N}(\Psi) \cdot \overline{\Psi} dx}{\|\mathfrak{N}(\Psi)\|^2_{(L_2(\Omega_\varepsilon))^3}}, \quad \Psi \in \mathfrak{X}_0. \tag{6.3.28}$$

Let $\nu_1 \leq \nu_2 \leq \nu_3$ denote the eigenvalues of the spectral problem induced by (6.3.28). According to (6.1.7) and Theorem 6.1 (we recall that in our case $q = \text{Const } \varepsilon^{-2}$)

$$\lambda_4 \geq \text{Const } \varepsilon^2,$$

and

$$\frac{\nu_j}{1 + C\varepsilon^{-2}\nu_j} \leq \lambda_j \leq \nu_j, \ j = 1,2,3. \tag{6.3.29}$$

Now, we need the asymptotic representation for ν_j. First, represent the spectral problem, corresponding to (6.3.28), in coordinate form.

Let

$$\Psi(\mathbf{x}) = \sum_{j=1}^{3} b_j \Psi_j(\mathbf{x}). \tag{6.3.30}$$

From Theorem 4.9 and the formulae for $\mathbf{u}^{(-1)}$, $\mathbf{u}^{(0)}$ from Section 6.1, the vector function $\mathfrak{N}(\Psi)$ admits the asymptotic representation

$$\mathfrak{N}(\Psi) = \varepsilon^{-4}(\alpha \mathbf{e}^{(1)} + \beta \mathbf{e}^{(2)} + \gamma \mathbf{e}^{(3)} \times \mathbf{x}) + \mathbf{w}, \tag{6.3.31}$$

where the remainder satisfies

$$\|\mathbf{w}\|_{(L_2(\Omega_\varepsilon))^3} \leq C\varepsilon^{-3}\|\Psi\|_{(L_2(\Omega_\varepsilon))^3}. \tag{6.3.32}$$

The constants α, β, γ solve the linear system

$$\mathcal{A}\begin{pmatrix} \alpha \\ \beta \\ \gamma \end{pmatrix} = \begin{pmatrix} h_1 \\ h_2 \\ h_3 \end{pmatrix},$$

with

$$h_j = \int_\Omega \mathbf{e}^{(j)} \cdot \Psi dx, \ j = 1,2, \tag{6.3.33}$$

$$h_3 = \int_\Omega (\mathbf{e}^{(3)} \times \mathbf{x}) \cdot \mathbf{\Psi} dx = \int_\Omega (\mathbf{x} \times \mathbf{\Psi})_3 dx. \qquad (6.3.34)$$

Let us introduce the matrix $\mathbf{N} = \{N_{jk}\}_{j,k=1}^3$ with elements

$$N_{jk} = \int_\Omega \mathbf{e}^{(j)} \cdot \mathbf{\Psi}^{(k)} dx \text{ for } j = 1,2, \ k = 1,2,3, \qquad (6.3.35)$$

$$N_{3k} = \int_\Omega (\mathbf{e}^{(3)} \times \mathbf{x}) \cdot \mathbf{\Psi}^{(k)} dx \text{ for } k = 1,2,3. \qquad (6.3.36)$$

Using the definition of $\mathbf{\Psi}^{(j)}$, $j = 1,2,3$, one can verify by direct calculations that the matrix $\mathbf{N}$ is symmetric. It follows from the linear independence of $\mathbf{\Psi}^{(1)}, \mathbf{\Psi}^{(2)}, \mathbf{\Psi}^{(3)}$ and the orthogonality conditions (6.3.5) that $\mathbf{N}$ is non-singular. Now, the equalities (6.3.33) and (6.3.34) can be written in the form

$$h_j = \sum_{k=1}^3 N_{jk} b_k, \ j = 1,2,3,$$

where b_k are the constant coefficients from (6.3.30). Therefore,

$$(\alpha, \beta, \gamma)^t = \mathcal{A}^{-1} \mathbf{N} \mathbf{b},$$

where $\mathbf{b} = (b_1, b_2, b_3)^t$. Hence,

$$\int_{\Omega_\varepsilon} \mathfrak{N}(\mathbf{\Psi}) \cdot \mathbf{\Psi} dx = (\mathcal{A}^{-1} \mathbf{N} \mathbf{b}, \mathbf{N} \mathbf{b})_{\mathbb{C}^3} \varepsilon^{-4} + O(\varepsilon^{-3}).$$

One can verify that

$$\|\mathfrak{N}(\mathbf{\Psi})\|_{(L_2(\Omega_\varepsilon))^3}^2 = \varepsilon^{-8} \left(\mathbf{J}(\alpha,\beta,\gamma)^t, (\alpha,\beta,\gamma)^t \right)_{\mathbb{C}^3} + O(\varepsilon^{-7}),$$

where $\mathbf{J}$ is defined in (6.3.25). Therefore,

$$\|\mathfrak{N}(\mathbf{\Psi})\|_{(L_2(\Omega_\varepsilon))^3}^2 = \varepsilon^{-8} (\mathbf{J}\mathcal{A}^{-1} \mathbf{N} \mathbf{b}, \mathcal{A}^{-1} \mathbf{N} \mathbf{b})_{\mathbb{C}^3} + O(\varepsilon^{-7}).$$

Now, the spectral problem, corresponding to (6.3.28), takes the form

$$\left(\mathbf{N} \mathcal{A}^{-1} \mathbf{N} + O(\varepsilon) \right) \mathbf{b} = \nu \varepsilon^{-4} \left(\mathbf{N} \mathcal{A}^{-1} \mathbf{J} \mathcal{A}^{-1} \mathbf{N} + O(\varepsilon) \right) \mathbf{b}, \qquad (6.3.37)$$

where the perturbation matrices are symmetric.

Let
$$\mathbf{q} = \mathcal{A}^{-1}\mathbf{N}\mathbf{b}.$$
The spectral problem (6.3.37) can be rewritten as
$$(\mathcal{A} + O(\varepsilon))\mathbf{q} = \nu\varepsilon^{-4}(\mathbf{J} + O(\varepsilon))\mathbf{q},$$
with the perturbation operators being symmetric. Hence, the eigenvalues ν_1, ν_2, ν_3 admit the asymptotic approximation
$$\nu_j = \varepsilon^4 \rho_j + O(\varepsilon^5).$$
This equality together with (6.3.29) yield the formula (6.3.27). □

6.4 Spectral problem for an inhomogeneous elastic multi-structure

6.4.1 The spectral problem

We formulate the spectral problem in the form

$$-\mathbf{L}(\partial_x)\mathbf{u} = \lambda\mathbf{u} \text{ in } \Omega \qquad (6.4.1)$$
$$-\mathbf{L}(\partial_x)\mathbf{u} = \varepsilon^2\lambda\mathbf{u} \text{ in } \Pi_\varepsilon \qquad (6.4.2)$$
$$\sigma^{(n)}(\mathbf{u}; \mathbf{x}) = 0 \text{ on } \partial\Omega\backslash S_\varepsilon \qquad (6.4.3)$$
$$\mathbf{u}(\mathbf{x}) = 0 \text{ on } S_\varepsilon, \qquad (6.4.4)$$

The parameter ε^2 in (6.4.2) corresponds to the density of the thin rods.

The objective is to obtain asymptotic formulae for the first six eigenvalues of the problem (6.4.1)–(6.4.4). The semi-linear form $a(\cdot,\cdot)$ is given by (6.3.4) and the space $\mathcal{X}$ is equipped with the norm

$$\|\mathbf{u}\|_{\mathcal{X}} = (\|\mathbf{u}\|^2_{(L_2(\Omega))^3} + \varepsilon^2\|\mathbf{u}\|^2_{(L_2(\Pi_\varepsilon))^3})^{1/2}.$$

The subspace $\mathbf{H}$ consists of vector functions from $\mathcal{H}$ which vanish on S_ε. The space $\mathcal{X}_0$ is defined as a linear hull of vector functions $\mathbf{a} + \mathbf{b} \times \mathbf{x}$, $\mathbf{a}, \mathbf{b} \in \mathbb{C}^3$, and $\mathbf{H}_0$ consists of the solutions of the problem (4.1.1)–(4.1.3) with $\mathbf{F} = \mathbf{a} + \mathbf{b} \times \mathbf{x}$, $\mathbf{P} = 0$ and $\mathbf{\Phi}^{(j)} = \mathbf{0}, j = 1, \ldots, K$.

Following the proofs of Lemmas 6.4 and 6.5 subject to evident changes we obtain

LEMMA 6.6

If $\mathbf{u} \in H$ satisfies
$$\int_{\Omega_\varepsilon} \mathbf{u} \cdot (\mathbf{a} + \mathbf{b} \times \mathbf{x}) dx = 0,$$
for all $\mathbf{a}, \mathbf{b} \in \mathbb{C}^3$, then
$$\|\mathbf{u}\|_{\tilde{x}}^2 \leq C a(\mathbf{u}, \mathbf{u})$$
where C does not depend on ε.

This assertion implies the estimate (6.1.5) with q independent of ε.

6.4.2 Asymptotic formulae for eigenvalues

Let $\rho_1 \leq \rho_2 \leq \rho_3$ be the first eigenvalues of the algebraic spectral problem (6.3.24) and $\sigma_1 \leq \sigma_2 \leq \sigma_3$ be the first eigenvalues of the problem

$$\mathcal{B}\eta = \sigma(\mathbf{S} - \mathbf{R}\mathbf{J}^{-1}\mathbf{R}^*)\eta \qquad (6.4.5)$$

where $\mathcal{B}$ is defined by (4.2.60) and

$$\mathbf{S} = \begin{pmatrix} \mathrm{mes}_3 \Omega & 0 & 0 \\ 0 & I_1 & -Q_3 \\ 0 & -Q_3 & I_2 \end{pmatrix} \qquad (6.4.6)$$

$$\mathbf{R} = \begin{pmatrix} 0 & 0 & 0 \\ 0 & -\kappa & -Q_2 \\ \kappa & 0 & -Q_1 \end{pmatrix}, \qquad (6.4.7)$$

with the same I_j, Q_j and κ as in (6.3.6), (6.3.7). Before presenting the main theorem of this section let us prove that the operator on the right-hand side of (6.4.5) is positive.

LEMMA 6.7

The operator $\mathbf{S} - \mathbf{R}\mathbf{J}^{-1}\mathbf{R}^*$ is positive.

Proof

Consider the 6×6 matrix

$$\mathbf{M} = \begin{pmatrix} \mathbf{J} & \mathbf{R}^* \\ \mathbf{R} & \mathbf{S} \end{pmatrix}.$$

One can directly verify that

$$\left(\mathbf{M}\begin{pmatrix} \boldsymbol{\xi} \\ \boldsymbol{\eta} \end{pmatrix}, \begin{pmatrix} \boldsymbol{\xi} \\ \boldsymbol{\eta} \end{pmatrix}\right)_{\mathbb{C}^6}$$
$$= \|\xi_1 \mathbf{e}^{(1)} + \xi_2 \mathbf{e}^{(2)} + \xi_3 \mathbf{e}^{(3)} \times \mathbf{x} + \eta_1 \mathbf{e}^{(3)} + \eta_2 \mathbf{e}^{(1)} \times \mathbf{x} + \eta_3 \mathbf{e}^{(2)} \times \mathbf{x}\|_{(L_2(\Omega))^3}^2.$$

Hence $\mathbf{M}$ is positive. Therefore,

$$(\mathbf{J}\boldsymbol{\xi} + \mathbf{R}^*\boldsymbol{\eta}, \boldsymbol{\xi})_{\mathbb{C}^3} + (\mathbf{S}\boldsymbol{\eta}, \boldsymbol{\eta})_{\mathbb{C}^3} + (\mathbf{R}\boldsymbol{\xi}, \boldsymbol{\eta}) > 0, \qquad (6.4.8)$$

provided $\boldsymbol{\xi}$ and $\boldsymbol{\eta}$ do not vanish simultaneously. Choosing $\boldsymbol{\xi} = -\mathbf{J}^{-1}\mathbf{R}^*\boldsymbol{\eta}$ we arrive at the required assertion. $\square$

THEOREM 6.4

The first six eigenvalues $\lambda_1 \leq \lambda_2 \leq \cdots \leq \lambda_6$ of the spectral problem (6.4.1)–(6.4.4) admit the asymptotic representation

$$\lambda_j = \varepsilon^4 \rho_j + O(\varepsilon^5), \quad j = 1, 2, 3,$$
$$\lambda_{3+j} = \varepsilon^2 \sigma_j + O(\varepsilon^3), \quad j = 1, 2, 3.$$

The remaining eigenvalues are bounded below by a positive constant C independent of ε.

Proof

We introduce a basis $\boldsymbol{\psi}_1, \boldsymbol{\psi}_2, \boldsymbol{\psi}_3, \boldsymbol{\phi}_1, \boldsymbol{\phi}_2, \boldsymbol{\phi}_3$ in the space of the rigid-body displacements in such a way that

$$\int_\Omega \mathbf{e}^{(3)} \cdot \boldsymbol{\psi}^{(j)} d\mathbf{x} = \int_\Omega (\mathbf{e}^{(1)} \times \mathbf{x}) \cdot \boldsymbol{\psi}^{(j)} d\mathbf{x} = \int_\Omega (\mathbf{e}^{(2)} \times \mathbf{x}) \cdot \boldsymbol{\psi}^{(j)} d\mathbf{x} = 0, \quad (6.4.9)$$

and

$$\int_\Omega \mathbf{e}^{(1)} \cdot \boldsymbol{\phi}^{(j)} d\mathbf{x} = \int_\Omega \mathbf{e}^{(2)} \cdot \boldsymbol{\phi}^{(j)} d\mathbf{x} = \int_\Omega (\mathbf{e}^{(3)} \times \mathbf{x}) \cdot \boldsymbol{\phi}^{(j)} d\mathbf{x} = 0, \quad (6.4.10)$$

for $j = 1, 2, 3$.

The right-hand sides $\mathbf{F} = \mathbf{V}$, $\mathbf{P} = 0$, $\mathbf{\Phi}^{(k)} = 0$, $k = 1, \ldots, K$, with $V \in \mathfrak{X}_0$ satisfy the conditions of Theorem 4.9 and, therefore, the solution of (4.1.1)–(4.1.3), denoted by $\mathfrak{N}(\mathbf{V})$, is represented by the asymptotic series of the form (4.3.5).

The ratio (6.1.7) of quadratic forms is equal to

$$\frac{2\mathcal{E}(\mathfrak{N}(\mathbf{V}); \Omega_\varepsilon)}{\|\mathfrak{N}(\mathbf{V})\|_\mathcal{H}^2}, \quad \mathbf{V} \in X_0,$$

which is the same as

$$\frac{\int_{\Omega_\varepsilon} \mathfrak{N}(\mathbf{V}) \cdot \bar{\mathbf{V}} d\mathbf{x}}{\|\mathfrak{N}(\mathbf{V})\|_\mathcal{H}^2}, \quad \mathbf{V} \in X_0. \tag{6.4.11}$$

Let $\nu_1 \leq \nu_2 \leq \nu_3 \leq \nu_4 \leq \nu_5 \leq \nu_6$ denote the eigenvalues of the spectral problem corresponding to (6.4.11); then by Theorem 6.1

$$\lambda_7 \geq \text{const}$$

and

$$\frac{\nu_j}{1 + C\nu_j} \leq \lambda_j \leq \nu_j, \quad j = 1, 2, \ldots, 6 \tag{6.4.12}$$

with C independent of ε. Thus it is enough to obtain asymptotic formulae for ν_j.

Let us represent the spectral problem corresponding to (6.4.11) in a coordinate form. Let

$$\mathbf{V}(\mathbf{x}) = \sum_{j=1}^{3} (a_j \phi_j(\mathbf{x}) + b_j \psi_j(\mathbf{x})).$$

Then from (6.4.9), (6.4.10)

$$\mathfrak{N}(\mathbf{V}) = \varepsilon^{-4}(\alpha \mathbf{e}^{(1)} + \beta \mathbf{e}^{(2)} + \gamma(\mathbf{e}^{(3)} \times \mathbf{x}))$$
$$+ \varepsilon^{-2}(d\mathbf{e}^{(3)} + p\mathbf{e}^{(1)} \times \mathbf{x} + q\mathbf{e}^{(2)} \times \mathbf{x} + \alpha_1 \mathbf{e}^{(1)} + \beta_1 \mathbf{e}^{(2)}$$
$$+ \gamma_1(\mathbf{e}^{(3)} \times \mathbf{x})) + \mathbf{w}(\mathbf{x}).$$

Here the remainder $\mathbf{w}$ admits the estimate

$$\|\mathbf{w}\|_\mathcal{H} \leq C\varepsilon^{-2}(|\alpha| + |\beta| + |\gamma|) + C_1 \varepsilon^{-1},$$

where the constants C, C_1 do not depend on V. The coefficients α, β, γ satisfy

INHOMOGENEOUS ELASTIC MULTI-STRUCTURE

$$\mathcal{A}\begin{pmatrix}\alpha\\ \beta\\ \gamma\end{pmatrix} = \begin{pmatrix}h_1\\ h_2\\ h_3\end{pmatrix}$$

with

$$h_j = \int_\emptyset a\mathbf{e}^{(j)}\cdot\mathbf{V}d\mathbf{x}, \quad j=1,2$$

$$h_3 = \int_\Omega (\mathbf{e}^{(3)}\times\mathbf{x})\cdot\mathbf{V}d\mathbf{x}.$$

The coefficients d, p, q solve the matrix equation

$$\mathcal{B}\begin{pmatrix}d\\ p\\ q\end{pmatrix} = \begin{pmatrix}h_4\\ h_5\\ h_6\end{pmatrix} - \mathcal{D}\begin{pmatrix}\alpha\\ \beta\\ \gamma\end{pmatrix},$$

where

$$h_4 = \int_\Omega \mathbf{e}^{(3)}\cdot\mathbf{V}d\mathbf{x}$$

$$h_{4+i} = \int_\Omega (\mathbf{e}^{(i)}\times\mathbf{x})\cdot\mathbf{V}d\mathbf{x}, \quad i=1,2,$$

and the matrices $\mathcal{A}$, $\mathcal{B}$ and $\mathcal{D}$ are given by (4.2.59)–(4.2.61).

Let the matrices $\mathbf{N} = \{N_{ij}\}_{i,j=1}$ and $\mathbf{M} = \{M_{ij}\}_{i,j=1}^{3}$ be specified by

$$N_{ij} = \int \mathbf{e}^{(i)}\cdot\boldsymbol{\psi}_j d\mathbf{x}, \quad i=1,2$$

$$N_{3j} = \int (\mathbf{e}^{(3)}\times\mathbf{x})\cdot\boldsymbol{\psi}_j d\mathbf{x}$$

$$M_{1j} = \int \mathbf{e}^{(3)}\cdot\boldsymbol{\phi}_j d\mathbf{x}$$

$$M_{ij} = \int (\mathbf{e}^{(i)}\times\mathbf{x})\cdot\boldsymbol{\phi}_j d\mathbf{x}, \quad i=2,3.$$

Then

$$\begin{pmatrix}\alpha\\ \beta\\ \gamma\end{pmatrix} = \mathcal{A}^{-1}\mathbf{N}\begin{pmatrix}B_1\\ B_2\\ B_3\end{pmatrix}$$

and

$$\begin{pmatrix}d\\ p\\ q\end{pmatrix} = \mathcal{B}^{-1}\mathbf{M}\begin{pmatrix}a_1\\ a_2\\ a_3\end{pmatrix} - \mathcal{B}^{-1}\mathcal{D}\mathcal{A}^{-1}\mathbf{N}\begin{pmatrix}b_1\\ b_2\\ b_3\end{pmatrix}.$$

Using this relation and orthogonality conditions (6.4.9), (6.4.10) we obtain

$$\int_{\Omega_\varepsilon} \mathfrak{N}(\mathbf{V}) \cdot \bar{\mathbf{V}} dx = \varepsilon^{-4}(\mathcal{A}^{-1}\mathbf{Nb}, \mathbf{Nb})_{\mathbb{C}^3} + \varepsilon^{-2}(\mathcal{B}^{-1}\mathbf{Ma}, \mathbf{Ma})_{\mathbb{C}^3}$$
$$+ \varepsilon^{-2}[(\mathbf{Gb}, \mathbf{a})_{\mathbb{C}^3} + (\mathbf{b}, \mathbf{Ga})_{\mathbb{C}^3}] + O(\varepsilon^{-2}\|\mathbf{b}\|^2) + O(\varepsilon^{-1}\|\mathbf{a}\|^2),$$

where $\mathbf{b} = (b_1, b_2, b_3)^t$, $\mathbf{a} = (a_1, a_2, a_3)^t$ and $\mathbf{G}$ is a 3×3 matrix, independent of ε. The explicit form of $\mathbf{G}$ is not required. Furthermore, one can verify directly that

$$\|\mathfrak{N}(\mathbf{V})\|^2_{\mathcal{H}} = \varepsilon^{-8}\left(\mathbf{J}\begin{pmatrix}\alpha\\\beta\\\gamma\end{pmatrix},\begin{pmatrix}\alpha\\\beta\\\gamma\end{pmatrix}\right)_{\mathbb{C}^3} + \varepsilon^{-4}\left(\mathbf{S}\begin{pmatrix}d\\p\\q\end{pmatrix},\begin{pmatrix}d\\p\\q\end{pmatrix}\right)_{\mathbb{C}^3}$$
$$+ \varepsilon^{-6}\left[\left(\mathcal{R}\begin{pmatrix}\alpha\\\beta\\\gamma\end{pmatrix},\begin{pmatrix}d\\p\\q\end{pmatrix}\right)_{\mathbb{C}^3} + \left(\begin{pmatrix}\alpha\\\beta\\\gamma\end{pmatrix},\mathcal{R}\begin{pmatrix}d\\p\\q\end{pmatrix}\right)_{\mathbb{C}^3}\right] + \ldots,$$

where $\mathbf{J}, \mathbf{S}$ and $\mathcal{R}$ are the matrices given by (6.3.25), (6.4.6) and (6.4.7). Thus,

$$\|\mathfrak{N}(\mathbf{V})\|^2_{\mathcal{H}} = \varepsilon^{-8}(\mathbf{J}\mathcal{A}^{-1}\mathbf{Nb}, \mathcal{A}^{-1}\mathbf{Nb})_{\mathbb{C}^3}$$
$$+ \varepsilon^{-4}(\mathbf{S}\mathcal{B}^{-1}\mathbf{Ma}, \mathcal{B}^{-1}\mathbf{Ma})_{\mathbb{C}^3}$$
$$+ \varepsilon^{-6}[(\mathcal{R}\mathcal{A}^{-1}\mathbf{Nb}, \mathcal{B}^{-1}\mathbf{Ma})_{\mathbb{C}^3} + (\mathcal{A}^{-1}\mathbf{Nb}, \mathcal{R}\mathcal{B}^{-1}\mathbf{Ma})_{\mathbb{C}^3}]$$
$$+ \text{lower-order terms}.$$

Now we arrive at the spectral problem in variables $\boldsymbol{\xi} = \mathcal{A}^{-1}\mathbf{Nb}$, and $\boldsymbol{\eta} = \mathcal{B}^{-1}\mathbf{Ma}$:

$$\begin{pmatrix} \varepsilon^{-4}\mathcal{A} & \varepsilon^{-2}\mathcal{T}^* \\ \varepsilon^{-2}\mathcal{T} & \varepsilon^{-2}\mathcal{B} \end{pmatrix}\begin{pmatrix}\boldsymbol{\xi}\\\boldsymbol{\eta}\end{pmatrix} = \nu \begin{pmatrix} \varepsilon^{-8}\mathbf{J} & \varepsilon^{-6}\mathcal{R}^* \\ \varepsilon^{-6}\mathcal{R} & \varepsilon^{-4}\mathbf{S} \end{pmatrix}\begin{pmatrix}\boldsymbol{\xi}\\\boldsymbol{\eta}\end{pmatrix},$$

where $\mathcal{T}$ is the 3×3 matrix

$$\mathcal{T} = \mathcal{B}\mathbf{M}^{-1*}\mathbf{C}\mathbf{N}^{-1}\mathcal{A}.$$

We did not write the lower-order terms here because they do not contribute to the final result. First, we are looking for eigenvalues of order $O(\varepsilon^4)$. If $\nu = \varepsilon^4\mu$ then

$$\mathcal{T}^*\boldsymbol{\xi} + \mathcal{B}\boldsymbol{\eta} = \mu[\mathcal{R}\boldsymbol{\xi} + \varepsilon^2\mathbf{S}\boldsymbol{\eta}]$$

and

$$\mathcal{A}\xi + \varepsilon^2 \mathcal{T}\eta = \mu(\mathbf{J}\xi + \varepsilon^2 \mathcal{R}\eta).$$

Hence
$$\eta = (\mathcal{B} - \varepsilon^2 \mu \mathbf{S})^{-1}(\mu\mathcal{R}\xi - \mathcal{T}^*\xi)$$
and
$$\mathcal{A}\xi + \varepsilon^2 \mathcal{T}(\mathcal{B} - \varepsilon\mu \mathbf{S})^{-1}(\mu\mathcal{R}\xi - \mathcal{T}^*\xi)$$
$$= \mu[\mathbf{J}\xi + \varepsilon^2 \mathcal{R}(\mathcal{B} - \varepsilon^2\mu\mathbf{S})^{-1}(\mu\mathcal{R} - \mathcal{T}^*)\xi].$$

It is clear that
$$\nu_j = \rho_j \varepsilon^4 + O(\varepsilon^5), \quad j = 1, 2. \tag{6.4.13}$$

Second, let $\nu = \varepsilon^2 \mu$ where μ has order $O(1)$. Then
$$\mathcal{A}\xi + \varepsilon^2 \mathcal{T}\eta = \mu(\varepsilon^{-2}\mathbf{J}\xi + \mathcal{R}\eta),$$
$$\mathcal{T}^*\xi + \mathcal{B}\eta = \mu(\varepsilon^{-2}\mathcal{R}\xi + \mathbf{S}\eta).$$

Hence
$$\xi = (\mathcal{A} - \mu\varepsilon^{-2}\mathbf{J})^{-1}(\mu\mathcal{R} - \varepsilon^2\mathcal{T})\eta$$
and
$$\mathcal{B}\eta + \mathcal{T}^*(\mathcal{A} - \mu\varepsilon^{-2}\mathbf{J})^{-1}(\mu\mathcal{R} - \varepsilon^2\mathcal{T})\eta$$
$$= \mu[\mathbf{S}\eta + \varepsilon^{-2}\mathcal{R}(\mathcal{A} - \mu\varepsilon^{-2}\mathbf{J})^{-1}(\mu\mathcal{R} - \varepsilon^2\mathcal{T})\eta].$$

This implies that
$$\nu_{3+j} = \varepsilon^2 \sigma_j + O(\varepsilon^3), \quad j = 1, 2, 3. \tag{6.4.14}$$

The formulae (6.4.13), (6.4.14) together with (6.4.12) lead to the required asymptotic representations. □

BIBLIOGRAPHICAL REMARKS

Chapter 1. Modulo our presentation, Sections 1.1 and 1.2 contain well-known material on asymptotic techniques for ordinary differential equations. We refer to the recent book by O'Malley (1991), where a concise survey of the historical development of singular perturbation methods is given. Mathematical foundations for the asymptotic analysis of singularly perturbed ordinary differential equations have been laid down by Vyshik and Lusternik (1957, 1960). A good source concerning the subject is the book by Nayfeh (1973).

For the examples of boundary value problems considered in Sections 1.3–1.5 we refer to the book by Maz'ya, Nazarov and Plamenevskii (1991, 1992). Numerical illustrations of asymptotic solutions can be found in Ward and Keller (1993) and Ward, Henshaw and Keller (1993). Practical applications of asymptotic analysis for problems posed in the interior or exterior of thin domains are considered in Keller (1993) and Ahluwalia and Keller (1986). Junctions between thin bodies treated in Section 1.6 were studied in detail by Bakhvalov and Panasenko (1989), Le Dret (1989, 1991), Panasenko (1991, 1994, 1996) and Panasenko and Paulin (1993). Many other examples of boundary value problems in parameter dependent domains solved by the method of compound asymptotic expansions are given in the book by Maz'ya, Nazarov and Plamenevskii (1991, 1992), which also includes a comprehensive bibliography.

Chapter 2 is an expanded version of Sections 1–9 of Kozlov, Maz'ya and Movchan (1994). Similar multi-structures were studied by Nazarov (1996). In connection with our study of model problems for unbounded domains we mention the book by Kozlov, Maz'ya and Rossmann (1997), where a systematic exposition of the theory of elliptic boundary value problems in domains with point singularities is presented. Properties of Sobolev spaces (extension, trace and embedding theorems) for domains singularly depending on small or large parameters are studied in Maz'ya and Poborchi (1997).

Chapter 3. For the basics of mathematical elasticity see, for example, Westergaard (1952), Lurie (1970), Gurtin (1981), Marsden and Hughes (1983) and Atkin and Fox (1980). Various forms of Korn's inequality can be found in Oleinik, Shamaev and Yosifian (1992), Oleinik (1996),

Nazarov (1992), Nazarov (1997a) and Ovtchinnikov and Xanthis (1997). Inequality (4.4.2) is due to Nazarov (1992).

The material of Section 3.4.3 has applications in the Saint-Venant theory of elastic beams: see, for example, Lurie (1970). Power-exponential solutions of operator differential equations and their connection with the spectral properties of corresponding operator pencils are discussed in Markus (1988), Shkalikov and Shkred (1991), Nazarov and Plamenevskii (1994), Kozlov and Maz'ya (1999).

Chapters 4, 5. Most of the results on solutions of elasticity problems for multi-structures, obtained by 1997, are presented or referred to in the book by Ciarlet (1997). For a description of the engineering theory of pile structures we refer to Asplund (1966).

The problem presented in Chapter 4 was solved in our article Kozlov, Maz'ya and Movchan (1995). Non-degenerate multi-structures considered in Chapter 5 were analysed in Kozlov, Maz'ya and Movchan (1996).

The work of Sanchez-Palencia (1988) includes the analysis of boundary value problems, where the Neumann boundary conditions are posed in a small region of the surface. Asymptotic results on elastic fields for a body contacting with a set of rigid punches of small contact area are obtained in Argatov and Nazarov (1994); the same authors developed a model of junctions for a rigid rod and a set of elastic bodies (see Argatov and Nazarov (1996)). The leading term of the solution for non-degenerate elastic multi-structures was outlined in Nazarov (1997b).

Chapter 6. Presentation is based on the work by Kozlov, Maz'ya and Movchan (1994, 1995, 1996). Asymptotic expansions for the eigenvalues of classical boundary value problems for the Laplacian for domains with small holes can be found in Maz'ya, Nazarov and Plamenevskii (1991, 1992) where earlier references are also given. Eigenvalue problems for junctions between elastic structures were studied in the work by Bourquin and Ciarlet (1989), where the authors used a 'limit' variational formulation. Asymptotic formulae for the eigenvalues of the Neumann problem for the union of a spherical layer and thin cylinders were obtained in Nazarov and Plamenevskii (1990). Problems of elastodynamics for multi-structures involving junctions of a three-dimensional elastic body and an elastic plate were studied by Raoult (1992); results on the dynamics of elastic multi-plate structures are discussed in Le Dret (1991); for additional references on time-dependent models of multi-structures see Section 2.9 of Ciarlet (1997).

BIBLIOGRAPHY

Agmon, S., Douglis, A. and Nirenberg, L. (1959) Estimates near the boundary for solutions of elliptic partial differential equations satisfying general boundary conditions. *Int. Commun. Pure Appl. Math.*, **12**, 623–727.

Ahluwalia, D.S. and Keller, J.B. (1986) Scattering by a slender body. *J. Acoust. Soc. Am.*, **80**, No. 6, 1782–1792.

Argatov, I.I. and Nazarov, S.A. (1993) Junction problem of shashlik (skewer) type. *C. R. Acad. Sci., Paris. Sér. I, Math.*, **316**, No. 2, 1329–1334.

Argatov, I.I. and Nazarov, S.A. (1994) Asymptotic solution of the problem of an elastic body resting on several small supports. *J. Appl. Math. Mech.*, **58**, No. 2, 303–311.

Argatov, I.I. and Nazarov, S.A. (1996) Asymptotic analysis of problems on junctions of domains of different limit dimensions. A body pierced by a thin rod. *Izv. Mat.*, **60**, No. 1, 1–37.

Åslund, J. (1999) *Asymptotic Analysis of a Mixed Boundary Value Problem for the Laplacian in a 2D-3D Multi-Structure*, Licentiate Thesis, LIU-TEK-LIC-1999, Linköping.

Asplund, S.O. (1966) *Structural Mechanics: Classical and Matrix Methods.* Prentice Hall, Englewood Cliffs, NJ.

Atkin, R.J. and Fox, N. (1980) *An introduction to the theory of elasticity.* Longman, London.

Bakhvalov, N. and Panasenko, G.P. (1989) *Homogenization: averaging processes in periodic media.* Kluwer, Dordrecht (translated from Russian: *Osrednenie processov v periodicheskih sredah*, Nauka, Moscow, 1984).

Birman, M.S. and Solomjak, M.S. (1987), *Spectral theory of self-adjoint operators in Hilbert space.* D. Reidel, Dordrecht.

Bourquin, F. and Ciarlet, P.G. (1989) Modelling and justification of eigenvalue problems for junctions between elastic structures. *J. Funct. Anal.*, **87**, No. 2, 392–427.

Cherepanov, G.P. (1974) *Mechanics of brittle fracture.* Nauka, Moscow.

Ciarlet, P.G. (1990) *Plates and junctions in elastic multi-structures.* Masson, Paris.

Ciarlet, P.G. (1997) *Mathematical elasticity. Volume II: theory of plates.* North-Holland, Amsterdam.

Ciarlet, P.G. and Destuynder, P. (1979) A justification of the two-dimensional plate model. *J. Mèc.*, **18**, 315–344.

Cole, J.D. (1968) *Perturbation methods in applied mathematics.* Blaisdell, Waltham.

Dautray, R. and Lions, J.-L. (1988) *Mathematical analysis and numerical methods for science and technology,* Vol. 2. Springer-Verlag, Berlin, Heidelberg, New York.

Fichera, G. (1984) Existence theorems in elasticity. *Mechanics of Solids,* Vol. II. Springer-Verlag, New York–Berlin, 347–389.

Gurtin, M.E. (1981) *An introduction to continuum mechanics.* Academic Press, New York, London.

Hadamard, J. (1908) *Sur le problème d'analyse relatif à l'équilibre des plaques élastiques encastrées.* Mémoire couronné en 1907 par l'Académie: Prix Vaillant, Mémoires présentés par divers savants à l'Académie des Sciences, **33**, No. 4; see also *Œuvres de Jacques Hadamard,* Centre National de la Recherche Scientifique, Paris, 1968, Vol. 2, 515–629.

Hörmander, L. (1983) *The analysis of linear partial differential operators I. Distribution theory and Fourier analysis.* Springer-Verlag, Berlin, Heidelberg, New York, Tokyo.

Il'in, A.M. (1992) *Matching of asymptotic expansions of solutions of boundary value problems,* Translations of Mathematical Monographs, 102. American Mathematical Society, Providence, RI.

Keller, J.B. (1993) Stresses in narrow regions. *J. Appl. Mech. Trans. ASME,* **60**, No. 4, 1054–1056.

Kozlov, V.A. and Maz'ya, V.G. (1999) *Differential equations with operator coefficients,* Springer monographs in Mathematics. Springer-Verlag, New York–Berlin.

Kozlov, V.A., Maz'ya, V.G. and Movchan, A.B. (1994) Asymptotic analysis of a mixed boundary value problem in a multi-structure. *Asymptotic Anal.,* **8**, 105–143.

Kozlov, V.A., Maz'ya, V.G. and Movchan, A.B. (1995) Asymptotic representation of elastic fields in a multi-structure. *Asymptotic Anal.,* **11**, 343-415.

Kozlov, V.A., Maz'ya, V.G. and Movchan, A.B. (1996) Fields in nondegenerate 1D-3D elastic multi-structures. *LiTH-MAT-R-96-14,* Preprint, Linköping University.

Kozlov, V.A., Maz'ya, V.G. and Rossmann, J. (1997) *Elliptic boundary value problems in domains with point singularities,* Mathematical Surveys and Monographs, Vol. 52. American Mathematical Society, Providence, RI.

Le Dret, H. (1989) Modelling of the junction between two rods. *J. Math. Pures Appl.*, **68**, 365–397.

Le Dret, H. (1991) *Problèmes variationnels dans les multi-domaines: Modélisation des jonctions et applications*, Recherches en Mathématiques Appliquées, 19. Masson, Paris.

Le Dret, H. (1994) Elastodynamics for multiplate structures. *Nonlinear partial differential equations and their applications, Collège de France Seminar*, **XI**, 151–180.

Lions, J.L. and Magenes, E. (1968) *Problèmes aux limites non homogènes et applications*. Dunod, Paris.

Lurie, A.I. (1970) *Theory of elasticity*. Nauka, Moscow.

Marchenko, V.A. and Khruslov, E.Ya. (1974) *Boundary value problems in domains with a fine-grained boundary*. Naukova Dumka, Kiev.

Markus, A.S. (1988) *Introduction to spectral theory of polynomial operator pencils*, Translation of Mathematical Monographs, Vol. 71. American Mathematical Society, Providence, RI.

Marsden, J.E. and Hughes, T.J.R. (1983) *Mathematical foundations of elasticity*. Prentice Hall, Englewood Cliffs, NJ.

Maz'ya, V.G. (1985) *Sobolev Spaces*. Springer-Verlag, Berlin.

Maz'ya, V.G., Nazarov, S.A. and Plamenevskii, B.A. (1991, 1992) *Asymptotische Theorie elliptischer Randwertaufgaben in singulär gestörten Gebieten*. Akademie-Verlag, Berlin, B.1, 1991; B.2, 1992.

Maz'ya, V.G. and Poborchi, S.V. (1997) *Differentiable functions on bad domains*. World Scientific, Singapore.

Mindlin, R.D. (1936) Force at a point in the interior of a semi-infinite solid. *Physics*, **7**, 195–202.

Nayfeh, A.H. (1973) *Perturbation methods*. Wiley, New York and Chichester.

Nazarov, S.A. (1983) The structure of solutions of elliptic boundary value problems in slender domains. *Vestn. Leningrad Univ. Math.*, **15**, 99–104.

Nazarov, S.A. (1992) Korn inequalities which are asymptotically exact for thin domains. *Vestn. St. Petersburg Univ. Math.*, **25**, No. 2, 18–22.

Nazarov, S.A. (1996) Junctions of singularly degenerating domains with different limit dimensions. *Int. J. Math. Sci.* **80**, No. 5, 1989–2034.

Nazarov, S.A. (1997a), Korn's inequalities for junctions of spatial bodies and thin rods. *Math. Methods Appl. Sci.*, **20**, No. 3, 219–243.

Nazarov, S.A. (1997b), Asymptotic behaviour of solutions of a problem of elasticity theory for a three-dimensional body with thin extensions. *Dokl. Akad. Nauk*, **352**, No. 4, 458–461.

Nazarov, S.A. and Plamenevskii, B.A. (1990) Asymptotics of spectrum of a Neumann problem in singularly perturbed thin domains. *Algebra*

Anal., **2**, No. 2, 85–111 (in Russian).
Nazarov, S.A. and Plamenevskii, B.A. (1994) *Elliptic problems in domains with piecewise smooth boundaries.* Walter de Gruyter, Berlin.
Oleinik, O.A. (1996) *Some asymptotic problems in the theory of partial differential equations.* Cambridge University Press, Cambridge.
Oleinik, O.A., Shamaev, A.S. and Yosifian, G.A. (1992) *Mathematical problems in elasticity and homogenization.* North-Holland, Amsterdam.
O'Malley, R.E. (1991) *Singular perturbation methods for ordinary differential equations.* Springer-Verlag, New York–London.
Ovtchinnikov, E.E. and Xanthis, L.S. (1997) The Korn's type inequality in subspaces and thin elastic structures. *Proc. R. Soc. London A*, **453**, 2003–2016.
Panasenko, G.P. (1991) Multi-component homogenization of processes in strongly nonhomogeneous structures. *Math. USSR Sborn.*, **69**, No. 1, 143–153.
Panasenko, G.P. and Saint Jean Paulin, J. (1993) An asymptotic analysis of junctions of non-homogeneous elastic rods: boundary layers and asymptotic expansions. *Zh. Vychisl. Mat. Mat. Phys.*, **33**, No. 11, 1693-1721.
Panasenko, G.P. (1994) Asymptotic analysis of bar systems. I. *Russian J. Math. Phys.*, **2**, No. 3, 325–352.
Panasenko, G.P. (1996) Asymptotic analysis of bar systems. II. *Russian J. Math. Phys.*, **4**, No. 1, 87–116.
Pólya, G. and Szegö, G. (1951) *Isoperimetric inequalities in mathematical physics.* Princeton University Press, Princeton, NJ.
Raoult, A. (1992) Asymptotic modelling of the elastodynamics of a multistructure. *Asymptotic Anal.*, **6**, 73–108.
Rudin, W. (1991) *Functional Analysis,* Second Edition. McGraw-Hill, New York.
Sanchez-Palencia, E. (1980) *Non-Homogeneous Media and Vibration Theory,* Lecture Notes in Physics, Vol. 127. Springer-Verlag, Heidelberg.
Sanchez-Palencia, E. (1988) Forces appliquées à une petite région de surface d'un corps élastique. Applications aux jonctions *C. R. Acad. Sci., Paris. Sér. II, Méc.*, **307**, 689–694.
Shkalikov, A.A. and Shkred, A.B. (1991) A problem on steady oscillations of a transversely isotropic half-cylinder. *Mat. Sborn.*, **182**, No. 8, 1222–1246.
Van Dyke, M.D. (1964) *Perturbation methods in fluid mechanics.* Academic Press, New York.

Vyshik, M.I. and Lusternik, L.A. (1957) Regular perturbation and boundary layer for linear differential equations with a small parameter. *Usp. Mat. Nauk,* **12**, No. 5, 3–122.

Vyshik, M.I. and Lusternik, L.A. (1960) Solution of some perturbation problems for the case of matrices and selfadjoint and non-selfadjoint differential equations. *Usp. Mat. Nauk,* **15**, No. 3, 3–80.

Ward, M.J. and Keller, J.B. (1993) Strong localized perturbation of eigenvalue problems. *SIAM J. Appl. Math.,* **53**, No. 3, 770–798.

Ward, M.J., Henshaw, W.D. and Keller, J.B. (1993) Summing logarithmic expansions for singularly perturbed eigenvalue problems. *SIAM J. Appl. Math.,* **53**, No. 3, 799–828.

Westergaard, H.M. (1952) *Theory of Elasticity and Plasticity.* Dover, New York.

INDEX

Agmon, S., 142
Ahluwalia, D.S., 274
Argatov, I.I., 275
Åslund, J. 108
Asplund, S.O., 155, 176
asymptotic expansion, 1, 8, 89, 92, 180, 198
Atkin, R.J., 115, 274

Bakhvalov, N., xii, 274
Betti's formula, 117, 134
biorthogonality conditions, 133
Birman, M.S., 249, 250
boundary layer, 4, 8, 12, 16, 24, 35
boundary value problem,
    Dirichlet, 5,22
    mixed, 30, 39, 59, 155, 230
    Neumann, 10, 40, 63
Bourquin, F., 275
Boussinesq–Cerruti solution, 121, 163

Cauchy's relations, 115
Cherepanov, G.P., xi
Ciarlet, P.G., xii, 275
Cole, J.D., xi
cylinder
    semi-infinite, 57, 63, 71, 107, 225
    thin, 58, 105, 155, 184

Dautray, R., 62
Destuynder, P., xii
displacement, 115, 157
Douglis A., 142

eigenfunctions, 253, 256
eigenvalues, 250, 253, 263, 268

energy
    asymptotic formula, 20
    functional, 143
    space E, 41

Fichera, G., 117
Fox, N., 115, 274

Green's
    formula, 37, 70, 102
    matrix, 159, 183, 234
Gurtin, M.E., 115, 274

Henshaw, W.D., 274
Hooke's law, 115
Hörmander, L., 73
Hughes, T.J.R., 274

Il'in, A.M., xii

junction points, 58, 157, 159, 215

Keller, J.B., 274
Khruslov, E.Ya., xii
Korn's inequality, 143, 198, 258
Kozlov, V.A., 274, 275

Lamé system, 116
Laplacian, 11, 22, 30, 56
layer
    bottom, 72, 87, 167, 191
    junction, 44, 70, 83, 160, 164, 189
Le Dret, H., xii, 274, 275
Lions, J.-L., 59, 93, 95
lock forces and moments, 176
Lurie, A.I., 115, 126, 128, 274

Magenes E., 59, 93, 95
Marchenko, V.A., xii
Markus, A.S., 275

Marsden, J.E., 274
maximum principle, 24, 25, 29
Maz'ya, V.G., xii, 274, 275
Mindlin, R.D., 123
    solution, 123
Movchan, A.B., 274, 275
multi-structure, 56, 155, 248
    degenerate, 213
    inhomogeneous, 267
    non-degenerate, 213
Nayfeh, A.H., 274
Nazarov, S.A., xii, 274, 275
Neumann function, 61
Nirenberg, L., 142
Oleinik, O.A., 117, 127, 144, 157, 274
O'Malley, R.E., 274
orthogonality conditions, 11, 141, 172, 260, 261
Ovtchinnikov, E.E., 275
Panasenko, G.P., xii, 274
perturbation
    regular, 1, 2
    singular, 3, 5, 30
Pile structure, 174, 176, 177, 216, 219
    degenerate and non-degenerate, 221, 222
Plamenevskii, B.A., xii, 274, 275
Poborchi, S.V., 274
Poincaré inequality, 15, 42
Poisson ratio, 118, 122
Pólya, G., xi
polynomial solution, 130, 138
Raoult, A., 275
Riesz theorem, 44, 163, 169, 251

rigid-body displacements, 133, 145, 177, 210, 234
Rossmann, J., 274
Rudin, W., 251

Saint Jean Paulin, J., 274
Sanchez-Palencia, E., xii, 275
scaled variables, 4, 12, 35, 78, 180, 232
Shamaev, A.S., 117, 127, 144, 157, 274
Shkalikov, A.A., 275
Shkred, A.B., 275
skeleton of the multi-structure, 46, 77, 105,
small parameter, 2, 11, 30, 58, 156
Sobolev space, 54, 95
Solomjak, M.S., 249, 250
spectral problem, 253, 258, 267
    finite-dimensional, 251
strain, 115
stress, 116
stiffness coefficients, 135, 174
stiffness matrix, 179, 223
Szegö, G., xi

temperature, 3, 39, 56
thin rectangle, 30

Van Dyke, M.D., xi

Ward, M.J., 274
Westergaard, H.M., 115, 121, 274

Xanthis, L.S., 275

Yosifian, G.A., 117, 127, 144, 157, 274